**Principles and Applications of
Ubiquitous Sensing**

Principles and Applications of Ubiquitous Sensing

Waltenegus Dargie
Dresden University of Technology, Germany

Library of Congress Cataloging-in-Publication Data applied for

ISBN: 9781119091349

A catalogue record for this book is available from the British Library.

Set in 10/12pt Warnock by SPi Global, Chennai, India
Printed and bound in Malaysia by Vivar Printing Sdn Bhd

10 9 8 7 6 5 4 3 2 1

To Josh and Phebie, with love, lots of love

Contents

Preface

Sensors have always been essential elements of all intelligent systems. Biological systems, for example, are equipped with millions of simple and unobtrusive yet indispensable sensors. Indeed, the trillions of nerve cells populating the cerebral cortex of a human brain are essentially sensors, for they respond to and generate action potentials in the magnitude, frequency, and shape of which messages pertaining to emotional, psychological, and physiological conditions are encoded. All advanced technological systems likewise consist of a large number of sensors that are needed for them to function properly. As human beings strive to make their environments intelligent, interactive, and adaptive, they ubiquitously embed small, unobtrusive, self-organising, energy-efficient, wireless, and interactive sensors. Most of these sensors undoubtedly imitate biological sensors in their simplicity, to enable large-scale and easy deployment. And indeed most existing sensors have inherently simple and comprehensible construction. The understanding of the basic principles of these sensors is vital in the development of sensors as well as the software programs and algorithms that manage them and process the data coming from them. The purpose of this book is to accomplish this goal.

I have endeavoured to make the book easy to read and self-contained. I make few assumptions as to the background knowledge readers should have to comprehend the concepts presented in the book. Where there are electrical circuits, logical explanations are given, so that readers with little electrical background can understand them. If we regard the first two chapters as motivational, the rest of the book can be logically organised into three parts. The first – the sensing part – consists of Chapters 4–8 and deals with the different possibilities (or principles) of sensing a physical phenomenon (the measurand). The second part (Chapter 3 and Chapters 9–11) deals with constructing and integrating physical sensors (and energy harvesters). I separate Chapter 3 from the rest of the second part because I make frequent reference to signal conditioning in the first part. This chapter may also be less comprehensible to readers who have little electrical background. It is not, however, essential to understand this chapter in order to understand the rest of the book. The last and third part of the book (Chapter 12) deals with data processing. In my view, this is an important chapter for many readers who develop or employ physical sensors, as the outputs of essentially all physical sensors contain error, and to reduce error it is paramount to understand the nature and sources of error and how to combine evidence about it.

I have made every endeavour to present the state-of-the-art principles, technologies, and applications of sensors. Unless it was absolutely necessary, my review of literature has been focused on the progress made in the field in the past five years or so. It has been a great pleasure preparing the material as well as writing the book. It is my sincere hope that my readers also take the same pleasure in reading and studying it.

Waltenegus Dargie
April 5, 2016

About the Companion Website

This book is accompanied by a companion website:

www.wiley.com/go/dargie2017

There you will find valuable material designed to enhance your learning, including:

- Lecture slides.

List of Abbreviations

AD	Alzheimer's disease
ADC	Analogue-to-digital converter
AGS	Automatic generating system
AMR	Anisotropic magnetoresistive
AP	Action potential
APD	Action potential duration
AV	Atrioventricular (node)
CLT	Central limit theorem
CDF	Cumulative distribution function
CVD	Chemical vapour deposition
DO	Dissolved oxygen
DOF	Degree of freedom
DSP	Digital signal processor
ECG	Electrocardiogram
ECoG	Electrocorticogram
EEG	Electroencephalography
EMG	Electromyogram
GMR	Giant magnetoresistance
EOG	Electrooculogram
GF	Gauge factor
IC	Integrated circuits
ICD	Implantable cardioverter defibrillator
iid	Independent and identically distributed
LACE	Laser-assisted chemical etching
LFP	Local field potential
LIGA	Lithographie, Galvanik und Abformung
LPCVD	Low-pressure chemical vapour deposition
LVDT	Linear variable differential transformer
MCU	Microcontroller unit
MDOF	Multi degree of freedom
MISO	Master-in, slave-out
MMSE	Minimum mean square error estimation
MR	Magnetoresistance
PD	Parkinson's disease

pdf	Probability density function
PDF	Probability distribution function
PDMS	Polydimethylsiloxane
PECVD	Plasma-enhanced chemical vector deposition
PI	Polyimide
PM	Particulate matter
PMMA	Polymethyl methacrylate
PNAS	Proceedings of the National Academy of Sciences
PZT	Lead zirconate titanate ($Pb[Zr_x Ti_{1-x}]O_3$)
RAP	Resistance area
RFID	Radio frequency identification
RIE	Reactive ion etching
R-SMA	Right supplementary motor area
RTD	Resistance temperature detector
SAN	Sinoatrial node
SIMO	Slave-in, master-out
S-LIGA	Sacrificial LIGA
SQUID	Superconducting quantum interference device
SPI	Serial peripheral interface
TMR	Tunnelling magnetoresistance

1

Introduction

Most advanced biological and man-made physical systems require reliable sensing to function properly. The more sensors they integrate, the more complete and comprehensive is the information they can gather from their surroundings. For a long time, practical challenges have set limits to the number of sensors that can be embedded into physical systems, processes, or environments. Among these challenges are limited space, the difficulty and obtrusiveness of wiring, heat dissipation, and power supply. Miniaturisation of sensors has fundamentally changed the way we deal with these challenges. Furthermore, integrating processing and wireless communication capabilities into sensing systems has enabled not only dynamic programmability but also networking, so that data can be processed (aggregated, filtered, compressed) in a distributed manner or can be packed in packets and transferred to a different location where they can be processed by employing advanced signal-processing algorithms.

The past decade has witnessed an explosion of interest in wireless sensors and wireless sensor networks, for which there are a variety of applications. In civil engineering, these sensors and networks can be employed to monitor the integrity of infrastructure, such as pipelines, bridges, and buildings. In the medicine and healthcare domain, they have already proved to be indispensable, but they are also finding new applications in augmenting existing diagnosis and monitoring infrastructure and in enabling more independent and flexible lifestyles for patients. In agriculture and environmental science, wireless sensors and wireless sensor networks are useful for precision agriculture, for monitoring the quality, amount, and flow of water, and for studying wildlife without the need to interfere with it.

However, the usefulness of the applications that employ sensors depends on the depth of understanding pertaining to the design and operation of the sensors as well as the quality of the data-processing algorithms employed. The faith an application developer puts in a sensor should be based on a quantitative understanding of its reliability, accuracy, precision, sensing range, sensitivity, and lifetime as well as on the strength of the assumptions supporting the data-processing schemes. Otherwise, the relevance of the application will necessarily be limited to laboratory settings, or prototypes. On the other hand, a fundamental understanding of sensors and their design leads to innovative ideas and the identification of totally new application domains.

Interestingly, the basic concepts of sensors are straightforward to grasp and the electrical circuits required to realise a sensing system are relatively simple and comprehensible, for example, compared to the design of high-frequency communication systems. This is because, in most practical situations, sensors have to deal with low-frequency

Principles and Applications of Ubiquitous Sensing, First Edition. Waltenegus Dargie.
© 2017 John Wiley & Sons, Ltd. Published 2017 by John Wiley & Sons, Ltd.
Companion Website: www.wiley.com/go/dargie2017

signals that can be detected and processed by relatively simple electrical components. The purpose of this book is to acquaint the reader with:

- the fundamental principles of electrical, ultrasonic, optical, and magnetic sensing
- the broad range of issues pertaining to the design and manufacturing of microelectromechanical sensors (MEMS)
- the principles of energy harvesting and sensor integration
- the fundamental assumptions and methodologies pertaining to the processing of sensed data.

I have tried to make the book self-contained by discussing the necessary prerequisites within the book itself, so that the reader is not obliged to refer to other materials in order to understand the text.

1.1 System Overview

Figure 1.1 displays the most essential building blocks of a self-contained sensing system. These are the sensing system, the conditioning system, the processor, and the wireless transceiver. The figure is intended to give a complete picture, but we shall not be dealing with the wireless transceiver here. Whether or not these blocks are distinct from each other depends on many factors, such as the quality of the signal that can be sensed by the sensor, the targeted energy and space efficiency, and the ease with which the wireless sensor should be integrated into and interact with other systems. The processor and the wireless transceiver are usually connected with the rest of the system using standard buses that use standard or quasi-standard protocols. Hence, the main design issue is how to integrate the remaining building blocks. I shall briefly summarise the function of each building block and highlight the different trade-offs that influence its integration with the other blocks.

1.1.1 Sensing System

The process we wish to observe or monitor is called the measurand. It either releases some form of energy that describes its condition in some way, or external energy in the form of an electrical (radio-frequency), optical, or acoustic signal is applied to it, so that from the way it modifies some of the characteristics of the signal (magnitude, phase, frequency or a combination of these), its condition can be determined or inferred. The human body is a typical example of a measurand, because it is a remarkable signal generator. The human brain generates electromagnetic signals that can be sensed by electroencephalogram or magnetoencephalogram. Likewise, the human cardiovascular and muscular systems generate electric potentials that can be sensed by employing an electrocardiogram or electromyogram. In contrast, ultrasound systems release ultrasonic

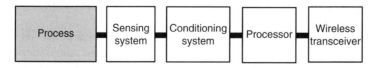

Figure 1.1 The main building blocks of a wireless sensor.

waves into a human organ and the spectrum of the reflected signal is analysed to determine the organ's condition. If a measurand's condition is determined from the signal it releases, then the sensing method is called "passive sensing". Otherwise, it is called active sensing. Passive sensing introduces less intrusion into the measurand compared to active sensing, but the amount of energy that can be collected through passive sensing is normally small.

1.1.2 Conditioning System

Regardless of the sensing mechanism employed, there are important conditions the sensor and the signal it produces should fulfil before useful features can be extracted from the signal. One of them is appropriate interfacing. When a sensor is employed to a measurand, apparently the measurand "perceives" the presence of the sensor, because the sensor draws some amount of power from it. This power must not affect the measurand's proper operation. A related issue to interfacing is impedance matching. If the impedance of the measurand as seen by the sensor is not matched by the sensor's own input impedance, maximum power does not transfer from the measurand to the sensor. Instead, power dissipation in the form of heat may be experienced at the interface, disturbing the measurand and reducing the efficiency of the sensor. Therefore, the impedance of the measurand (human body, water, air) should be taken into account when the sensor circuits are designed.

Even with the interfacing problem solved, the signal produced by the sensor may not accurately reflect the measurand's true condition for a number of reasons. Noise may be added to the sensed signal from the surrounding environment or from the internal circuits of the sensor itself. Likewise, some portion of the signal may be removed, suppressed, or distorted, because the sensor circuits act as filters. Therefore, the bandwidth of the desired signal and the bandwidth of the sensing circuits should be matched. It must be noted that in most real-world cases, the signal produced by a measurand contains a range of frequency components. The purpose of the conditioning system is to deal with all these issues. A conditioning component typically consists of a filter circuit and a differential amplifier, the order in which they appear usually depending on the nature of the measurand as well as the strength of the signal produced by the sensing system.

1.1.3 Analogue-to-digital Signal Conversion

This component is not directly shown in Figure 1.1 because it may be a part of the conditioning system or the processor or it may be a distinct entity. Regardless of its specific position, the analogue signal the sensor produces and the conditioning system pre-processes should first be converted to a digital bit stream before it can be further processed by a microcontroller or a digital signal processor (DSP). In some sensors, the analogue-to-digital converter (ADC) is an integral part of the conditioning system, while in others it is a separate block. Modern microcontrollers also integrate multiple general-purpose ADCs, to one of which the analogue signal coming from a conditioning circuit can directly be fed. Next to the transceiver and the processor, the ADC is the largest power consumer and hence care must be taken in choosing a suitable ADC. Several factors determine the choice of an ADC. For example, if the sensor signal is noisy, it is better not to use a powerful pre-amplifier lest the noise is amplified together with the useful signal. In this case it is better to use a high-resolution

ADC, so that an efficient DSP algorithm can eliminate the noise. But a high-resolution ADC consumes a large amount of power and generates a large amount of data, which require a sizeable resource to process, store, and communicate. If, on the other hand, there is a small noise component in the signal, then it is better first to amplify the signal and use a low-resolution ADC. If the ADC is not an integral component of the sensor or conditioning system, then it is possible to use the sensor for different applications which require different resolutions (accuracies), in which case separating the ADC from the conditioning stage enables the choice of suitable ADCs, independent of the sensing system.

1.1.4 Processor

The processor is a multi-purpose system, but as far as a wireless sensor is concerned, the level of data processing it can support is limited by factors such as available RAM, processor speed, battery capacity, the amount of heat dissipation that can be tolerated by the object or person, and the sensor's size. In wireless sensor networks and in wireless body-area networks, the processor is mainly responsible for low-level DSP (such as digital filtering and data compression) and for managing the various communication protocols which transfer the raw data to a nearby base station.

1.2 Example: A Wireless Electrocardiogram

A wireless electrocardiogram (ECG) measures the electric or action potentials that are generated in the heart and propagated through its electrical conduction system (a combination of nerve fibres, muscles, and tissues). These electrical potentials are responsible for creating and regulating the diastole–systole rhythms of the heart. Action potentials are produced at the sinoatrial (SA) node (located in the right atrium of the heart) and propagate through the atrial muscles to the atrioventricular (AV) node and further into the ventricular muscles of the heart through the His bundle, the left and the right bundle, and the Purkinje fibres (see Figure 1.2).

The propagating potential difference can be sensed by placing two or more electrodes on the skin at the right and the left sides of the heart (Figure 1.3). The magnitude of the pulses that can be picked up by the electrodes can reach up to 5 mV and their frequency varies between 0.05 Hz and 150 Hz. By analysing the shape, magnitude and the frequency of these pulses it is possible to determine several cardiac conditions.

Whilst the pulses themselves can be easily detected, the design of a safe and reliable electrocardiogram involves several components and DSP stages. Figure 1.4 displays the essential building blocks of an electrocardiogram. Between the electrodes and the rest of the sensing system there should be a protection mechanism to ensure that the system's operation does not interfere with the operation of the body. Both to prevent the ECG from overloading and interfering with the functions of the heart and to pick up as much voltage as possible, the ECG should have a high input impedance (because the body has a high output impedance, which has to be matched by the sensing system).

The electrodes capturing the action potentials of the heart also pick up electrical signals from their surroundings which have nothing to do with the action potentials of the heart and are therefore unwanted. The human skin itself produces a DC signal of up to

Figure 1.2 The generation and propagation of action potentials in the heart. (1) the sinoatrial (SA) node, (2) the atrioventricular (AV) node, (3) His bundle, (6 and 10) left and right bundle branches. Courtesy of J. Heuser (2007). Original image of the heart was by Patrick J. Lynch and C. Carl Jaffe, Yale University, Center for Advanced Instructional Media.

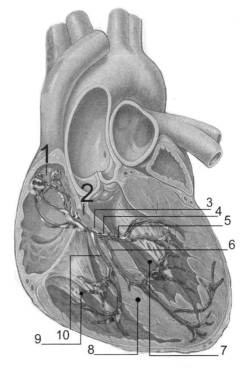

Figure 1.3 Two electrodes are used to measure cardiac action potentials.

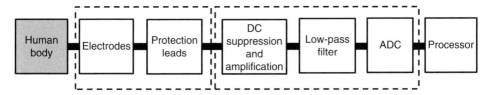

Figure 1.4 The essential building blocks of a wireless electrocardiogram.

300 mV, and other sources of noise include the power-line and the radio-frequency signals that are generated by nearby wireless and microwave devices as well as fluorescent lamps. Because of the small amplitude of the useful signal, it is not possible to separate all the unwanted signals from it right from the beginning. However, the pre-amplification and DC-suppression stages can remove the DC components by using relatively simple coupling capacitors. This same stage can amplify the rest to an appreciable level.

The pre-amplifier is typically a differential amplifier, the main purpose of which is to suppress all unwanted signals that have equal effects on all the electrodes. For example, noise that is generated by the power line will equally affect all the electrodes. Therefore, combining the outputs of the two electrodes in a differential mode (subtractive mode) suppresses the signals that are produced by the power line.

After the pre-amplification stage, an additional amplification is applied, followed by a low-pass filtering with a cut-off frequency of 150 Hz to remove all signals that have higher-frequency components. Then the analogue signal is converted to a digital signal and supplied to the processor. A DSP algorithm further processes the digital signal to improve the quality of the ECG measurements. For example, errors that can occur as a result of shaking or vibrating electrodes can be detected by a digital filter and corrected. In the case of a wireless ECG, the digital stream after the DSP stage can be packed into packets and transferred to a remote location where it can be further processed or displayed to a physician, who remotely monitors the patient.

Figure 1.5 displays a five-cord wireless ECG consisting of three distinct stages: the electrodes, the conditioning system, and the processor with a memory subsystem and a wireless transceiver. The memory subsystems enables data to be logged locally. Figure 1.6 highlights both the achievements and challenges of using a wireless ECG. Measurements were taken in our laboratory using a wireless ECG while a person was freely moving on a flat surface. Apparently, because of different movement-related artefacts, the measurements suffer from both long-term and short-term drift and signal distortions, which is why the various building blocks and signal processing algorithms are necessary. We return to this issue in subsequent chapters.

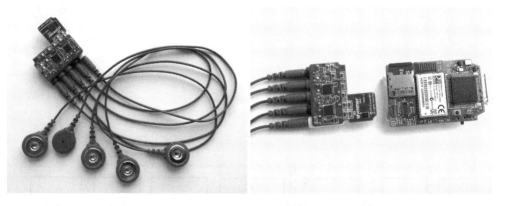

Figure 1.5 A wireless electrocardiogram with its processor and conditioning systems as well as electrodes as separate building blocks: Left, the electrodes and the conditioning system; right, the conditioning system and the processor.

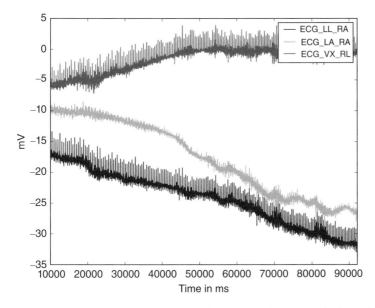

Figure 1.6 ECG measurements using the Shimmer wireless sensor platform while a person was freely moving on a flat surface. The measurements were taken from three different places: top trace, between the left leg and right arm; middle trace, left arm and right arm; bottom trace, neutral and right leg.

1.3 Organisation of the Book

The book is organised into several logical components. Chapter 2 provides an overview of emerging applications in the ubiquitous computing, wireless body-area network, and wireless sensor network communities. The typical features of these applications are the comprehensiveness of the sensing task, the intelligence and self-organising features embedded in the sensing systems, and the novelty of the applications themselves. The specific applications are selected to highlight the diversity of sensing techniques and the issues involved, or rather the challenges surrounding the design, deployment, and signal processing aspects of ubiquitous sensing.

Chapter 3 provides a brief introduction to signal conditioning and addresses the most essential aspects. Chapters 4–7 introduce the essential aspects of electrical, ultrasonic, optical, and magnetic sensing. These chapters cover essential aspects of most important regions of the wide spectrum of sensing.

Chapter 8 presents the most important aspects of medical sensing. I decided to give a separate treatment to this subject because of the growing number of medical applications in the communities listed at the beginning of this subsection.

Chapter 9 provides a comprehensive insight into the design and manufacturing of microelectromechanical sensors and demonstrates how the various sensing mechanisms (electrical, optical, magnetic, and so on) can be employed to develop practical sensors such as inertial, pressure, and fluid sensors.

Chapter 10 addresses an important issue in sensor deployment, namely energy harvesting. It describes the need for and the advantages of energy harvesting, discusses the choice of suitable sensing mechanisms, proposes a conceptual architecture, and presents various prototypes, highlighting their merits and demerits.

Chapter 11 addresses practical issues surrounding sensor integration. In most practical cases, a sensor is a part of a more complex system, the operation of which, unfortunately, may produce undesirable effects on the quality of sensing, such as radiation and thermal noise. The chapter recommends several integration strategies.

Finally, the last chapter, Chapter 12, addresses the data processing aspects of sensing, the main objective of which is minimising uncertainty. The chapter describes how the outputs of sensors can be regarded as random variables and discusses the different evidence-combining techniques used to reduce sensing error. To make the subject both interesting and useful, I give several examples and endeavour to take the reader step by step into the different stages of estimation.

2

Applications

Sensors ubiquitously surround human existence. Deeply embedded into the fabrics of everyday life in systems – consumer electronics, civil and medical infrastructure, cars, homes, business and office buildings, and smart phones – they gather a vast amount of useful information that can be used for various purposes. Providing an exhaustive review of the applications of sensors is beyond the scope of this chapter. Instead, it focuses on a handful of emerging applications that use ubiquitous (wireless) sensing and can be considered representative of both the opportunities and challenges associated with ubiquitous sensing. These applications are:

- civil infrastructure monitoring
- medical diagnosis and monitoring
- water-quality monitoring.

The idea is to illustrate the diversity of sensing, the novelty of the applications, and the various design issues and challenges that have to be dealt with when designing, developing, and deploying sensors.

2.1 Civil Infrastructure Monitoring

In civil infrastructure monitoring, accelerometers, gyroscopes, tilt sensors, and piezoresistive sensors have been widely employed to measure the response of bridges and buildings to ambient and forced excitations. Ambient excitations are produced by natural causes such as earthquakes, wind, or the movement of cars, whereas forced excitations are intentionally produced by impact hammers and shakers or by placing bumpers on bridges to determine the response of a structure. Tilt sensors, gyroscopes, and accelerometers can sense two- or three-dimensional angular accelerations of a structure whilst piezoresistive sensors can sense linear displacements. Likewise, in water, gas, and oil pipelines, pressure and acoustic sensors are employed to detect the magnitude, direction, and speed of pressure transients caused by bursts, leakages, or valve operations in pipelines. In a pipeline, when the flow of a fluid or gas is suddenly interrupted or modified due to a change in the structure of the pipe or a valve operation, a region of high pressure builds up immediately behind the direction of the flow and a region of low pressure in front of it, as a result of which the momentum of flow is suddenly transferred forward according to Newton's third law. The build up of high pressure propagates in the backward direction along the pipe in the form of a wave, typically travelling at a speed of $500 - 1400\,\mathrm{ms}^{-1}$, depending on the material and the wall thickness of the pipe.

Principles and Applications of Ubiquitous Sensing, First Edition. Waltenegus Dargie.
© 2017 John Wiley & Sons, Ltd. Published 2017 by John Wiley & Sons, Ltd.
Companion Website: www.wiley.com/go/dargie2017

Figure 2.1 The Golden Gate Bridge in San Francisco, USA.

2.1.1 Bridges and Buildings

The response of bridges and buildings to external excitations is usually known (or approximated) at the time of construction. These structures experience three-dimensional oscillations even when no force is applied to them. These types of oscillations are known as "free" or "natural" responses. When a force of known magnitude and duration is applied to them, the structures respond with predictable oscillations when they are healthy. A departure from the predictable oscillation indicates that some of the components of the structure are defective. Both the free and forced responses of a structure can be estimated by using single and multiple degree of freedom (DOF) models.

Let us examine the response of the bridge shown in Figure 2.1[1] using a single degree of freedom model. In this model, when an excitation is applied to a structure that has a

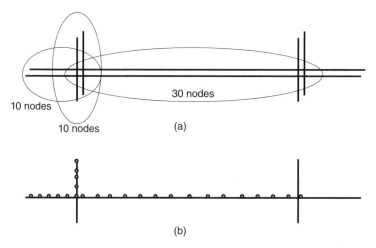

Figure 2.2 A wireless sensor network deployed on the Golden Gate Bridge to monitor its response to ambient and forced excitations: (a) nodes deployed on the deck and tower of the bridge; (b) the actual placement of the nodes on the bridge structure (Dargie and Poellabauer 2010).

---

1 Figure 2.2 shows how a wireless sensor network was deployed on the Golden Gate Bridge in San Francisco to monitor its response to ambient and forced excitations (Dargie and Poellabauer 2010; Kim et al. 2007).

Figure 2.3 Representation of a structure using a single degree of freedom model.

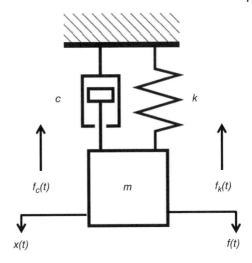

mass of m, it undergoes a time-varying oscillation, $x(t)$. Therefore, the structure can be modelled as a harmonic oscillator (spring) with a spring or stiffness constant of k. Due to the motion of the mechanical components of the structure and the resulting friction, some of the mechanical force will be converted to heat. The effect is that the oscillation decreases gradually. This phenomenon is called "damping" and it is a function of the velocity, which is the derivation with respect to time of the oscillation (displacement), $\dot{x}(t)$, and the damping coefficient, c, of the structure. The forces exerted on the structure can be illustrated by using Figure 2.3. Mathematically,

$$m\ddot{x}(t) + c\dot{x}(t) + kx(t) = 0, \quad x(0) = d_0, \quad \dot{x}(0) = v_0 \tag{2.1}$$

where $\ddot{x}(t)$ is the acceleration of the oscillation, and d_0 and v_0 refer to the initial displacement and initial velocity of the structure, respectively. You may recall that Eq. (2.1) is a second order differential equation. In general, the oscillation of the bridge can be expressed using Fourier series as a combination of infinite sinusoidal functions with infinite harmonic components which are properly weighted:

$$x(t) = \sum_{n=-\infty}^{\infty} a_n \cos(\omega_n t) + jb_n \sin(\omega_n t) \tag{2.2}$$

where $j = \sqrt{-1}$ is the imaginary unit[2] signifying that the two components, namely, $\cos(\omega t)$ and $\sin(\omega t)$, are orthogonal. In other words, if we integrate the inner product of these two components, the result will be zero. Assuming that the oscillation of the bridge can be approximated by just two sinusoidal components:

$$x(t) = a \cos(\omega t) + jb \sin(\omega t) \tag{2.3}$$

Alternatively, using Euler's formula:

$$\cos(\omega t) + j \sin(\omega t) = e^{j\omega t}$$

---

2 In mathematics and physics the imaginary unit is represent by the letter i, but in electrical engineering, the letter i is reserved for AC current, which is why the imaginary unit is represented by j.

and

$$\cos(\omega t) - j\sin(\omega t) = e^{-j\omega t}$$

we can rewrite Eq. (2.3) as follows:

$$x(t) = Ae^{j\omega t} + A^* e^{-j\omega t} \tag{2.4}$$

where $A = (a + jb)/2$ and A^* is the complex conjugate of A.[3] If the bridge oscillates with a constant magnitude and frequency (Figure 2.4, left), Eq. (2.4) is sufficient to describe $x(t)$, but due to the damping effect of the bridge, the amplitude of the oscillation usually decreases over time (Figure 2.4, right). So, Eq. (2.4) should be modified to:

$$\begin{aligned}
x(t) &= e^{-\sigma t}(Ae^{j\omega t} + A^* e^{-j\omega t}) \\
&= Ae^{-(\sigma - j\omega t)} + A^* e^{-(\sigma + j\omega t)} \\
&= Ae^{-\lambda t} + A^* e^{-\lambda^* t}
\end{aligned} \tag{2.5}$$

where $\lambda = \sigma - j\omega$ and $\lambda^* = \sigma + j\omega$. For the natural or free response, σ and ω are determined from m, c, and k while the constant a equals the initial displacement d_o and b depends on the initial displacement, the velocity, and the mass of the structure, as well as the damping and the stiffness coefficients. We shall demonstrate this by inserting Eq. (2.5) into Eq. (2.1) and by setting $f(t) = 0$ (considering the unforced response of the bridge):

$$\begin{aligned}
m\lambda^2 Ae^{\lambda t} + c\lambda Ae^{\lambda t} + kAe^{\lambda t} &= 0 \\
(m\lambda^2 + c\lambda + k)Ae^{\lambda t} &= 0
\end{aligned} \tag{2.6}$$

In Eq. (2.6), all quantities are independent of time except the exponential term. The equation can be satisfied either by setting $X = 0$ or by setting $m\lambda^2 + c\lambda + k = 0$. We are

---

3 Using Euler's formula, we can express

$$\cos(\omega t) = \frac{e^{j\omega t} + e^{-j\omega t}}{2}, \sin(\omega t) = \frac{e^{j\omega t} - e^{-j\omega t}}{j2}$$

Letting $X = a + jb$ and $X^* = a - jb$, it is possible to rewrite Eq. (2.3):

$$\begin{aligned}
x(t) &= a\cos(\omega t) + b\sin(\omega t) \\
&= \frac{X + X^*}{2}\left[\frac{e^{j\omega t} + e^{-j\omega t}}{2}\right] + \frac{X - X^*}{2j}\left[\frac{e^{j\omega t} - e^{-j\omega t}}{j2}\right] \\
&= \frac{1}{4}[Xe^{j\omega t} + Xe^{-j\omega t} + X^* e^{j\omega t} + X^* e^{-j\omega t}] - \\
&\quad \frac{1}{4}[X^* e^{j\omega t} - Xe^{-j\omega t} + X^* e^{j\omega t} - X^* e^{-j\omega t}] \\
&= \frac{X}{2}e^{j\omega t} + \frac{X^*}{2}e^{-j\omega t}
\end{aligned}$$

Letting $A = \frac{X}{2}, A^* = \frac{X^*}{2}$, then,

$$x(t) = Ae^{j\omega t} + A^* e^{-j\omega t}$$

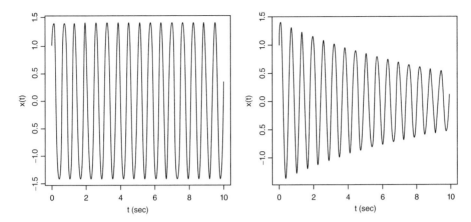

Figure 2.4 The response of a structure to an excitation: left, undamped oscillation; right, damped oscillation.

interested in the second term. As can be seen, this term is a quadratic equation and has two solutions:

$$\lambda_{1,2} = -\frac{c}{2m} \pm \sqrt{\left(\frac{c}{2m}\right)^2 - \frac{k}{m}} \tag{2.7}$$

The solution to a homogeneous second-order differential equation requires two independent initial conditions: the initial displacement and the initial velocity. These two initial conditions are used to determine the coefficients A and A^* by the two linearly independent solutions corresponding to λ_1 and λ_2. Suppose that the bridge oscillates forever without being damped. In this case $c = 0$. Hence Eq. (2.7) reduces to:

$$\lambda_{1,2} = \pm j\sqrt{\frac{k}{m}} = \pm j\omega_n \tag{2.8}$$

where ω_n is the natural frequency of the bridge. In other words, the bridge oscillates freely at this frequency and the displacement it undergoes in time can be expressed as:

$$x(t) = Ae^{j\omega_n t} + A^* e^{-j\omega_n t} = a\cos(\omega_n t) + b\sin(\omega_n t) \tag{2.9}$$

If the magnitude of oscillation gradually reduces in time, then c is no longer zero. If $\left(\frac{c}{2m}\right)^2 = \frac{k}{m}$ (equivalently, if $c = 2\sqrt{mk}$), then the square-root term of Eq. (2.7) is zero. In this case, the bridge's oscillation is termed "critically damped" and the damping coefficient is labelled as $c_c = 2\sqrt{mk}$. The solutions of the quadratic equation in Eq. (2.7) are:

$$\lambda_1 = \lambda_2 = \frac{-c}{2m} = -2\sqrt{\frac{mk}{2m}} = -\omega_n \tag{2.10}$$

In order to admit arbitrary initial displacement and velocity, the solution to the differential equation should be described as:

$$x(t) = x_1 e^{-\omega_n t} + x_2 t e^{-\omega_n t} \tag{2.11}$$

where x_1 and x_2 are determined from the initial displacement, d_0, and the initial velocity, v_0. Often, Eq. (2.7) is expressed in terms of the ratio of the three coefficients that characterise the quality of the bridge, namely its mass, m, and the spring and the damping coefficients, k and c. Let $\zeta = \frac{c}{2\sqrt{km}}$ and

$$\frac{c}{m} = c\frac{\sqrt{k}}{\sqrt{k}}\frac{1}{\sqrt{m}\sqrt{m}} = 2\frac{c}{2\sqrt{km}}\sqrt{\frac{k}{m}} = 2\zeta\omega_n \tag{2.12}$$

Now we can rewrite Eq. (2.1) as follows:

$$\frac{1}{m}f(t) = \ddot{x}(t) + \frac{c}{m}\dot{x}(t) + \frac{k}{m}x(t)$$
$$= \ddot{x}(t) + 2\zeta\omega_n\dot{x}(t) + \omega_n^2 x(t) \tag{2.13}$$

Likewise, the two roots of Eq. (2.7) can be expressed as:

$$\lambda_{1,2} = -\zeta\omega_n \pm \omega_n\sqrt{\zeta^2 - 1} \tag{2.14}$$

So far, we have considered the free response of the bridge without any external excitation applied to it. Now suppose we apply an external excitation $f(t)$ and that we wish to determine how it responds. To ease our analysis, suppose the excitation force is a sinusoidal force with frequency ω, $f(t) = F\cos(\omega t)$. Clearly, if $f(t)$ is persistent, then the bridge will oscillate with the frequency of the excitation force. In this case, we can express Eq. (2.1) as follows:

$$F\cos(\omega t) = m\omega^2(-a\cos(\omega t) - b\sin(\omega t))$$
$$c\omega(-a\sin(\omega t) + b\cos(\omega t)) \tag{2.15}$$
$$k(a\cos(\omega t) + b\sin(\omega t))$$

Equating the sine and the cosine terms will result in two separate equations:

$$(-m\omega^2 a + c\omega b + ka)\cos(\omega t) = F\cos(\omega t) \tag{2.16}$$
$$(-m\omega^2 b + c\omega a + ka)\sin(\omega t) = 0 \tag{2.17}$$

Or, in matrix form, we have:

$$\begin{bmatrix} k - m\omega^2 & c\omega \\ -c\omega & k - m\omega^2 \end{bmatrix}\begin{bmatrix} a \\ b \end{bmatrix} = \begin{bmatrix} F \\ 0 \end{bmatrix} \tag{2.18}$$

Subsequently, using a matrix manipulation, it is possible to determine a and b as functions of ω:

$$a(\omega) = \frac{c\omega}{(k - m\omega^2)^2 + (c\omega)^2}F \tag{2.19}$$

$$b(\omega) = \frac{(k - m\omega^2)^2}{(k - m\omega^2)^2 + (c\omega)^2}F \tag{2.20}$$

Equations (2.7), (2.19), and (2.20) describe the bridge's natural and forced responses. Each of the quantities are known at the time of construction and can be directly measured by employing sensors that capture displacement, acceleration, force, or pressure.

The single degree of freedom model is simple to analyse and comprehend, but it does not accurately represent the characteristics of a complex structure such as a bridge or a high building. In practice, a multi degree of freedom model (MDOF) is often used. A 2MDOF model is displayed in Figure 2.5, in which an excitation force applied to one

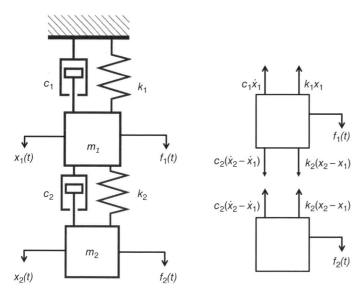

Figure 2.5 Representation of a structure using a two degree of freedom model.

part of a structure (for example, a deck) induces another force on another part of the structure (for example, a suspension cable or a tower). The equation of motion for such a system can be described as follows:

$$f_1(t) = m_1\ddot{x}_1(t) + (c_1 + c_2)\dot{x}_1(t) - c_2\dot{x}_2(t) + (k_1 + k_2)x_1(t) - k_2x_2(t) \tag{2.21}$$

$$f_2(t) = m_2\ddot{x}_2(t) + c_2\dot{x}_2 - c_2\dot{x}_1 + k_2x_1 - k_2x_1 \tag{2.22}$$

Or, in matrix form,

$$\begin{bmatrix} f_1(t) \\ f_2(t) \end{bmatrix} = \begin{bmatrix} m_1 & 0 \\ 0 & m_2 \end{bmatrix} \begin{bmatrix} \ddot{x}_1(t) \\ \ddot{x}_2(t) \end{bmatrix} + \begin{bmatrix} c_1 + c_2 & -c_2 \\ -c_2 & c_2 \end{bmatrix} \begin{bmatrix} \dot{x}_1(t) \\ \dot{x}_2(t) \end{bmatrix}$$

$$+ \begin{bmatrix} k_1 + k_2 & -k_2 \\ -k_2 & k_2 \end{bmatrix} \begin{bmatrix} x_1(t) \\ x_2(t) \end{bmatrix} \tag{2.23}$$

Unlike the force we considered in our analysis – $f(t) = F\cos(\omega t)$ – most real-world excitations have multiple frequency components. For example, a force produced by an impact hammer or a bumper can be approximated by the Dirac delta function:

$$\delta(t) = \begin{cases} 1 & x = 0 \\ 0 & \text{otherwise} \end{cases} \tag{2.24}$$

In the time domain, $\delta(t)$ has an infinite magnitude for a brief duration and then it disappears. In the frequency domain, this type of function has infinite frequency components. Fortunately, a healthy structure responds only to a selected number of the frequency components. Once again, the magnitude of oscillation at these frequencies can be estimated. Some of the most frequently employed techniques for identifying defects in a structure examine shifts in the frequency response of the structure and use techniques such as "peak picking" and "frequency domain decomposition". For further information, refer to Brincker et al. (2001); Cho et al. (2008); Hackmann et al. (2014); Mascarenas et al. (2007); Zimmerman et al. (2008).

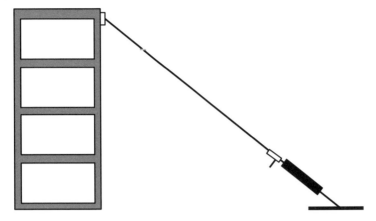

Figure 2.6 The horizontal oscillation (sway) of a long building can be measured by using a horizontally oriented single-dimensional accelerometer sensor. Alternatively, a piezoresistive sensor can be attached to the foot of the supporting cable to measure the linear horizontal displacement of the building.

Example 2.1 We wish to sense the horizontal natural sway (oscillation) of the building shown in Figure 2.6 using a single-dimensional accelerometer. Assuming that the fundamental oscillation frequency of the building is 10 Hz, what should be the minimum sampling frequency of the accelerometer if we are interested in up to the third harmonic frequency?

Regardless of the nature of the signal that can be sensed by the accelerometer, if the signal exhibits periodicity, it is possible to describe it using an infinite combination of properly weighted sinusoidal signals, the frequency of which is an integer multiple of the fundamental frequency. Moreover, because of the existence of periodicity, it suffices to consider the amplitude variation of a single period. Figure 2.7 illustrates the use of Fourier series to describe an arbitrary signal with simple sinusoidal functions. Hence, if the fundamental frequency is 10 Hz, then the third harmonic will be at 30 Hz. The relationship between the sampling frequency and the signal's frequency is illustrated in Figure 2.8. If, for example, the accelerometer output is sampled at 30 Hz, then what we obtain is a constant set of values that does not represent the true characteristic of the third harmonic (we shall have enough information about the fundamental frequency and the second harmonic, nevertheless). If the accelerometer output is sampled at twice the frequency of the third harmonic, then the output more or less resembles the third harmonic, but still there is a loss of information. As the sampling rate increases, then the digital (sampled) signal becomes more and more representative of the actual third harmonic, as can be seen in the figure. In reality, to deal with noise and interference that distort the signal considerably, the accelerometer should have to be sampled at a much higher rate than the Nyquist frequency, which is twice the frequency of the desired signal: for us the third harmonic. Readers wishing to obtain a more comprehensive knowledge about the signal processing aspect of multi-dimensional accelerometers can refer to the literature (Dargie 2009; Dargie and Denko 2010).

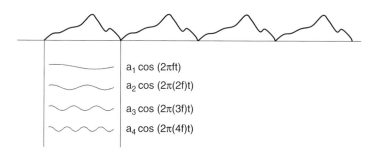

Figure 2.7 Describing a periodic signal by a combination of an infinite number of simple sinusoidal signals, the frequency of which is a linear multiplication of the fundamental frequency. Here, only the cosine terms are shown.

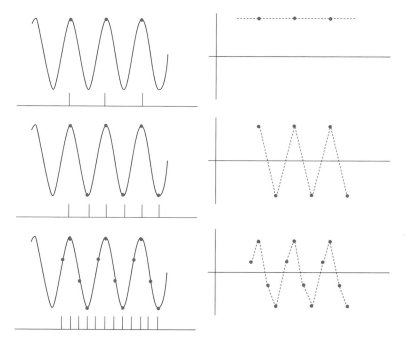

Figure 2.8 The sampling frequency in converting an analogue signal to a digital signal is related to the frequency of the signal. According to Nyquist, the minimum sampling frequency should be greater than or equal to twice the frequency of the signal. Top: the signal frequency and the sampling frequency are equal. Middle: the sampling frequency is twice the signal frequency. Bottom: the sampling frequency is four times the signal frequency.

2.1.2 Water Pipelines

Water pipelines critically affect living and work conditions around the world and considerably influence the quality, amount, and cost of water they transport. For example, according to Whittle et al. (2013), in May 2010, a huge water main in Boston ruptured, depriving two million residents of the greater Boston area of drinking water, and thereby causing a state of emergency. Since the backup water did not meet the US Environmental Protection Agency's standards, the public was forced to survive without a supply of drinking water for several days. Similarly, during the summer of 2009, the city of Los Angeles experienced a series of pipe breaks and leaks, which caused a significant disruption of water supply.

Due to the inherent dynamics of water transportation and several environmental and operational aspects, pipelines are often subject to leaks and bursts. According to the American Society of Civil Engineering (2013) the water infrastructure across the US is in a critical condition. A recent report from the US Environmental Protection Agency (2016) also indicates that more than 200,000 pipes break and 7000 km of pipeline requires replacement every year and, as a result, the country loses 10% of its aggregate water supply each year. The UK Office of Water Services (2016) classifies the condition of the existing water pipeline infrastructure in the UK as critical too. According to Lin et al. (2008), a BBC report revealed that 10,500 litres of water was lost every second in 2006 on account of Thames Water's leaky pipes. Citing a survey from the German Association of Energy and Water Industries, Whittle et al. (2013) indicate that losses in other European countries are comparable, with water losses in France and Italy reaching up to 30%.

Damage to pipelines typically occurs as a result of the cumulative effects of corrosion at the joints of pipes, structural fatigue, and sudden surges in water pressure. The speed at which damage is detected and localised is of paramount importance because they are associated with large water and energy losses, service disruption, and traffic delays. Conventionally, structural damage is detected using acoustic and pressure sensors during manual inspections. The sensors are placed at two joints or valves to pick up the sounds and vibrations induced by water as it escapes from pressurised pipes. In some cases, leak-noise correlators are also integrated with the acoustic sensors. Bursts in pipes cause a sudden change in the flow of water and produce pressure transients that propagate in the water along the pipeline. The transients travel away from the burst origin in both directions at the speed of sound. The position of a leak or burst is determined by analysing the cross-correlation and time shift of the signals captured by the acoustic and pressure sensors. Even though the accuracy of existing technologies is appreciable, they require a significant amount of manpower, making damage detection a tedious and time-consuming task. As a result, eliminating leakage is perhaps impossible and certainly expensive with these technologies alone.

Whittle et al. (2013) propose a distributed wireless system for monitoring the integrity of water pipelines and water quality. The system consists of a wireless sensor network, an event-detection and alarm system, and a decision-support system. The nodes that make up the wireless sensor network are reconfigurable and each of them integrates a host of sensors that can measure pressure, water flow, water temperature, water pH, water conductivity, and oxidation redox potential. Furthermore, each node integrates a GPS receiver, a 3G modem, and a microcontroller, among other things (see Figure 2.9). Each sensor node is battery powered but during the daytime it can harvest energy using a solar panel, which can be attached to a nearby lamppost. The hydraulic and water-quality sensors are mounted on an integrated multi-probe that is inserted into the water pipe via a gate valve.

Individual sensors are sampled at different intervals, in accordance with the expected rates of change of the quantities they monitor. For example, the hydrophone sensor is used to detect background leakage and is sampled at kilohertz rates, but only for a few seconds every hour. The pressure sensor, on the other hand, is continuously sampled

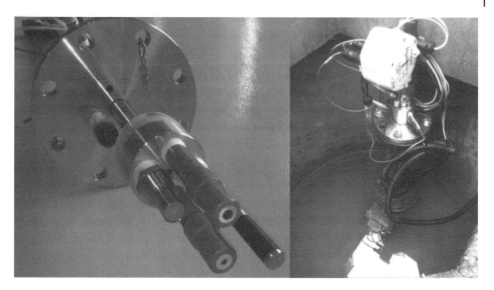

Figure 2.9 A wireless sensor node integrating several sensors to monitor the structural integrity of a water pipeline (left) and the deployment of a wireless sensor node at a gate valve (right).

at hundreds of hertz. The sensor data are either transmitted in 30 s windows or first processed locally using an event detection algorithm, which is an integral part of the alarm system. The sensors pertaining to water quality, flow, and battery life are typically sampled at 30 s intervals and transmitted every 30 min. The wireless sensor nodes can buffer several days of data in the event of a network outage and prioritise lower-rate data when network connections have low bandwidth.

The event-detection and alarm system is responsible for streaming data from the sensor network to a remote server where advanced data processing takes place. It integrates a set of algorithms that are applied to data streams in order to detect abnormal events and provide location estimates. In order to achieve scalability, abnormal events are detected in the data stream of each sensor node and are grouped together and used to estimate the location of the abnormal event. The system is also responsible for notifying subscribers about important events.

The decision-support system uses the sensor data streams provided by the alarm system to estimate water demand and supply in the entire system. The central element of this system is a real-time hydraulic model of the water distribution. The most recent pressure and flow data from sensor nodes are used, along with a baseline demand pattern and seasonal information to estimate the consumption for each demand zone in the network.

Example 2.2 Two ultrasonic sensors are placed at two known positions on a pipeline as shown in Figure 2.10. Suppose a leakage at position x results in the creation of a pressure transient at time t_0, which then propagates in both directions along the pipeline at the speed of sound. The first sensor detects the event at time t_1 and the second at time t_2. When they detect the event, both sensors notify the base station, which computes the exact location of damage. How can the base station determine this position?

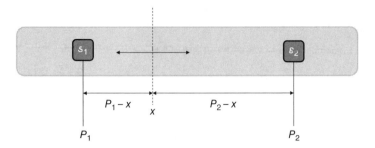

Figure 2.10 Two ultrasonic sensors which are placed in two different but known position can be used to determine the exact location of a damage in a pipeline.

By the time the pressure transient arrives at the first sensor, it has travelled for $t_1 - t_0$ seconds. Similarly, by the time it arrives at the second sensor, it has travelled for $t_2 - t_0$ seconds. The distances travelled can be expressed as:

$$(P_1 - x) = (t_1 - t_0) V$$
$$(P_2 - x) = (t_2 - t_0) V$$

where P_1 and P_2 are the positions of the first and the second sensor, respectively; t_1 and t_2, respectively, are the times the first and the second sensor detect the event; and V is the speed of sound. Since we have two equations for two unknown variables (x and t_0), we can determine both quantities. First, as the velocity of propagation is the same in both directions, we can express it as follows:

$$\frac{P_1 - x}{t_1 - t_0} = \frac{P_2 - x}{t_2 - t_0}$$

Simplifying and rearranging terms will yield:

$$P_1 t_2 - P_2 t_1 - (P_1 - P_2) t_0 = x (t_2 - t_1)$$

or

$$x = \frac{P_1 t_2 - P_2 t_1 - (P_1 - P_2) t_0}{(t_2 - t_1)}$$

Notice that, except t_0, all the other terms are known, so we can replace them by different constants:

$$x = \frac{C_1 - C_2 t_0}{C_3}$$

If we insert the above expression in one of the previous expressions (for example, in $(P_1 - x) = (t_1 - t_0) V$, we shall solve for t_0:

$$t_0 = \frac{C_3 P_1 - C_1 - C3 t_1 V}{C_2 - V}$$

With this,

$$x = \frac{C_1}{C_3} - \frac{C_2}{C_3} \left(\frac{C_3 P_1 - C_1 - C_3 t_1 V}{C_2 - V} \right)$$

Example 2.3 Example 2.2 implicitly assumes that some essential prerequisites or aspects of distributed sensing are fulfilled in order to determine the location and timing of the pressure transient. Can you list some of them?

The following aspects have been assumed to have already been addressed:

1) The two sensors are supposed to be time synchronised, otherwise, the time "perceived" by one of the sensors may differ from the time perceived by the other. Similarly, the time perceived by both sensors may differ from the time perceived by the end device, which computes x and t_0.
2) The physical properties of the pipeline are assumed to be homogeneous, so that v can be considered the same all over.
3) The wireless sensor nodes should have a mechanism or a local signal-processing sub-system to detect an event (the pressure transient), a timer to associate time with the event, and a communication subsystem to transmit the event and the timestamps to the base station.
4) A wireless communication between the two sensors and the base station simplifies the deployment and monitoring cost; a pipeline that is completely wired is both costly and difficult to maintain.

2.2 Medical Diagnosis and Monitoring

In the healthcare domain, wireless sensor networks have found several applications, including monitoring

- people with Parkinson's disease, epilepsy, asthma, and arthritis
- patients rehabilitating after hip and knee surgery (arthroplasty)
- elderly people, enabling them to lead an independent life.

Most of these applications are intended for use in a home setting. As a result we now have

- wireless pulse oximeters for measuring blood flow and heart rate (Li and Warren 2012; Mendelson et al. 2006; Watthanawisuth et al. 2010)
- a combination of oximeter and accelerometer for measuring physical activities and their impact on the heart (typically, overexertion in cardiac patients) (Buttussi and Chittaro 2010)
- a combination of accelerometer, gyroscope, and magnetic sensors for measuring joint flexion and extension as well as muscle tremor (Dejnabadi et al. 2005; Favre et al. 2008; O'Donovan et al. 2007; Takeda et al. 2009)
- wireless electrocardiograph (ECG) (Burns et al. 2010; Hopman et al. 2003, 2007; Istvan et al. 2007).

2.2.1 Parkinson's Disease

Parkinson's disease (PD) is a result of the deficiency of dopamine due to the degeneration of neurons in the substantia nigra pars compacta (a region of the midbrain; see Figure 2.11). In its early stages, the disease manifests itself through characteristic motor

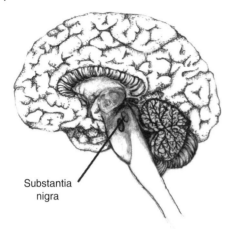

Figure 2.11 Dopamine receptors at the substantia nigra.

Substantia
nigra

features such as tremor, slowness of movements (bradykinesia), rigidity, and impaired postural balance. These features can be diagnosed by performing a set of standardised motor tasks specified by the Unified Parkinson's Disease Rating Scale. These include finger-to-nose movement, finger tapping, repeated hand movement, heel tapping, quiet sitting, and alternate hand movements. The treatment of the disease primarily entails augmentation or replacement of dopamine using bio-synthetic precursors that activate dopamine receptors.

Wireless accelerometer sensors can be employed both during the diagnosis and the treatment phases of PD. The sensors are typically placed on five different locations in the body: torso, ankles, and wrists (Ko et al. 2010). The aim is to correlate, cross-correlate, and compare the measurements from these locations with each other and with expected results. Once the disease is diagnosed, patients are prescribed with medication but the dosage and frequency are different from patient to patient. One of the challenges with most PD medication processes is that they are initially successful but patients eventually develop motor complications such as wearing off, the abrupt loss of efficacy at the end of each dosing interval, and dyskinesias (an involuntary and sometimes violent writhing) (Ko et al. 2010; Lorincz et al. 2009). The main objective of wireless sensing during treatment is identifying the occurrence of dyskinesia and its magnitude in patients who are freely leading their everyday lives, so that the appropriate medication and dosage can be estimated.

Researchers at different institutions have employed wireless accelerometers to diagnose and monitor PD. At the Memorial Hospital's Parkinson Day Center in Cambridge, Massachusetts, Weaver (2003) employs wireless 3D accelerometer sensors to monitor the presence and magnitude of dyskinesia during treatment. The sensors were sampled at 40 Hz whilst the patient was engaged in everyday activities. The sensor data were supplied to a neural network that classified the patient's state, while a doctor physically observed both the patient and the outcome of the neural network. The state of dyskinesia was correctly recognised 85% of the time by the system. With the inclusion of another neural network to declassify rapid movements that had no correlation with dyskinesia, the system improved its estimation accuracy to 91%.

Similarly, Patel et al. (2009) carried out a more comprehensive study by deploying uniaxial accelerometers on different parts of PD patients' bodies. The chief aim was to diagnose the symptoms pertaining to PD. They extracted time-domain features (rms values and data range) and frequency-domain features (dominant frequency and spectral-energy ratio) from the measurements of the accelerometer sensors. The authors report that their approach was able to recognise tremor with 3.4% estimation error, bradykinesia with 2.2% estimation error, and dyskinesia with 3.2% estimation error in real patients.

Example 2.4   One of the most important techniques in analysing movement-related data is to examine how the movement develops in time. Often there are quantitative relationships between precursors leading to the onset of an important event, the event itself (tremor, for instance having a specific duration), and the disappearance or decaying of the event. The relationships between these features can be established by examining the cross-correlation of the time series data. Show how this can be done in practice.

Figure 2.12 shows a plot of samples taken from a single-dimension accelerometer, which was attached to the elbow of a person. The accelerometer was sampled at 1 kHz rate for 1 s. One way the time-domain features are analysed to determine the significance of the movement is the cross-correlation. In order to obtain a cross-correlation, the following important steps are required:

1) Segmentation : In this step, the time series is divided into multiple, equal segments. The size of a segment is application specific. If the movement is a fast movement,

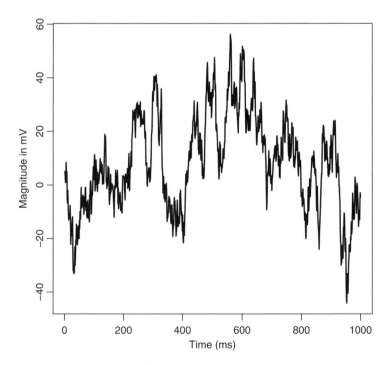

Figure 2.12 A time series of 1000 samples taken from a single-dimensional accelerometer sampled at a rate of 1 kHz.

the segments should be short, but this means the associated data-processing cost is large; if, on the other hand, the movement is slow, the size of the segment can be large. Another factor that determines the size of a segment is the duration of the interesting event or symptom we wish to capture. Symptoms that are by nature evanescent require short segments.

2) Overlapping segments: Overlapping segmentation provides useful insights into the relationship between past, present, and future contexts. Once again, the level of overlapping between segments depends on many factors and requires knowledge of the application. The larger the overlapping, the more correlated are the segments, but this also means more resource is required to process them.

3) Correlation and covariance matrices: Once the segments are produced, it is possible to compute vital statistics such as autocorrelation, cross-correlation, and covariance.

Figure 2.13 illustrates the segmentation strategy with two different overlapping regions. For the example in Figure 2.14, a segment size of 100 samples and 20% segment overlap was used. Since we have 100 samples in each segment, we have sufficient statistics to establish the correlation and covariance between the segments. For example, the covariance between any of the two segments can be computed as:

$$\text{Cov}(S^I, S^J) = \frac{1}{N^2} = \sum_{i=1}^{N} \sum_{j=1} (S_i^I - S_j^I)(S_i^J - S_j^J) \tag{2.25}$$

where S^I, S^J are two arbitrary segments and i and j refer to the individual samples in each segment. If we have n segments, then we have an $n \times n$ covariance matrix:

$$\text{Cov}(S^I, S^J) = \begin{bmatrix} \text{Cov}(S^1, S^1) & \text{Cov}(S^1, S^2) & \cdots & \text{Cov}(S^1, S^n) \\ \text{Cov}(S^2, S^1) & \text{Cov}(S^2, S^2) & \cdots & \text{Cov}(S^2, S^n) \\ \vdots & & & \\ \text{Cov}(S^n, S^1) & \text{Cov}(S^n, S^2) & \cdots & \text{Cov}(S^n, S^n) \end{bmatrix} \tag{2.26}$$

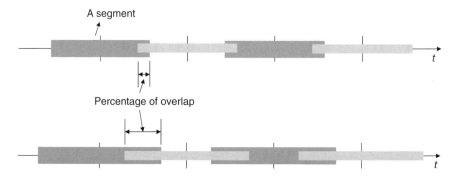

Figure 2.13 Segmentation of the time series into *n* overlapping regions. The degree of overlap depends on many factors, such as the speed at which the motion changes, the expected duration of the features (symptoms), and the expected storage and processing cost.

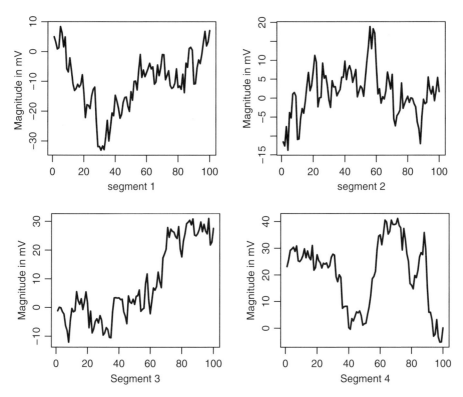

Figure 2.14 The values of the first four segments constructed from the time series of Figure 2.12 with a 20% overlap between the segments. The first segment is constructed by taking the first 100 samples, the second segment by taking 100 samples from locations 81 to 180, the third segment by taking 100 samples from locations 161 to 260, the fourth segment by taking 100 samples from locations 241 to 340, and so on.

For our example, taking only the first four segments, the covariance matrix is:

$$\text{Cov}(4 \times 4) = \begin{bmatrix} 92.7 & -17.6 & 55.4 & 18.5 \\ -17.6 & 39.2 & -18.4 & -19.6 \\ 55.4 & -18.4 & 174 & -9.2 \\ 18.5 & -19.6 & -9.2 & 163.1 \end{bmatrix}$$

2.2.2 Alzheimer's Disease

Alzheimer's disease (AD) is one of the most common neurodegenerative disorders that afflicts the ageing population. The main cause of the disease is an abnormal activation of glial cells in the brain, especially the astrocytes and microglia, which damages the neurons and leads to progressive neurodegeneration (Beach et al. 2012; Byrne et al. 2014). The normal role of glial cells is to protect the neurons and to ensure that the brain operates smoothly. When an injury or a change in the brain occurs, the glial cells mount a desirable inflammation response to fight off the the injury and restore the brain to its proper functioning. Whereas a controlled inflammatory response is beneficial, sometimes it gets out of balance and the inflammation becomes too strong or does not shut

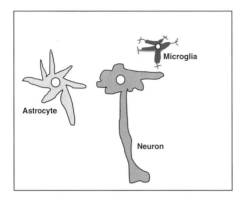 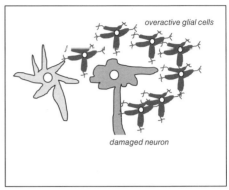

Figure 2.15 Comparison of the presence of glial cells in a healthy brain and in a neurodegenerative brain: left: normal brain; right, neurodegenerative brain.

off on time. In neurodegenerative diseases, the glial cells are over-activated and produce damaging inflammatory molecules called proinflammatory cytokines. This results in the death of nerve cells and the acceleration of the progression of the disease. Figure 2.15 compares the activation of normal and abnormal glial cells in the brain.

Clinically, AD is manifested as early memory deficit and progressive cognitive and functional disorientation. Several researchers have asserted that AD patients are prone to fall due to difficulty with balancing (Mace and Rabins 2011; Mawuenyega et al. 2010; Suttanon et al. 2012). Hence, early assessment of the ability of patients to balance themselves in different situations is important for timely detection of the existence of the disease.

Wang et al. (2013) employed inertial sensors consisting of 3D accelerometers and uni-axial and bi-axial gyroscopes to diagnose AD. The data from these sensors were supplied to a cognitive assessment algorithm. The prototype was tested with 16 AD patients and 25 healthy subjects. Cognitive assessments were carried out while the test subjects wore the inertial sensors on their waist and performed specific movements. The data from the inertial sensors were used to determine the anterior–posterior (AP) and medial–lateral (ML) sway speed, since the body's centre of mass can be determined from these variables. If a subject has difficulty with balancing, his/her AP and ML sway speeds are bigger. In the prototype, the parameters and features pertaining to movements and balance were computed using the cognitive assessment algorithm.

2.2.3 Sleep Apnea and Medical Journalling

A sleep apnea is a sleep disorder characterised by the existence of intermittent pauses (apnea) in breathing or shallow breaths (hypoapnea) during sleep. Each pause (apnea) may last from a few seconds to several minutes and may occur 30 times or more per hour. The disease is treatable, but may remain undetected for a long time. Persons suffering from sleep apnea exhibit symptoms such as loud and chronic snoring, and choking and gasping during sleep. They may also exhibit headache, restlessness, irritability, and insomnia when awake. Failure to deal with a sleep apnea leads to diabetes, high blood pressure, heart disease, stroke, and weight gain (Dempsey et al. 2010; Gottlieb et al. 2010; Redline et al. 2010).

Clinically, sleep apnea is classified into three categories: obstructive, central, and complex. The most common of these is obstructive sleep apnea and occurs when the collapsible walls of a soft tissue in the throat relaxes during sleep, obstructing breathing. In central sleep apnea, the brain's respiratory control centres are imbalanced during sleep and unable to signal the muscles that control breathing. As a consequence, the neurological feedback mechanism that monitors the level of carbon dioxide in the blood does not react quickly enough to maintain an even respiratory rate. Patients with complex (or mixed) apnea have a combination of both types, and the condition is usually correlated with heart failure.

The diagnosis of sleep apnea can take place at home or in a clinical environment using the techniques of polysomnography. At home, patients are required to record in detail their biological as well as physiological experiences during the night and the hours immediately after waking up. This enables doctors to identify internal and external causes of sleep disturbance and the magnitude of apnea if present. Patients are also encouraged to video or audio record their sleep activity. In a clinical setting, patients are brought to a "sleep house", where biopotential measurements are taken from different parts of the upper body while the patients are asleep. The biopotentials indicate brain activities, eye movements, muscle activity, and heart rhythm, as shown in Figure 2.16.

Oliver and Flores-Mangas (2006) employed a lightweight wireless, reprogrammable pulse oximeter to diagnose sleep apnea in a home setting. The oximeter is made up of an optical transmitter and an optical receiver. The transmitter emits a red light or an infrared signal into the blood close to the skin (a finger tip or an ear lobe) and the receiver receives the optical signal that is not absorbed by the blood. The transmitted signal undergoes both absorption and a shift in wavelength when it passes through the blood, and the amount of signal absorbed and the shift in wavelength depend on the percentage of hemoglobin in the blood saturated with oxygen and the flow of blood. Knowledge of the received light and its wavelength provides sufficient information about

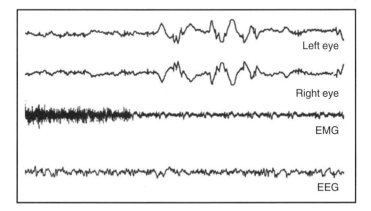

Figure 2.16 Measurement of biopotentials from a sleeping patient. The electroencephalogram (EEG) measures the electrical activities of the brain. EEG electrodes are attached to the scalp. The electrooculogram measures the electrical potential difference between the cornea (which has a positive potential) and the retina (which has a negative potential). The rate of change of the potential difference depends on how fast the eyes move during sleep. Likewise, muscular activities are obtained by measuring the biopotential of facial muscles using electromyogram or EMG.

the level of oxygen in the blood (SpO_2) as well as the expansion of blood vessels, both in real time. From these two parameters, it is also possible to determine heart rate and blood flow. The received light is processed by an integrated microcontroller to identify additional useful features such as the fluctuation of oxygen in the blood. This simple technique can measure heart rates ranging from 18–300 beats per minute and oxygen saturation ranging from 70–100%. The sensor is sampled with two different sampling frequencies. The first, at 3 Hz, is used to measure the concentration of haemoglobin and heart rate whereas the second, at 75 Hz, measures change in the volume of blood (plethysmograph).

The researchers developed two types of algorithms to detect sleep apnea. One of them is a multi-threshold time series analysis technique personalised to individual baselines. According to this technique the existence of sleep apnea is marked by a known pattern of oxygen desaturation. First the desaturation steadily falls below a baseline down to a specific lower level and begins to recover towards the baseline up to a level approximately 25% higher than the lowest level, after which no further recovery is observed. The second technique relies on the spectral characteristics of nocturnal oximetry and heart-rate variability.

The prototype was tested on 20 volunteers who were between 25 and 65 years of age, 80% of them being male. 70% of the volunteers either suspected or knew that they were suffering from a sleep apnea. The volunteers recorded their sleeping condition for an entire night and the data were fed to the two algorithms, both of which, according to the report of the researchers, were able to identify with 100% accuracy all three cases of known obstructive sleep apnea, and one case of severe and two cases of mild obstructive sleep apnea among the volunteers who suspected they might be suffering from the condition but had not undergone any medical diagnosis.

The device can be integrated into a wireless personal area network, which can augment medical journalling (Redline et al. 2012; Stankovic et al. 2012). The idea is to objectively identify potential environmental factors that cause or exacerbate sleep apnea, such as temperature, humidity, barometric pressure, and sunlight. The wireless network can consist of sensors that can be deployed on and around patients and measure environmental, biomedical, and physiological parameters during sleep. Likewise, changes in body functions (such as range of motion, pain, fatigue, headache, and irritability) can be seamlessly measured when patients are awake. Then doctors can establish any correlation between environmental and biological measurements. The approach not only enriches doctors' knowledge about the patients, but also helps them to filter subjective and inconsistent reports from patients.

2.2.4 Asthma

Asthma is a chronic respiratory disease caused by chronic inflammation of the lung's. According to a 2016 survey by the WHO, around 235 million people suffer from asthma worldwide (WHO 2016). In the USA alone, nearly 1 in 15 young children suffer from the disease, however, asthma prevalence is higher among females, children, and people living in low-income countries. The study estimates that the annual cost for the treatment of asthma is $56 billion in the US and €17.7 billion in Europe. The estimated annual death rate due to asthma is 250,000 worldwide (Akinbami et al. 2011, 2012; Partridge 2007).

The actual cause of asthma is not well defined, but researchers correlate the condition with heredity, exposure to air pollution (indoor as well as outdoors), and poor working environments (such as exposure to chemicals and/or health hazardous materials). Moreover, factors such as air pollution, excessive exercise, emotional upsets, chemicals, perfumes, hormones, and viral infections are taken as common asthma triggers, even though the specific triggers vary from person to person. Symptoms of asthma include episodic periods of wheezing, chest tightness, shortness of breath, and coughing. The treatment of asthma depends on the degree of severity as well as the outcome of a clinical diagnosis involving patient medical history, breathing tests, exercise, chest X-rays, and allergy tests. Generally, asthma patients must carefully manage the disease through the use of medications as well as by avoiding conditions that can exacerbate the disease, such as allergens, irritants, and physical exercise when pollution levels are high (Jackson et al. 2011).

Seto et al. (2009) proposed an open source platform for monitoring the activities and environments of asthmatic children. It consists of three layers: a body-sensor layer, a personal-network layer, and a global-network layer. The body-sensor layer consists of a 3D accelerometer, a 2D gyroscope, a GPS receiver, and a mass particle counter. The latter measures a variety of pollutants such as dust, fly ash, diesel exhaust particles, wood smoke, and sulfate aerosols. The motion sensors are used to detect acceleration, which can be correlated with surge, sway, and heave. The personal-network layer is responsible for collecting preprocessed data from the body-sensor layer, for performing higher-level signal processing (to detect the pitch and roll of the motion-sensor orientation in space), and for forwarding the output to the global-network layer. The aggregated data pertaining to motion, location, and particulate matter (PM) concentrations at the personal-network layer form the basis of a comprehensive asthma management system. The global-network layer runs a web service in order to manage the information coming from the middle layer. Thus asthmatics can have a global view of their air pollution exposure patterns and history and correlate periods of high exposure with symptoms and medication.

Likewise, Fu et al. (2009) propose an asthma management system based on distributed wearable sensors. The system provides reminders, early warnings, and instructions to patients, so that they can reduce the likelihood of an asthma attack. The system consists of body sensors, environmental sensors, and a central control unit. The body sensors are piezoceramic microphones with cylindrical air chambers to acquire body sounds. The environment sensors consist of an optical sensor (for monitoring airborne particles), temperature and humidity sensors, an accelerometer, and an SpO_2 sensor. The accelerometer measures the activity level of patients to contextualise analysis. The SpO_2 sensor measures the peripheral capillary blood oxygen saturation, which is a key indicator of advanced pulmonary symptoms. The various sensors and the control unit make up the body-area network. The raw sensor data are processed using a smartphone (that serves as a central control unit), which runs a signal-processing module and a disease-specific rule-based expert system. The expert system estimates a patient's level of pulmonary disease exacerbation by "scoring" current and recent patient history.

Kuryloski et al. (2009) employ a physiological sensor node to measure respiratory minute volume (the volume of gas inhaled or exhaled by a person's lungs per minute). The physiological sensor contains four electrodes connected to the surface of the ribcage

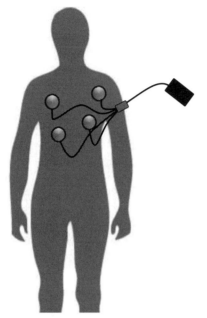

Figure 2.17 The use of wireless pneumography to measure the quality of respiration in asthmatic patients.

(see Figure 2.17). Electrical impedance pneumography is a common technique to measure respiration rate (Sanchez et al. 2013; Seppä et al. 2010). In this technique, a small amount of high-frequency AC current is applied to the thorax through the drive electrodes. The AC current produces a potential difference between the receiver electrodes (the tissue resistance between the electrodes is related to the potential difference). The potential difference is directly related to the change in the conductivity of the thorax due to breathing. During inspiration, the volume of gas in the chest increases with respect to the volume of fluid. This reduces the conductivity of the thorax, resulting in an increase in the electrical impedance. During expiration, the electrical impedance decreases due to the reverse phenomenon. There is an approximately a linear relationship between the changes in the electrical impedance and the volume of respired air. The change in the potential difference can be used to determine breathing rate.

The authors implemented a web service to aggregate sensor data pertaining to physical activity, location, and PM concentration. The web service is categorised into four different applications. The first is a personal health record that supports asthma patients to effectively plan their activity by managing their activity path and intensity of exercise in relation to their symptoms and pollution exposure history. The second is an electronic medical record, which stores comprehensive patient records during physical activity and pollution exposure; this enables doctors to monitor patients remotely and to make treatment-related decisions. The third is a participatory sensing in which data from many asthmatic patients are stored in a database to help health departments and city planners to improve the design of cities and to reduce environmental pollution. Lastly, the fourth application is intended as a health research system in which the stored data from different asthma patients can be investigated by researchers to examine the relationship between asthma symptoms and environmental triggers.

The prototype was tested on six adults (5 males and 1 female). The candidates were asked to perform a series of prescribed walks on a 2.4 km path route. The path route included sections of uphill, downhill, flat, and busy roadways as well as a commercial area. During the walks, data were logged from an accelerometer, GPS, and particle mass counter. The motion and GPS data were logged at 40 Hz and 1 Hz, respectively. The sensed data were combined and processed to analyse individual physical activity patterns at certain geographical locations.

2.2.5 Gastroparesis

Gastroparesis is a medical condition in which the contraction of muscles in the stomach or intestine do not function properly, preventing the normal processing of food in these organs. There can be many causes to gastroparesis: uncontrolled diabetes, PD, multiple sclerosis, deposits of protein fibres in tissues and organs, and narcotics and antidepressants, but the primary cause is damage to the vagus nerve, which regulates the digestive system. Its typical symptoms are nausea, vomiting, and constipation (Camilleri et al. 2011, 2013).

Conventionally, the diagnosis of gastroparesis involves nuclear medicine but additional approaches such as consideration of antro-duodenal motility, electrogastrograms, upper gastrointestinal endoscopy, and computer tomography, can also be used. During a nuclear medicine test, a patient is given a meal containing a small amount of radioactive material. Then a scanner is placed over the stomach for several hours to monitor the amount of radioactivity in the stomach. In patients with gastroparesis, the stomach takes a longer time than usual (more than several hours) to process the food. This approach can be effective to detect gastroparesis but it is tedious and invasive. In an antro-duodenal motility study, the pressure that is generated by the contraction of the muscles of the stomach and the first portion of the small intestine is measured using pressure sensors. The pressure sensors are housed in a tube, which is then inserted through the nose, the oesophagus, and the stomach into the duodenum. The pressure sensors measure the strength of the contraction of the muscles of the stomach and the small intestine before and after a meal. In most gastroparesis patients, either infrequent or very weak contractions are detected, depending on whether the damage is to the nerve or the muscles. When the damage is to the nerve, infrequent contractions are observed and when the damage is to the muscles, very weak contractions are detected.

An electrogastrogram (see Figure 2.18) is similar to an electrocardiogram; the difference is that it measures the biopotentials that are generated and propagate inside the stomach and intestine muscles (as opposed to the cardiac muscles). Several electrodes are attached to the abdomen of patients for several hours. In normal individuals there is a regular electrical rhythm and the magnitude of the biopotentials increases after a meal. But in most gastroparesis patients, the rhythm is not regular and no increase in the magnitude of the biopotentials is observed. Sometimes, an upper gastrointestinal endoscopy test is necessary to exclude conditions that may produce similar symptoms.

More recently, SmartPill Corporation has developed a cylindrical wireless sensor node (a wireless motility capsule) for diagnosing gastroparesis (Rao et al. 2009; Sarosiek et al. 2010). The node consists of sensors for temperature, pressure, and pH, a microcontroller, and a wireless transmitter inside a rigid polyurethane shell containing a battery that lasts for a minimum of 120 h. It has a length of 26.8 mm and a diameter of 11.7 mm (see

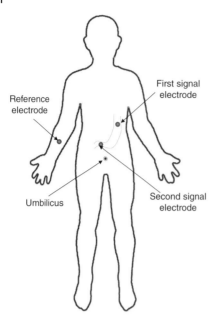

Figure 2.18 The placement of electrodes in the use of an electrogastrogram for diagnosing gastroparesis.

Reference electrode

First signal electrode

Umbilicus

Second signal electrode

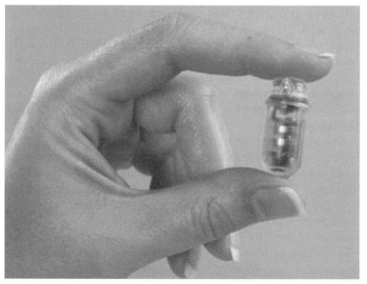

Figure 2.19 A wireless sensor node (wireless mobility capsule) which integrates temperature, pressure, and pH sensors, a microcontroller, and a wireless transmitter to diagnose gastroparesis.

Figure 2.19). The node is capable of real-time sensing and transmission. The temperature sensor can measure body temperatures ranging from 25–49 °C; the pressure sensor measures the pressure of its immediate surrounding ranging from 0–350 mmHg, and the pH sensor measures body pH in a range between 0.05 and 9.0. The wireless transmitter uses the 434 MHz carrier frequency for transmitting the sensed data from within the body to an external base station, which then logs the data to a nearby computer.

The node is a single-use, orally ingested, non-digestible device. Once it is swallowed, it undergoes the same route digestible foods take, so that from its travel time in the body, it is possible to accurately measure gastric emptying time, small bowel transit time, colonic transit time, and whole gut transit time. The node also measures intraluminal pressure, which is useful for determining pressure patterns in the regions of the gastrointestinal tract.

The diagnosis of gastroparesis using the wireless motility capsule typically takes place as follows (Saad and Hasler 2011). First, in the presence of a doctor, a patient undergoes an overnight fast and discontinues medication that can potentially alter gastric pH and gastrointestinal motility. Second, the patient is given a standardised 260 kcal nutrient bar consisting of 17% protein, 66% carbohydrate, 2% fat, and 3% fiber and 50 m of water. The transit times observed after the ingestion of the nutrient bar are comparable with those observed in healthy controls and patients with dysmotility after the consumption of a low-fat egg-substitute meal. Third, immediately following the ingestion of the nutrient bar, the patient ingests the wireless sensor node. At this stage the patient can leave the doctor and freely carry out everyday activities, but the external base station should either be attached to the patient's waist or placed within a 5 m distance for the next 3 – 5 days. The patient will be advised to avoid strenuous exercise, which can strain the stomach and intestinal muscles and thereby generate false readings. Moreover, in order to obtain an accurate measurement of gastric emptying time, the patient has to abstain from eating for 6 h. When transmission with the wireless sensor node is interrupted or the base station detects an abrupt temperature drop, then it confirms the passage of the node out of the body.

A combination of pH and temperature readings are used to calculate emptying times. The gastric emptying time is calculated as the time from the ingestion of the sensor node to its departure from the stomach. Immediately following ingestion of the node, an abrupt rise in temperature reading from ambient temperature to normal body temperature (approximately 36 °C) is expected. The exit of the node from the stomach is marked by an abrupt rise in pH ($\geq$ 2 pH units), which corresponds to the passage of the node from the acidic environment of the stomach to the more alkaline environment of the proximal duodenum. Likewise, the small bowel transit time is defined as the time from the entrance of the wireless sensor node into the duodenum to its passage through the ileocecal valve and into the cecum. The passage into the cecum is defined by "a sustained (> 10 min in duration) pH drop of at least 1 pH unit, which occurs as the node leaves the alkaline environment of the terminal ileum and enters the more acidic environment of the cecum. If the pH drop in the ileocecal junction is not evident, abrupt changes in pressure wave frequency or amplitude may provide supportive evidence of the transition from the small bowel to the colon" (Saad and Hasler 2011). The colonic transit time is defined as the time from the entry of the node into the cecum to its passage from the body and the whole gut transit time is defined as the time from the ingestion of the wireless sensor node to its exit from the body.

Several experiments and actual diagnoses confirm that the wireless sensor node is an effective tool for diagnosing gastroparesis. For example, a study conducted on 87 healthy adults and 61 adults with gastroparesis indicates that the detection of gastric emptying time with the wireless sensor node was more accurate than a scintigraphic gastric emptying study (Saad and Hasler 2011). A similar study conducted on 158 adults who had tested positive for chronic functional constipation demonstrated overall device agreement of 87% between the wireless sensor node and a widely accepted procedure for determining delayed versus normal colonic transit (Saad and Hasler 2011).

2.3 Water-quality Monitoring

Because of its indispensable function, monitoring the quality of water is vital for various purposes. In most urban areas of the world, pollution is the main challenge. In countries where there is scarcity of water, such as Africa, Australia, the Middle East, and southern European countries, the efficient distribution of water is the main challenge. In countries like South Africa, where there is a shortage of water and huge demand and perhaps inefficient utilisation of water by the mining industry, the challenge is manifold, including the efficient distribution of water, underground mixing of clean and used water, the latter potentially contaminated by heavy metals, and uneven distribution of underground water. As I am editing this chapter, a water contamination crisis unfolding in Flint, Michigan, United States, has exposed more than 10,000 children to heavy metals and potentially caused an outbreak of the Legionnaires' disease that killed 10 and seriously affected another 77 people (Bellinger 2016; Hanna-Attisha et al. 2016; Hurley 2016). The problem was a result of more than two years of unattended leaching of lead from ageing and corroded pipes into the water they transport. This ultimately caused an extremely elevated level of heavy metal contamination and created a terrifying public health danger.

The concentration of hydrogen ions and oxygen in water determines the quality of water for supporting human and aquatic life. The same is true for the amount of mineral and salt impurities in water. For example, the concentration of impurities in drinking water should not exceed 500 parts per million (ppm); for agriculture, it should not exceed 1200 ppm. High-technology manufacturers often require impurity-free water. Similarly, the clarity of water or the amount of particulates floating in it (such as debris, sand, silt, and clay) affect the amount of sunlight that can reach aquatic plants, which in turn affects their reproduction quality.

The concentration of hydrogen ions in water, pH, is a measure of its acidity (p stands for potential). The pH ranges from 0 (very acidic) to 14 (very basic) and 7 is neutral. In most water the pH range is between 5.5 and 8.5. Changes in the pH affect how chemicals dissolve in water and the survival of aquatic organisms. High acidity (pH < 4) is deadly to fish, for example. Dissolved oxygen (DO) is a measure of the concentration of oxygen in water. Healthy water has high amounts of DO, whereas water in swamps has a low amount of DO. Temperature, the amount and speed of flow, the diversity and quantity of aquatic plants, pollution, and the composition of a stream bottom affect the amount of DO.

The amount of particulates in water due to mineral and salt impurities are referred to as total dissolved solids, which is measured in parts per million and quantifies the amount of impurities in 1 million units of water. Likewise, the clarity of water and the amount of sunshine it permits to pass through it is determined by the turbidity of the water, which measures the amount of particulates floating in it. This is also referred to as "total suspended solids".

A large range of sensors is available to monitor water quality. Often these sensors measure the electrical properties of the water being monitored. For example, the electrical potential of water is proportional to the hydrogen ions it contains. With a proper mechanism to compensate for the effect of temperature, it is possible to map the water potential to a pH level. DO sensors use an oxygen-permeable membrane to set up a current, the magnitude of which indicates the oxygen level. Typical DO sensors can measure dissolved oxygen from saturation (approximately 8 ppm) to a few particles per billion. Similarly, electrical conductivity sensors are used to measure the ability of water to conduct electricity. Whereas pure water is a poor conductor, a water's conductivity improves proportionally when dissolved salts are present. The amount of dissolved solids is expressed in mhos/cm or in mS/cm.

The turbidity of water can be measured by optical sensors. A light beam passing through water can be partly absorbed by it and the rest can be received by an underground optical receiver. From the portion of light that is received by the receiver, it is possible to estimate the portion of light that has been absorbed by particulates. Alternatively the portion of reflected light can be measured to avoid deploying underwater sensors.

Jiang et al. (2009) list four types of existing approaches for water-quality monitoring and highlight their inadequacy for large-scale and online monitoring. One of these approaches relies on laboratory tests after water samples are manually collected from different locations and different water bodies (reservoirs, rivers, lakes, streams, water catchments, and surface collection dumps) at different points in time. This approach is not scalable, and it is slow, tedious, and expensive. The second approach is both automatic and continuous and consists of monitoring and control centres and monitoring substations. Each substation is responsible for a particular body of water, while the monitoring stations are responsible for aggregating the data from the substations. The control station is responsible for controlling, configuring, and managing all the stations. This architecture is expensive both to set up and maintain. The third approach relies on remote sensing technologies that capture and analyse the spectrum of electromagnetic waves radiated, reflected, or scattered from water bodies. While it is scalable and capable of monitoring an extensive body of water, it is inaccurate and slow. The fourth approach monitors aquatic organisms and their activity to indirectly reason about the quality of the water in which they live.

Jiang et al. (2009) propose wireless sensor networks for water quality monitoring. Their system architecture incorporates several wireless sensor networks, each of which monitors a particular body of water. The base station of each network collects data from the network and transmits it via a GPRS link to a remote monitoring centre. Each node inside a wireless sensor network integrates a temperature and a pH sensor as well as a ZigBee (Zigbee Alliance 2006) compatible transceiver and a microcontroller. Both the temperature and the pH sensors measure the potential of the water with respect to a reference voltage and produce a weak voltage. This weak voltage is first converted to current (4-20 mA), which is then amplified and converted back to a relatively high voltage (between -1.5 V and 1.5 V) by differential circuits. Then the analogue signal is directly fed to the microcontroller, which converts the analogue signal to a digital signal using one of its internal ADCs, and processes it to determine the pH level to which the voltage refers. A local software program coordinates the sampling of the sensors and the timing of aggregation and processing of their outputs.

The researchers set up a prototype wireless sensor network consisting of five nodes and deployed it in an artificial lake at Hangzhou Dianzi University in China for one month (November 2008) and extracted useful data from the network. Figure 2.20 shows

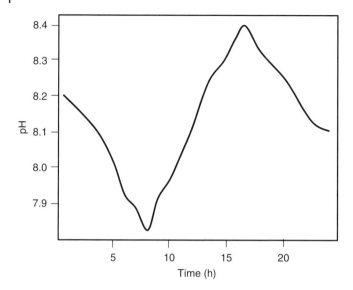

Figure 2.20 The change of pH in an artificial lake during a single day.

the pH variation in the lake for a single day taken from one of the nodes. The researchers noted that the maximum pH value occurred between 3 and 5 pm. This was because the phytoplankton in the lake reached its peak photosynthesis rate during this period. During this peak, plenty of carbon dioxide (CO_2) was changed to sugar ($C_6H_{12}O_6$), which lowers the content of carbonic acide (H_2CO_3) and the acidity of the lake and increases the pH. This phenomenon is expressed as:

$$6CO_2 + 12H_2O \longrightarrow C_6H_{12}O_6 + 6H_2O + 6O_2 \tag{2.27}$$

$$CO_2 + H_2O \longrightarrow H_2CO_3 \tag{2.28}$$

Similarly, Peter Corke and his team at CSIRO, Australia, deployed a wireless sensor network into an area which was 2000 km away from their lab in Brisbane (Corke et al. 2010; Le Dinh et al. 2007). Their aim was to monitor the salinity, water-table level, and water-extraction rate at a number of bores within the Burdekin irrigated sugar-cane growing district, a coastal region with a total monitored area of 2 km by 3 km. Often overextraction of water by the irrigation system led to salt water intrusion into the aquifer. The researchers reported that the network was deployed in 2006 and operated for 1.5 years, delivering more than 1 million water quality readings and requiring only two maintenance visits. Figure 2.21 displays one of the wireless sensor nodes the researchers designed and deployed at one of the pump sites in the Burdekin.

The network consisted of nine nodes, in a tree topology with an average single-hop distance of 800 m (see Figure 2.22). Each node integrated commercially available water-conductivity and water-flow rate sensors, a microcontroller, and a wireless transceiver which used the 915 MHz frequency to communicate at a rate of $50-100$ ks^{-1} at a transmission range of 1000 m. The sensors were interfaced with the microcontroller using a 16-bit ADC and a SPY serial bus.

Tokekar et al. (2010) deployed a robotic sensor network on one of Minnesota's lakes to monitor the activity and distribution of the carp population in the lake. Carp are

Figure 2.21 The wireless sensor node deployed at one of the pump sites in the Burdekin: (A) overview of the wireless sensor node; (B) sensor node housing; (C) bore containing water-level sensor; (D) pump; (E) flow meter; (F) EC sensor; (G) tank. EC, electrical conductivity. Source: Le Dinh et al. (2007).

an oily freshwater species which spend some or all of their lives in fresh water and can seriously affect the quality of water due to the harmful nutrients they release while bottom-feeding. Locations such as shallow wetlands, where carp migrate and reproduce, are of particular interest. The researchers argue that deploying and maintaining stationary sensors in such environments is a formidable challenge. For example, Minnesota has more than 10,000 lakes with different sizes and some of these are interconnected, forming complex ecosystems.

The system they propose consists of a mobile robot carrying an on-board sensing system to monitor the pH level of the lake. At the same time, the robot interacts with several wireless mobile transmitters that are surgically inserted into the bodies of some of the fish. The fish were first caught and the radio transmitters were implanted

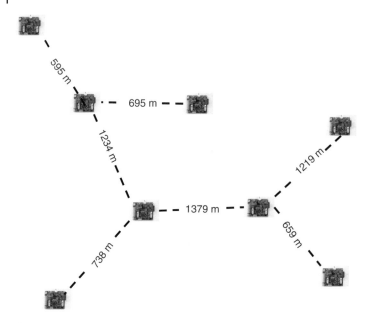

Figure 2.22 A tree topology wireless sensor network for monitoring water quality and water extraction rates at a number of bores within the Burdekin irrigated sugar cane growing district in Australia.

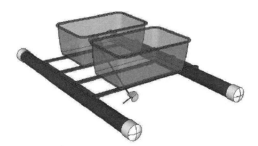

Figure 2.23 A mobile robot for collecting data from wireless transmitters implanted in carp living in one of Minnesota's lakes. Source: Tokekar et al. (2010).

under their skin in a laboratory and then they were reintroduced into the lake. Each transmitter emits short regular pulses (with unique radio frequency identification or RFID) and can be detected within a 50 m radius under ideal conditions.

The researchers built a lightweight and sturdy catamaran raft using polyvinyl chloride pipes and wood (see Figure 2.23) which was capable of carrying up to 20 kg of equipment: the sensing and controlling devices. The robot was driven by a 3-inch diameter, [4] three-blade propeller attached to a 12 V DC motor through a flexible shaft. A 12 V, 18 Ah sealed lead-acid battery was used as the main power source for steering and propulsion. The battery provided 5 h of continuous operation. The on-board sensing systems gathered data on the whereabouts of fish and the water quality, whilst the control system was responsible for driving the propeller and steering the robot. The short regular pulses received from the fish were logged onto an on-board laptop computer together with data from a compass and a GPS receiver to record location and

---

4 Approximately 7.6 cm.

cruising directions. A route-planner algorithm, which was a part of the control module, commanded the cruising speed and direction of the robot. The system was deployed on Lake Keller in Maplewood, Minnesota to trace the activity of 22 fish, which were tagged with wireless transmitters. The robot was able to travel approximately 1 km and to interact with the fish.

References

Akinbami OJ, Moorman JE, Liu X *et al.* 2011 *Asthma prevalence, health care use, and mortality: United States, 2005–2009.* US Department of Health and Human Services, Centers for Disease Control and Prevention, National Center for Health Statistics, Washington, DC.

Akinbami OJ *et al.* 2012 Trends in asthma prevalence, health care use, and mortality in the United States, 2001–2010. *NCHS Data Brief*, **94**, 1–8.

American Society of Civil Engineering 2013 *A 2013 report card for America's Infrastructure.* URL: http://www.infrastructurereportcard.org/a/#p/drinking-water/overview

Beach TG, Monsell SE, Phillips LE and Kukull W 2012 Accuracy of the clinical diagnosis of Alzheimer disease at National Institute on Aging Alzheimer Disease Centers, 2005–2010. *Journal of Neuropathology & Experimental Neurology*, **71** (4), 266–273.

Bellinger DC 2016 Lead contamination in Flint–an abject failure to protect public health. *New England Journal of Medicine*, **374** (12), 1101–1103.

Brincker R, Zhang L and Andersen P 2001 Modal identification of output-only systems using frequency domain decomposition. *Smart materials and structures*, **10** (3), 441.

Burns A, Greene B, McGrath M, O'Shea T, Kuris B, Ayer S, Stroiescu F and Cionca V 2010 Shimmer –a wireless sensor platform for noninvasive biomedical research. *IEEE Sensor Journal*, **10** (9), 1527–1534.

Buttussi F and Chittaro L 2010 Smarter phones for healthier lifestyles: An adaptive fitness game. *IEEE Pervasive Computing*, **9** (4), 51–57.

Byrne JH, Heidelberger R and Waxham MN 2014 *From Molecules to Networks: An Introduction to Cellular and Molecular Neuroscience.* Academic Press.

Camilleri M, Bharucha AE and Farrugia G 2011 Epidemiology, mechanisms, and management of diabetic gastroparesis. *Clinical Gastroenterology and Hepatology*, **9** (1), 5–12.

Camilleri M, Parkman HP, Shafi MA, Abell TL and Gerson L 2013 Clinical guideline: management of gastroparesis. *The American Journal of Gastroenterology*, **108** (1), 18–37.

Cho S, Yun CB, Lynch JP, Zimmerman AT, Spencer Jr BF and Nagayama T 2008 Smart wireless sensor technology for structural health monitoring of civil structures. *Steel Structures*, **8** (4), 267–275.

Corke P, Wark T, Jurdak R, Hu W, Valencia P and Moore D 2010 Environmental wireless sensor networks. *Proceedings of the IEEE*, **98** (11), 1903–1917.

Dargie W 2009 Analysis of time and frequency domain features of accelerometer measurements. *Computer Communications and Networks, 2009. ICCCN 2009. Proceedings of 18th Internatonal Conference on*, pp. 1–6.

Dargie W and Denko MK 2010 Analysis of error-agnostic time-and frequency-domain features extracted from measurements of 3-D accelerometer sensors. *Systems Journal, IEEE*, **4** (1), 26–33.

Dargie W and Poellabauer C 2010 *Fundamentals of Wireless Sensor Networks: Theory and Practice.* John Wiley & Sons.

Dejnabadi H, Jolles BM and Aminian K 2005 A new approach to accurate measurement of uniaxial joint angles based on a combination of accelerometers and gyroscopes. *Biomedical Engineering, IEEE Transactions on,* **52** (8), 1478–1484.

Dempsey JA, Veasey SC, Morgan BJ and O'Donnell CP 2010 Pathophysiology of sleep apnea. *Physiological Reviews,* **90** (1), 47–112.

Environmental Protection Agency 2016 *Water supply and use in the United States.* URL: https://www3.epa.gov/watersense/docs/ws_supply508.pdf.

Favre J, Jolles B, Aissaoui R and Aminian K 2008 Ambulatory measurement of 3D knee joint angle. *Journal of Biomechanics,* **41** (5), 1029–1035.

Fu Y, Ayyagari D and Colquitt N 2009 Pulmonary disease management system with distributed wearable sensors. *Engineering in Medicine and Biology Society, 2009. EMBC 2009. Annual International Conference of the IEEE,* pp. 773–776.

Gottlieb DJ, Yenokyan G, Newman AB, O'Connor GT, Punjabi NM, Quan SF, Redline S, Resnick HE, Tong EK, Diener-West M *et al.* 2010 Prospective study of obstructive sleep apnea and incident coronary heart disease and heart failure the sleep heart health study. *Circulation,* **122** (4), 352–360.

Hackmann G, Guo W, Yan G, Sun Z, Lu C and Dyke S 2014 Cyber-physical codesign of distributed structural health monitoring with wireless sensor networks. *Parallel and Distributed Systems, IEEE Transactions on,* **25** (1), 63–72.

Hanna-Attisha M, LaChance J, Sadler RC and Champney Schnepp A 2016 Elevated blood lead levels in children associated with the flint drinking water crisis: a spatial analysis of risk and public health response. *American Journal of Public Health* (0), e1–e8.

Hopman NC, Williams DL and Lodato F 2003 Wireless electrocardiograph system and method. US Patent 6,611,705.

Hopman NC, Williams DL and Lodato F 2007 Wireless electrocardiograph system and method. US Patent 7,272,428.

Hurley D 2016 Lead crisis in Flint exposes continuing risk to children nationwide: What neurologists should know and what they can do about it. *Neurology Today,* **16** (5), 1–14.

Istvan R, Gregory B, Solovay K, Chastain DP, Gundlach JD, Hopman NC, Williams DL, Lodato F and Salem M 2007 Wireless ECG system. US Patent 7,197,357.

Jackson DJ, Sykes A, Mallia P and Johnston SL 2011 Asthma exacerbations: origin, effect, and prevention. *Journal of Allergy and Clinical Immunology,* **128** (6), 1165–1174.

Jiang P, Xia H, He Z and Wang Z 2009 Design of a water environment monitoring system based on wireless sensor networks. *Sensors,* **9** (8), 6411–6434.

Kim S, Pakzad S, Culler D, Demmel J, Fenves G, Glaser S and Turon M 2007 Health monitoring of civil infrastructures using wireless sensor networks *Information Processing in Sensor Networks, 2007. IPSN 2007. 6th International Symposium on,* pp. 254–263.

Ko J, Lu C, Srivastava MB, Stankovic JA, Terzis A and Welsh M 2010 Wireless sensor networks for healthcare. *Proceedings of the IEEE,* **98** (11), 1947–1960.

Kuryloski P, Giani A, Giannantonio R, Gilani K, Gravina R, Seppa VP, Seto E, Shia V, Wang C, Yan P, Yang A, Hyttinen J, Sastry S, Wicker S and Bajcsy R 2009 Dexternet: An open platform for heterogeneous body sensor networks and its applications. *Wearable and*

Implantable Body Sensor Networks, 2009. BSN 2009. Sixth International Workshop on, pp. 92–97.

Le Dinh T, Hu W, Sikka P, Corke P, Overs L and Brosnan S 2007 Design and deployment of a remote robust sensor network: Experiences from an outdoor water quality monitoring network. *Local Computer Networks, 2007. LCN 2007. 32nd IEEE Conference on*, pp. 799–806.

Li K and Warren S 2012 A wireless reflectance pulse oximeter with digital baseline control for unfiltered photoplethysmograms. *Biomedical Circuits and Systems, IEEE Transactions on*, **6** (3), 269–278.

Lin M, Wu Y and Wassell I 2008 Wireless sensor network: Water distribution monitoring system. *Radio and Wireless Symposium, 2008 IEEE*, pp. 775–778.

Lorincz K, Chen Br, Challen GW, Chowdhury AR, Patel S, Bonato P and Welsh M 2009 Mercury: a wearable sensor network platform for high-fidelity motion analysis. *Proceedings of the 7th ACM Conference on Embedded Networked Sensor Systems*, pp. 183–196.

Mace NL and Rabins PV 2011 *The 36-Hour Day: A Family Guide to Caring for People who have Alzheimer Disease, Related Dementias, and Memory Loss*. JHU Press.

Mascarenas DL, Todd MD, Park G and Farrar CR 2007 Development of an impedance-based wireless sensor node for structural health monitoring. *Smart Materials and Structures*, **16** (6), 2137.

Mawuenyega KG, Sigurdson W, Ovod V, Munsell L, Kasten T, Morris JC, Yarasheski KE and Bateman RJ 2010 Decreased clearance of CNS β-amyloid in Alzheimer's disease. *Science*, **330** (6012), 1774–1774.

Mendelson Y, Duckworth R and Comtois G 2006 A wearable reflectance pulse oximeter for remote physiological monitoring *Engineering in Medicine and Biology Society, 2006. EMBS'06. 28th Annual International Conference of the IEEE*, pp. 912–915.

Metje N, Chapman DN, Cheneler D, Ward M and Thomas AM 2011 Smart pipes–instrumented water pipes, can this be made a reality? *Sensors*, **11** (8), 7455–7475.

O'Donovan KJ, Kamnik R, O'Keeffe DT and Lyons GM 2007 An inertial and magnetic sensor based technique for joint angle measurement. *Journal of Biomechanics*, **40** (12), 2604–2611.

Ofwat 2016 *The development of the water industry in England and Wales*. URL: http://www .ofwat.gov.uk/wp-content/uploads/2015/11/rpt_com_devwatindust270106.pdf.

Oliver N and Flores-Mangas F 2006 Healthgear: A real-time wearable system for monitoring and analyzing physiological signals. *Proceedings of the International Workshop on Wearable and Implantable Body Sensor Networks*, pp. 61–64.

Partridge M 2007 Examining the unmet need in adults with severe asthma. *European Respiratory Review*, **16** (104), 67–72.

Patel S, Lorincz K, Hughes R, Huggins N, Growdon J, Standaert D, Akay M, Dy J, Welsh M and Bonato P 2009 Monitoring motor fluctuations in patients with Parkinson's disease using wearable sensors. *Information Technology in Biomedicine, IEEE Transactions on*, **13** (6), 864–873.

Rao SS, Kuo B, McCallum RW, Chey WD, DiBaise JK, Hasler WL, Koch KL, Lackner JM, Miller C, Saad R *et al.* 2009 Investigation of colonic and whole-gut transit with wireless motility capsule and radiopaque markers in constipation. *Clinical Gastroenterology and Hepatology,* **7** (5), 537–544.

Redline S, Tosteson T, Tishler PV, Carskadon MA and Milliman RP 2012 The familial aggregation of obstructive sleep apnea. *American Journal of Respiratory and Critical Care Medicine,* **151** (3), 682–687.

Redline S, Yenokyan G, Gottlieb DJ, Shahar E, O'Connor GT, Resnick HE, Diener-West M, Sanders MH,Wolf PA, Geraghty EM *et al.* (2010) Obstructive sleep apnea–hypopnea and incident stroke: the sleep heart health study. *American Journal of Respiratory and Critical Care Medicine,* **182** (2), 269–277.

Saad RJ and Hasler WL 2011 A technical review and clinical assessment of the wireless motility capsule. *Journal of Gastroenterology and Hepatology (NY),* **7** (12), 795–804.

Sanchez B, Louarroudi E, Jorge E, Cinca J, Bragos R and Pintelon R 2013 A new measuring and identification approach for time-varying bioimpedance using multisine electrical impedance spectroscopy. *Physiological Measurement,* **34** (3), 339.

Sarosiek I, Selover K, Katz L, Semler J, Wilding G, Lackner J, Sitrin M, Kuo B, Chey W, Hasler W *et al.* 2010 The assessment of regional gut transit times in healthy controls and patients with gastroparesis using wireless motility technology. *Alimentary Pharmacology & Therapeutics* **31** (2), 313–322.

Seppä VP, Viik J and Hyttinen J 2010 Assessment of pulmonary flow using impedance pneumography. *Biomedical Engineering, IEEE Transactions on,* **57** (9), 2277–2285.

Seto EY, Giani A, Shia V, Wang C, Yan P, Yang AY, Jerrett M and Bajcsy R 2009 A wireless body sensor network for the prevention and management of asthma. *Industrial Embedded Systems, 2009. SIES'09. IEEE International Symposium on,* pp. 120–123.

Stankovic JA, Cao Q, Doan T, Fang L, He Z, Kiran R, Lin S, Son S, Stoleru R and Wood A 2012 *American Journal Phys Med Rehabil.* 2012 Jan;91(1):12–23.

Suttanon P, Hill KD, Said CM, LoGiudice D, Lautenschlager NT and Dodd KJ 2012 Balance and mobility dysfunction and falls risk in older people with mild to moderate Alzheimer disease. *American Journal of Physical Medicine & Rehabilitation,* **91** (1), 12–23.

Takeda R, Tadano S, Todoh M, Morikawa M, Nakayasu M and Yoshinari S 2009 Gait analysis using gravitational acceleration measured by wearable sensors. *Journal of Biomechanics,* **42** (3), 223–233.

Tokekar P, Bhadauria D, Studenski A and Isler V 2010 A robotic system for monitoring carp in Minnesota lakes. *Journal of Field Robotics,* **27** (6), 779–789.

Wang WH, Chung PC, Hsu YL, Pai MC and Lin CW 2013 Inertial-sensor-based balance analysis system for patients with Alzheimer's disease *Technologies and Applications of Artificial Intelligence (TAAI), 2013 Conference on,* pp. 128–133.

Watthanawisuth N, Lomas T, Wisitsoraat A and Tuantranont A 2010 Wireless wearable pulse oximeter for health monitoring using Zigbee wireless sensor network. *Electrical Engineering/Electronics Computer Telecommunications and Information Technology (ECTI-CON), 2010 International Conference on,* pp. 575–579.

Weaver JA 2003 *Wearable health monitor to aid Parkinson's disease treatment.* Master's thesis, Massachusetts Institute of Technology.

Whittle A, Allen M, Preis A and Iqbal M 2013 Sensor networks for monitoring and control of water distribution systems. *6th International Conference on Structural Health Monitoring of Intelligent Infrastructure (SHMII 2013).*

WHO 2016 *Asthma: Key facts.*

Zimmerman AT, Shiraishi M, Swartz RA and Lynch JP 2008 Automated modal parameter estimation by parallel processing within wireless monitoring systems. *Journal of Infrastructure Systems*, **14** (1), 102–113.

Zigbee Alliance 2006 Zigbee specification.

3

Conditioning Circuits

In producing an electrical signal that describes the condition of a measurand, the sensing system acts as a voltage or a current source. In some situations, the measurand itself acts as a voltage or a current source. An example is the human body, which produces action potentials (Kimura 2013; Pennisi 2005). Whenever a difference in the concentration of ions (or electrical charges) between two points exists, there will be a potential difference, which we measure in volts. If we set up an external circuit connecting these two points, charged particles (ions, electrons) begin to flow from the higher potential region to the lower potential region (see Figure 3.1). The amount of charge flow (current) depends on both the potential difference and the impedance of the external circuit.

Figure 3.2 displays an electrocardiogram attached to a human body. Figure 3.3a displays the abstraction of these two systems, the human body as a generator of potential difference and the ECG as a sensing system as a whole. The two systems are connected with each other with two electrodes. Figure 3.3b displays the electrical representation of the two systems. The human body (for example, the skin) acts as an internal impedance resisting the flow of electrons. This impedance is represented by Z_S and the impedance seen by the current that flows out of the body and into the sensing system is the input impedance of the sensing system. Incidentally, this impedance is equal to the load impedance of the sensing system.

3.1 Voltage and Current Sources

The potential source can be considered as a voltage or a current source, depending on the magnitude of its internal impedance compared to the load impedance. If it has a series internal impedance that is much smaller than the load impedance (ideally zero), then the voltage across the load is almost always equal to the source voltage, even though the load impedance varies considerably. In this case, the source is considered to be a voltage source. Figure 3.4a displays an ideal voltage source with a zero internal resistance and Figure 3.4b displays a constant load voltage for a variable load impedance. However, no voltage source with a zero impedance can be found in reality. Instead, it has a small internal impedance and the load voltage varies as the load impedance varies, as can be seen in Figure 3.5.

A current source outputs a constant current for a variable load impedance. Unlike a voltage source, it has a much larger internal impedance, so that the current it produces and delivers is not affected by variations in the load impedance.

Principles and Applications of Ubiquitous Sensing, First Edition. Waltenegus Dargie.
© 2017 John Wiley & Sons, Ltd. Published 2017 by John Wiley & Sons, Ltd.
Companion Website: www.wiley.com/go/dargie2017

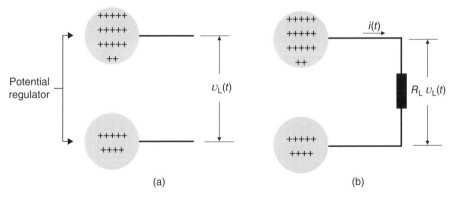

Figure 3.1 Electrical potential difference and current: (a) a difference in the concentration of ions between two locations creates a potential difference between them; (b) when a load is connected between these two locations, a complete electrical circuit is established and the ion concentration difference forces a flow of charged particles. The amount of current that flows in the circuit depends both on the potential difference and the load resistance.

Figure 3.2 An electrocardiogram attached to a human body to sense active potentials of the heart.

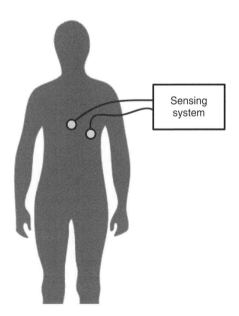

3.2 Transfer Function

The transfer function (or gain) of a linear, time-invariant system is the ratio of the output voltage to the input voltage. It is expressed as a function of the frequency of the input signal because we assume that the system's behaviour is not affected by time but it is affected by the frequency of the input signal (in other words, the system behaves differently for different frequencies). This is typically the case with the steady-state characteristics of most existing systems. Often the transfer function is sufficient to

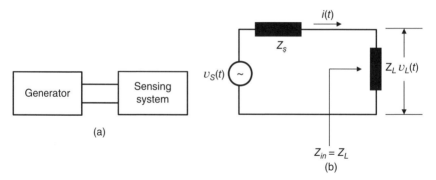

(a)

(b)

Figure 3.3 The human body and an ECG: (a) abstraction of the human body as a generator of action potentials and the sensing system as the load of the generator; (b) electrical representation of the generator and its load. In this simple representation, the resistance experienced (or "seen") by the current that enters into the sensing system is equal to the load impedance of the sensing system.

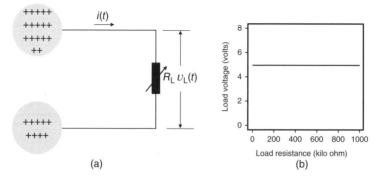

(a)

(b)

Figure 3.4 (a) A variable load resistance connected to an ideal voltage source. (b) The output of an ideal voltage source is fixed despite the variation of the load resistance (in other words, an ideal voltage source produces whatever current is necessary to maintain the voltage across the load. Similarly, an ideal current source produces whatever voltage is necessary to maintain the current across the load).

describe the electrical behaviour of a system:

$$H(j\omega) = \frac{v_{\mathrm{L}}(j\omega)}{v_{\mathrm{S}}(j\omega)} \tag{3.1}$$

where $\omega = 2\pi f$ is the radial frequency of the source voltage and the imaginary component j signifies the phase difference between the source and the load voltages. In other words, there may be a delay between the time an input is applied to the system and the time the system reacts to the input by producing a corresponding output. This is typically the case when the system consists of inductive and capacitive elements. The magnitude component of a transfer function is also referred to as gain. In Figures 3.4 and 3.5 we assumed that the source impedance and the load impedance are frequency invariant; in other words, they are not affected by the change in the frequency of the source voltage. This can be true if the source and the load impedance consist only of resistive components ($Z_{\mathrm{S}} = R_{\mathrm{S}}, Z_{\mathrm{L}} = R_{\mathrm{L}}$).

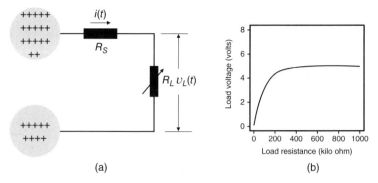

(a)   (b)

Figure 3.5 (a) A variable load resistance connected to a real voltage source which has an internal resistance. (b) In a real voltage source the voltage produced by a source depends on the load resistance and the internal resistance of the source itself.

However, this is not a valid assumption for most of the sensing systems we are concerned with. For example, the frequency of the action potentials generated by the SA node in the heart vary from $0.5 - 150$ Hz (Webster 2009). Because the conditioning system contains inductive and capacitive elements (the impedance of electrical circuits containing capacitive and inductive elements depends on the frequency of the source voltage), the amount of power that can be transferred from the human body to the ECG front-end is a function of the frequency of the action potentials (Nemati et al. 2012; Webster 2009).

The impedance produced by a capacitor is known as a capacitive reactance; it is inversely proportional to the frequency of the source voltage. In other words, a capacitor is approximated by an open circuit for a DC signal while for a high-frequency signal it is approximated as a short circuit. Furthermore, a capacitor opposes a sudden change in the voltage across it (because a capacitor is inherently a charge accumulator). In order to describe these two properties, the reactance of a capacitor ($X_C(j\omega)$) is mathematically described as:

$$X_C(j\omega) = \frac{1}{j\omega C} \tag{3.2}$$

where C is the capacitance of the capacitor and j describes the opposition of the capacitor to a change in voltage. Similarly, the impedance introduced by an inductor is known as an inductive reactance ($X_L(j\omega)$). It is directly proportional to the frequency of the source voltage and its opposition is to a change in current. Hence, it is approximated by a short circuit for a DC signal but by an open circuit for a high-frequency signal (recall that a high-frequency current that passes through a coil produces a high magnetic flux, the tendency of which is to oppose the current that produces it). Mathematically, the inductive reactance is expressed as:

$$X_L(j\omega) = j\omega L \tag{3.3}$$

where L is the inductance of the inductor.

With this in mind, now suppose the source impedance of Figure 3.3 can be approximated by a resistive element and the load impedance by a capacitive element (see

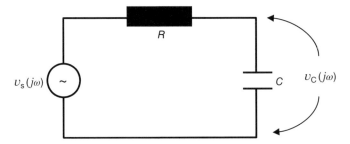

Figure 3.6 A first-order low-pass system approximating the relationship between a measurand and a sensing system.

Figure 3.6). Then, the output voltage can be determined by voltage division:

$$v_L(j\omega) = \frac{1/j\omega C}{1/j\omega C + R}v_S(j\omega) = \frac{1}{1 + j\omega RC}v_S(j\omega) \tag{3.4}$$

As a result, the transfer function of the system is:

$$H(j\omega) = \frac{1}{1 + j\omega RC} \tag{3.5}$$

The speed at which the system responds to the applied source voltage depends on the time constant $\tau = RC$. As can be seen in Eq. (3.5), the load voltage is maximum when $\omega = 0$. On the contrary, when the frequency of the source voltage approaches infinity ($\omega \to \infty$), the load voltage approaches zero. In other words, the system is a low-pass system. A second-order low-pass system (Figure 3.7) better approximates the relationship between the measurand and the sensing system. In order to determine the transfer function of this system, we should apply Thévenin's theorem. The steps are illustrated in Figure 3.7, where we first disconnected the load from the rest of the system to determine the voltage across the two end terminals (the Thévenin voltage), which is the voltage drop across the capacitor. Then we short circuit the source voltage to determine the Thévenin impedance, which is the combination of $(R_1 \| C_1) + R_2$. Consequently,

$$V_{th}(j\omega) = \frac{1/j\omega C_1}{1/j\omega C_1 + R_1}v_S(j\omega) = \frac{1}{1 + j\omega R_1 C_1}v_S(j\omega) \tag{3.6}$$

Similarly,

$$z_{th}(j\omega) = \frac{(1/j\omega C_1 \times R_1)}{(1/j\omega C_1 + R_1)} + R_2 = \frac{R_1 + R_2 + j\omega C_1 R_1 R_2}{1 + j\omega C_1 R_1} \tag{3.7}$$

The Thévenin equivalent circuit of the second-order low-pass system is shown in Figure 3.8. Finally, the transfer function of the second-order low-pass system can be determined by applying voltage division on Figure 3.8:

$$H(j\omega) = \left(\frac{1}{1 + j\omega C_1 R_1}\right)\left(\frac{1 + j\omega C_1 R_1}{(1 - \omega^2 C_1 C_2 R_1 R_2) + j\omega(C_1 R_1 + C_2 R_1 + C_2 R_2)}\right) \tag{3.8}$$

Figure 3.9 compares the transfer functions of the first-and second-order low-pass systems. As can be seen, the gain of the first-order low-pass filter is high compared to the gain of the second-order low-pass filter, but it varies gently as the frequency of the source voltage increases, which means its attenuation capacity of high-frequency

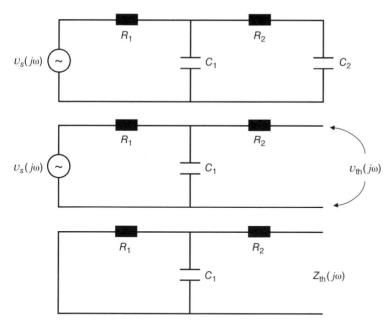

Figure 3.7 A second-order low-pass system approximating the relationship between a measurand and a sensing system.

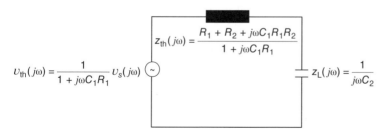

Figure 3.8 The Thévenin equivalent circuit of a second-order low-pass system.

components is weak. In contrast, the gain of the second-order filter reduces rapidly as the frequency of the source voltage increases, resulting in a superior attenuation of high-frequency components. A third-order low-pass filter is even more severe in rejecting high-frequency components but it also attenuates the desired signal and produces delay.

If, instead, we approximate Figure 3.3 by an RL system as in Figure 3.10, the transfer function of the system is given as:

$$H(j\omega) = \frac{j\omega L}{R + j\omega L} \tag{3.9}$$

As can be seen, at low frequency, $H(j\omega) = 0$, but as ω becomes large, the denominator of Eq. (3.9) is dominated by the $j\omega L$ term, as a result of which the transfer function approaches unity. In other words, the system is a high-pass system. Figure 3.3 can also be approximated by a circuit combining a resistive, an inductive, and a capacitive element,

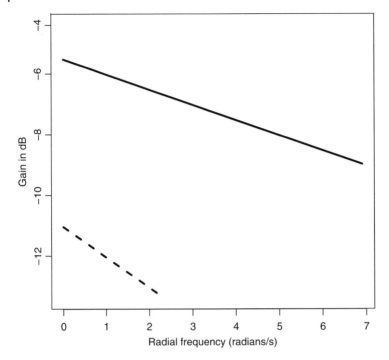

Figure 3.9 Comparison of the gain (transfer function) of a first- and a second-order low-pass systems in a log–log scale. The gain of the second order system (dashed line) drastically diminishes for high frequencies while the gain gently diminishes for the first-order system (solid line). $C = 100\ \mu F$ and $R = 100\ M\Omega$ for the first system and $C_1 = 100\ \mu F, R_1 = R_2 = 100\ M\Omega$ for the second system.

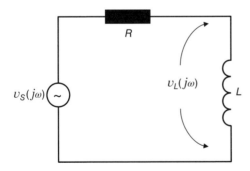

Figure 3.10 A first-order high-pass system approximating the relationship between a measurand and a sensing system.

as in Figure 3.11. The transfer function of this circuit can be determined in two steps. First, we determine the impedance of the load, which is a parallel circuit:

$$z_L(j\omega) = \frac{1/j\omega C \times j\omega L}{1/j\omega C + j\omega L} = \frac{j\omega L}{1 - \omega^2 LC} \tag{3.10}$$

With the load impedance determined, the transfer function is expressed as:

$$H(j\omega) = \frac{z_L(j\omega)}{z_L(j\omega) + R} = \frac{j\omega L}{R(1 - \omega^2 LC) + j\omega L} \tag{3.11}$$

Figure 3.11 A first-order band-pass system approximating the relationship between a measurand and a sensing system.

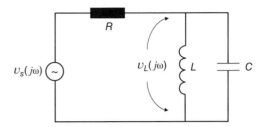

Eq. (3.11) approaches zero as ω approaches zero or infinity, suggesting that Figure 3.11 is a band-pass system. The transfer function will be a maximum when $R(1 - \omega^2 LC) = 0$. For this to happen the radial frequency of the source voltage should be equal to:

$$\omega_p = \frac{1}{\sqrt{LC}} \qquad (3.12)$$

where ω_p is called the resonance frequency, in radians per second. But what does ω_p signify? If the system is supplied with a signal having a radial frequency of ω_p, then this signal will reach the load without undergoing significant attenuation.

3.3 Impedance Matching

Even though voltage and current sources allow the variation of a load impedance, the amount of power that can be transferred from the source to the load depends on the relationship between the source and load impedance (Kong et al. 2010; Li et al. 2007; Liu et al. 2010). It so happens that maximum power is transferred to the load when the internal impedance of the source is equal to the load impedance. Consider Figure 3.3b, where the generator and the sensing systems are connected in series. The power drawn by the sensing system (load impedance) is equal to the load voltage multiplied by the current that passes through the load:

$$p = i \times v_L \qquad (3.13)$$

Where $i = v_L/Z_L$. Because of the series connection, the load voltage is given as:

$$v_L = \frac{Z_L}{Z_L + Z_S} v_S \qquad (3.14)$$

Hence, Eq. (3.13) can be rewritten as:

$$p = \left(v_S \frac{Z_L}{Z_L + Z_S} \frac{1}{Z_L} \right) \left(\frac{Z_L}{Z_L + Z_S} v_S \right) = v_S^2 \frac{Z_L}{(Z_L + Z_S)^2} \qquad (3.15)$$

As Z_L approaches zero, p in Eq. (3.15) approaches zero. Similarly, as Z_L approaches infinity, p approaches zero. To find the load impedance that maximises p, we have to differentiate Eq. (3.15) with respect to Z_L and set it to zero. Recalling that:

$$\frac{d}{dx} \left(\frac{f(x)}{g(x)} \right) = \frac{\dot{f}(x)g(x) - \dot{g}(x)f(x)}{g^2(x)} \qquad (3.16)$$

hence,

$$\frac{\partial p}{\partial Z_L} - v_S^2 \frac{\partial}{\partial Z_L}\left(\frac{Z_L}{(Z_L + Z_S)^2}\right) = \frac{(Z_1 + Z_S)^2 - 2(Z_1 + Z_S)Z_1}{(Z_1 + Z_S)^4} = 0 \tag{3.17}$$

Equation (3.17) yields $Z_L = Z_S$, which means that maximum power is transferred to the load when the load impedance is equal to the source impedance. Another reason for impedance matching is to prevent reflected power from building standing waves, which may damage the voltage source (in the case of biomedical devices, such as an electro-cardiogram, the body). In the absence of impedance matching, there will be reflected power from the load to the source, thereby building standing waves on the transmission circuits. Depending on the phase difference between the forward and reflected powers, the voltage produced by the reflected power may be add to the source voltage either constructively or destructively. If it is added destructively, a large amount of current may flow into the source, which may damage it (Ueno et al. 2007).

Example 3.1 We wish to design an impedance-matching circuit that should isolate a 5 kΩ voltage source from a 50 Ω sensor front-end (load), as shown in Figure 3.12.

The first step in designing impedance matching circuits (also known as matching networks in electronics) is to equate $Z_{in} = Z_0$. If we place Z_M between the load and the source as shown in the figure, the input impedance seen by the load impedance can be determined by applying Thévenin's theorem. So, when we short circuit the voltage source, the input impedance will be a parallel combination of R_S and Z_M. Hence,

$$Z_{in} = Z_L = \frac{Z_M R_S}{Z_M + R_S} = \frac{50\,k\Omega Z_M}{50\,k\Omega + Z_M} = 50\Omega \tag{3.18}$$

The simplest solution is to make Z_M a resistive component, which is insensitive to the frequency of the signal delivered by the source. In this case, we have:

$$50\,\Omega(50\,k\Omega + R_M) = 50\,k\Omega R_M \tag{3.19}$$

By rearranging terms we obtain:

$$R_M = \frac{50\,\Omega \times 50\,k\Omega}{50\,k\Omega - 50\,\Omega} \tag{3.20}$$

The problem with a resistive component is that there will be a voltage drop across it and, as a result, a power dissipation. This, of course, is a big disadvantage because the signal we get from the measurand is typically very small and further attenuating the signal (reducing its power) by the resistive component reduces its signal quality. What will happen if we exchange the resistive component with an LC circuit (which, ideally, should store energy rather than dissipate it), as shown in Figure 3.13 (top)? The input impedance of the Thévenin equivalent circuit is expressed as:

$$Z_{in} = \frac{R_S \frac{1}{j\omega C}}{R_S + \frac{1}{j\omega C}} + j\omega L = \frac{R_S}{1 + j\omega R_S C} + j\omega L = R_L \tag{3.21}$$

Simplifying Eq. (3.21) leads to:

$$R_L + j\omega R_L R_S C = R_S + j\omega L - \omega^2 R_S LC \tag{3.22}$$

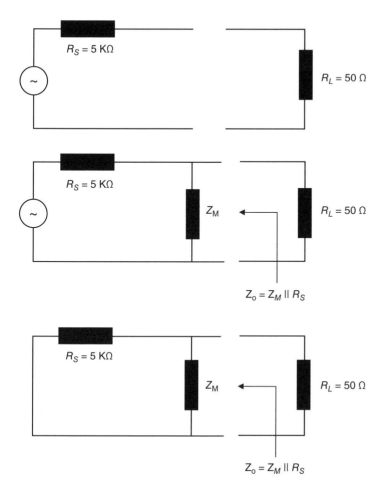

Figure 3.12 A resistive impedance-matching circuit for isolating a voltage source from a low-resistance load.

For the equality in Eq. (3.22) to hold, the real part of the left-hand term should be equal to the real part of the right-hand expression and the imaginary part of the left-hand expression should be equal to the imaginary part of the right-hand expression. Therefore, from the real terms we have:

$$R_L = R_S(1 - \omega^2 LC)$$
$$\frac{R_L}{R_S} = (1 - \omega^2 LC)$$
$$\omega^2 LC = 1 - \frac{R_L}{R_S} \tag{3.23}$$

Likewise, from the imaginary terms we have:

$$L = R_L R_S C \tag{3.24}$$

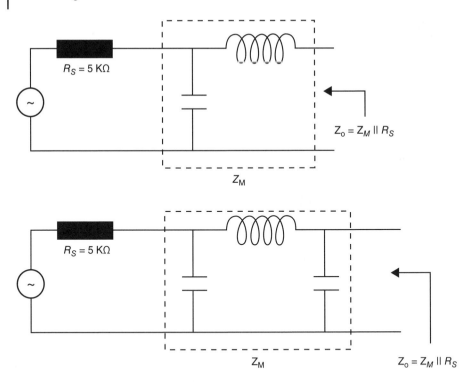

Figure 3.13 An LC impedance-matching circuit for isolating a voltage source from a low-resistance load: top, L-network; bottom, π-network.

If we insert Eq. (3.24) into Eq. (3.23), then we have:

$$C = \pm \sqrt{\left(1 - \frac{R_L}{R_S}\right) \frac{1}{\omega^2 R_L R_S}} \tag{3.25}$$

Notice that the final term, Eq. (3.25), is expressed in terms of R_L, R_S, and $\omega = 2\pi f$, all of which we already have. Once C is determined, then it is possible to determine L as well. In general, however, the LC match circuit is inefficient if the signal's frequency changes considerably. For example, for a wireless ECG or a 3D accelerometer, the frequency of the signal coming from the source (representing, for example, the heart beats and movements of different people doing different exercises) may vary considerably. In order to make sure that the matching circuit aptly deals with wide variations of input-signal frequency, more complex matching networks can be employed, for examples a π-network (Figure 3.13 (bottom)), a T-network, or a lattice network. Good and practical references can be found in the literature (Becciolini 1974; Bosco et al. 2003; Yang and Yacoub 2006).

Example 3.2 Suppose we model or approximate the heart, the interface between the heart and the electrodes (which include tissues, the skin, and the air interface between the skin and the electrodes), and the electrodes' resistance during an ECG measurement by a combination of a source voltage with an internal resistance, an L-impedance matching circuit, and a resistive load, respectively, as shown in Figure 3.14. Determine the transfer function of the set up.

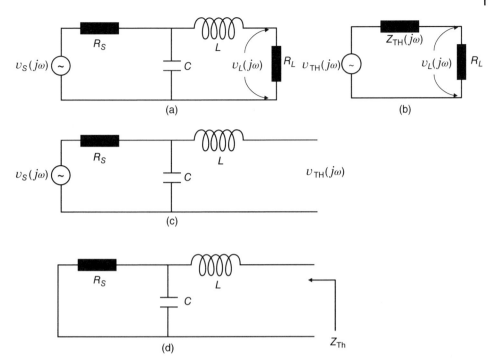

Figure 3.14 Approximating an ECG set up by a source voltage with an internal resistance, a match circuit, and a resistive load: (a) complete circuit; (b) Thévenin's equivalent circuit; (c) determination of Thévenin's voltage; (d) determination of Thévenin's impedance.

Since the transfer function establishes the relationship between the input and output voltages ($v_S(j\omega)$ and $v_L(j\omega)$, respectively) in the frequency domain, we should first establish the output voltage. One way to do this is to determine the Thévenin equivalent circuit and use a voltage divider between the load and the Thévenin impedance, as shown in Figure 3.14b. To do this, first we remove the resistive load and determine the Thévenin voltage, which is simply the voltage across the capacitor:

$$v_{th}(j\omega) = \frac{1/j\omega C}{1/j\omega C + R_S} v_S(j\omega) = \frac{1}{1 + j\omega R_S C} v_S(j\omega) \tag{3.26}$$

Similarly, the Thévenin impedance can be determined by short circuiting the voltage source and determining the impedance of the equivalent circuit. This is the same impedance we labelled as Z_{in} previously (in Eq. (3.21)). Hence:

$$Z_{th}(j\omega) = \frac{(R_S - \omega^2 R_S LC) + j\omega L}{1 + j\omega R_S C} \tag{3.27}$$

Consequently, the output (or load) voltage can be determined using voltage division:

$$v_L(j\omega) = \frac{R_L}{R_L + Z_{th}(j\omega)} v_{th}(j\omega) = \left(\frac{1}{1 + j\omega R_S C} \right) \left(\frac{R_L}{R_L + Z_{th}} \right) v_S(j\omega) \tag{3.28}$$

After simplifying Eq. (3.28), we obtain:

$$H(j\omega) = \left(\frac{R_L + j\omega R_S R_L C}{(R_S + R_L - \omega^2 R_S LC) + j\omega(L + R_S R_L C)} \right) \left(\frac{1}{(1 + j\omega R_S C)} \right) \tag{3.29}$$

3.4 Filters

The desired signal that is produced by a measurand rarely reaches the sensing system in its cleanest form. Instead, it will be mixed with undesired signals that the system internally produces or picks up from its surroundings. In some situations, the sensor output may slowly drift as a result of unintended displacement of sensors from their original positions, loose electrodes, shake, and vibration. Electrical filters are required to suppress the effects of noise artefacts. The choice of the appropriate types of filters depends on the spectrum of both the desired and the undesired signals as well as the spectral overlap between them. In most cases, the signals we wish to sense are low frequency (baseband) whereas the undesired signals can have both low-frequency and high-frequency components. The high-frequency components can be removed by using low-pass filters but other conditioning systems should be put in place to deal with the low-frequency components.

Figure 3.15 shows how movement artefacts create a drift (a gradual shift of the reference line) on the measurements of a wireless ECG. An interesting aspect of the drift is that it is a gradual process compared to the frequency of the ECG signal, so that it can be

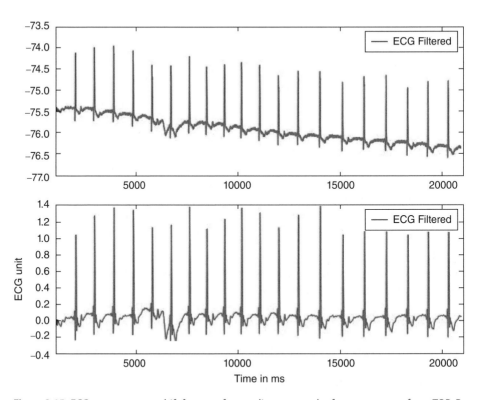

Figure 3.15 ECG measurements drift from a reference line as a result of movement artefacts. TOP: Raw ECG measurements (measured in mV). Bottom: The effect of drift is suppressed by applying a digital filter. Compared to the frequency of the ECG signal, the drift frequency is very small which enables the use of a high-pass filter.

considered as a low-frequency noise signal that is superimposed with the desired ECG signal. Therefore, we apply a digital high-pass filter to correct the effect of drift.

Broadly speaking, electrical filters can be passive or active and digital or analogue; for further reading, refer to Haykin et al. (1989); Jackson (2013); Oppenheim et al. (1989); Parks and Burrus (1987) and Su (2012). Active filters consist of active components such as transistors and operational amplifiers together with inductive, capacitive, and resistive elements. Active filters combine filtering with amplification but require a separate supply voltage and are susceptible to producing internal noise. The gain of an active filter for the desired frequency domain is typically very much greater than unity whereas for the undesired frequency domain it is considerably lower. Passive filters, on the other hand, consist of passive elements (resistive, inductive, resistive) only, and do not require an external supply voltage. The gain of the transfer function of a passive filter is less than one. Analogue filters are electrical circuits and are employed at the early stages of signal processing, so they belong to the conditioning subsystem. Digital filters are often algorithms applied on digital samples, and hence are applied at the later stages of signal processing. There are, however, exceptions to this; digital filters can also be hardware implementations consisting of inverters, adders, multipliers, and delay circuits.

In general, digital filters are more complex than analogue filters. The frequency beyond which the gain of a filter drops below −3 dB of the the maximum gain is called the cut-off or corner frequency. This frequency has to be carefully selected so that no useful signal is suppressed and no unwanted signal is admitted into the next stage.

In Figure 3.6 we regarded the sensing system as a whole as a first-order low-pass system (first-order, because there is a single RC circuit, as a result of which the transfer function of the system is first-order). Characterising the system as a whole should not be confused with intentionally inserting a low-pass filter between the source and the load in order to suppress noise. The insertion of a third entity between the source and the load changes the transfer function of the system. As an example, consider Figure 3.16 where we have a first-order low-pass filter. The output voltage is the voltage across the parallel $C_F R_L$ circuit and can be determined by voltage division:

$$V_L(j\omega) = \frac{Z_{C_F\|R_L}(j\omega)}{Z_{C_F\|R_L}(j\omega) + R_S + R_F} v_S(j\omega) = \frac{R_L}{R_L + (R_S + R_F)(1 + j\omega R_L C)} v_S(j\omega) \quad (3.30)$$

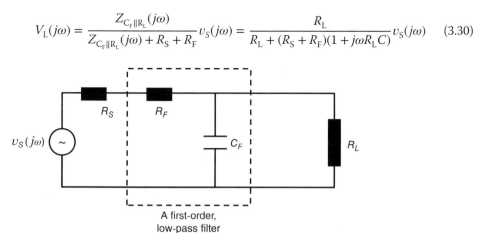

A first-order,
low-pass filter

Figure 3.16 A first-order low-pass filter inserted between the source (the measurand) and the load (the sensing load) to suppress high-frequency noise.

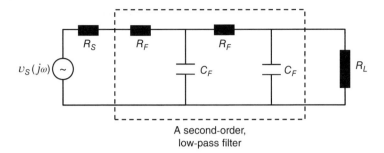

A second-order,
low-pass filter

Figure 3.17 A second-order low-pass filter inserted between the source (the measurand) and the load (the sensing load) to suppress high-frequency noise.

Therefore, the transfer function of the low-pass filter is:

$$H(j\omega) = \frac{R_L}{R_L + (R_S + R_F)(1 + j\omega R_L C_F)} = \frac{R_L}{R_R + j\omega R_I C_F} \tag{3.31}$$

where $R_R = R_S + R_F + R_L$ is the real part of the denominator of the transfer function and $R_I = R_L(R_S + R_F)$. Figure 3.16 is easy to realise, consumes a small amount of power, and occupies a small area in the sensing system. However, it is weak in suppressing unwanted high frequency noise and interfering signals. The suppression (rejection) capacity of the low-pass filter can be improved by adding another RC block as shown in Figure 3.17, making the filter a second-order low-pass filter.

In order to compute the transfer function of the second-order low-pass filter, we can simplify Figure 3.18 by combining the source and the filter resistance as $R_{in} = R_S + R_F$ and the filter capacitive impedance and the load impedance $(Z_L(j\omega) = X_C \| R_L)$, because the voltage across the load impedance and the capacitive impedance is the same. This is shown in Figure 3.18. With the simplification, Figure 3.18 becomes similar to Figure 3.7. Therefore, we can use the same procedure to express the load voltage in terms of the source voltage. Hence,

$$v_{th}(j\omega) = \frac{1/j\omega C_F}{1/j\omega C_F + R_{in}} v_S(j\omega) = \frac{1}{1 + j\omega R_{in} C_F} v_S(j\omega) \tag{3.32}$$

Similarly,

$$z_{th}(j\omega) = \frac{(1/j\omega C_F \times R_{in})}{(1/j\omega C_F + R_{in})} + R_F = \frac{R_{in} + R_F + j\omega C_F R_{in} R_F}{1 + j\omega C_F R_{in}} \tag{3.33}$$

The Thévenin equivalent circuit of the second-order low-pass system is shown in Figure 3.19. Consequently, the load voltage, $v_L(j\omega)$ can be be determined by applying voltage division on Figure 3.19:

$$v_L(j\omega) = \frac{Z_L(j\omega)}{Z_L(j\omega) + z_{th}(j\omega)} v_S(j\omega) \tag{3.34}$$

Finally, the transfer function of the second-order low-pass filter is:

$$H(j\omega) = \left(\frac{1}{1 + j\omega C_F R_{in}} \right) \left(\frac{R_L + j\omega C_F R_{in} R_L}{(R_R - \omega^2 C_F^2 R_m) + j\omega C_F (2R_{in} R_L + R_L R_F)} \right) \tag{3.35}$$

where $R_p = R_{in} + R_F + R_L$ and where $R_m = R_{in} R_F R_L$.

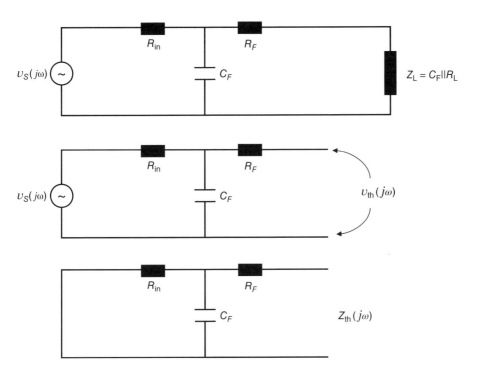

Figure 3.18 Top: A simplified version of the second-order low-pass filter. Middle: The load impedance is removed to compute the Thévenin voltage. Bottom: The load impedance is removed to compute the Thévenin impedance.

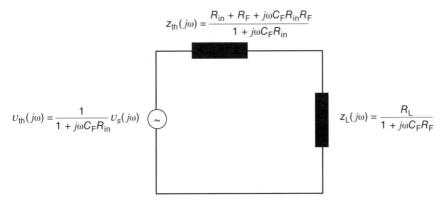

Figure 3.19 The Thévenin equivalent circuit of a second-order low-pass system.

Similarly, a high-pass and a band-pass filter (shown in Figure 3.20) can be inserted into the conditioning subsystem to suppress low-frequency components (with the high-pass filter) and both low- and high-frequency components (with the band-pass filter). Once again, the quality of suppression depends on the order of the filters. Most existing filters are second-order filters because of the good trade-off between suppression quality and design complexity.

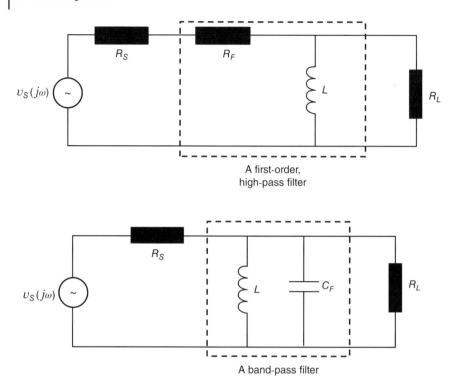

A first-order,
high-pass filter

A band-pass filter

Figure 3.20 A first-order high-pass filter (top) and a band-pass filter (bottom).

Example 3.3 We wish to design a low-pass filter that also serves as an impedance-matching network for interfacing the electrodes of a two-lead wireless ECG with a voltage (front-end) amplifier at 150 Hz. Suppose we can model the electrical circuit up to the output of the two electrodes as a voltage source having an impedance of $R_S - jXC_S$ ohms and the load impedance as $R_L - jXC_L$ ohms. The desired set up is shown in Figure 3.21. Assuming that all the source and load components are known, determine the filter components.

In order to ensure that maximum power is transferred at 150 Hz, the load impedance should be matched with the output impedance of the rest of the system as shown at the bottom of Figure 3.21. For this to happen, the real component of Z_O should be equal to the real component of Z_L and the imaginary component of Z_O should be equal to the imaginary component of Z_L. Hence:

$$Z_O = -jX_F \| (R_S + R_F - jXC_S) = \frac{-XC_SXC_F - jXC_F(R_S + R_F)}{(R_S + R_F) - j(XC_S + XC_F)} \qquad (3.36)$$

Since it is difficult to separate the real from the imaginary part in Eq. (3.36), we should multiply both the numerator and denominator by the conjugate of the denominator: $(R_S + R_F) + j(XC_S + XC_F)$. Consequently, the real part is given as:

$$\Re\{Z_O\} = \frac{(R_S + R_F)(XC_SXC_F)(1 + XC_F^2)}{(R_S + R_F)^2 + (XC_S + XC_F)^2} = R_L \qquad (3.37)$$

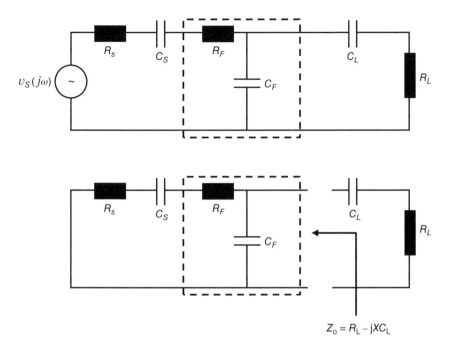

Figure 3.21 A low-pass filter optimised as an impedance matching network interfacing a two-lead wireless ECG with a voltage amplifier (top). The output impedance of the sensing system as seen by the front-end amplifier matched to the load impedance (bottom).

Likewise, the imaginary part is given as:

$$\Im\{Z_O\} = \frac{(XC_S + XC_F)(XC_S XC_F) - (R_S + R_F)^2 XC_F}{(R_S + R_F)^2 + (XC_S + XC_F)^2} = XC_L \tag{3.38}$$

We have now two equations for two unknowns. Therefore, by equating the last two equations, it is possible to determine R_F and C_F.

3.5 Amplification

The signal that can be produced by interfacing the sensing system with the measurand is normally very small in magnitude (of the order of a few millivolts). The signal first has to be amplified to an appreciable level before it can be further processed. A simple amplifier with gain A boosts the magnitude of the input signal by a factor of A (Figure 3.22). During the amplification process, there may be a phase difference between the input and the output voltages, depending on the ports to which the input voltage is supplied and from which the output voltage is taken. A non-inverting amplifier does not change the phase of the signal it amplifies and therefore the input and the output voltages are in phase (they vary over time in the same direction; see Figure 3.23). In an inverting amplifier, however, the input and the output voltages are out of phase by 180°. In other words, they vary over time in opposite directions (Figure 3.24).

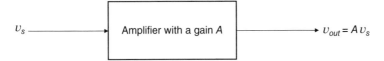

Figure 3.22 An open-loop amplifier that magnifies the input by a factor of *A*.

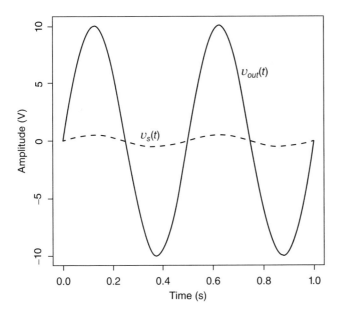

Figure 3.23 A non-inverter amplifier: the source and output voltages are in phase.

The amplification stage may involve several amplifiers connected to each other in tandem or having conditioning circuits (such as high-pass and low-pass filters) in between. Having two or more amplifiers in tandem enables their joint optimisation, which can then produce a collective amplification gain, but a noise inserted in the beginning will also be amplified in all the subsequent stages. Inserting filters in the intermediate stages has the advantage of controlling the amplification process and giving systematic suppression of undesired signals.

In general, the design process of practical amplifiers takes several factors into account. Fortunately, the signals that come from the sensing system are low-frequency and band-limited. These two aspects considerably simplify the design process. Nevertheless, factors such as impedance matching, sensitivity to internal and external noise (such as thermal noise), stability, and common mode rejection have to be properly addressed. Figure 3.25 displays the way the input and the output impedances of an amplifier should be matched to the output impedance of the preceding and the input impedance of the subsequent stages for maximum power to be transferred from the preceding to the last stage (from the sensing system to the low-pass filter, in this case).

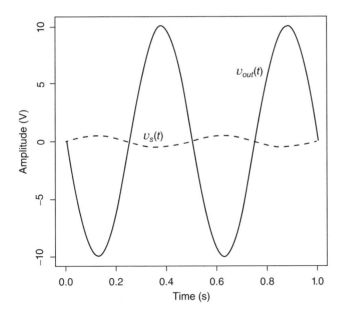

Figure 3.24 An inverter amplifier: the input and the output voltages are $180°$ out of phase.

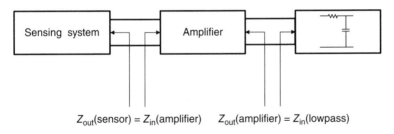

Figure 3.25 Impedance matching in amplifier design.

3.5.1 Closed-loop Amplifiers

The amplifier model we considered in Figure 3.22 is an open-loop amplifier. It is rarely used in practical situations because it is unstable. Because of variations in the internal impedance of the components from which amplifiers are made (transistors and operational amplifiers), their gain may not be constant for all operation frequencies and magnitudes of the input voltage. Similarly, offset voltages and thermal noise can influence the amplifier gain. A feedback loop taken from the output and supplied to the input frees the overall gain of the amplifier from all these factors. Consider Figure 3.26. The output of the feedback equals:

$$v_{out} = A v_{in} \tag{3.39}$$

But,

$$v_{in} = v_S + v_f = v_S + \beta v_{out} = v_S + \beta A v_{in} \tag{3.40}$$

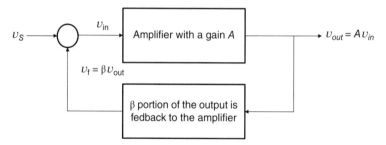

Figure 3.26 A closed-loop (feedback) amplifier.

Reorganising the terms in Eq. (3.40) yields:

$$v_{in} = \frac{v_S}{(1 - \beta A)} \tag{3.41}$$

Finally, combining Equations (3.39) and (3.41), the overall gain of the closed-loop amplifier is expressed as:

$$A_v = \frac{v_{out}}{v_S} = \frac{A}{(1 - \beta A)} \tag{3.42}$$

The open loop gain A is a large quantity. If β is chosen, such that $\beta A \gg 1$, then Eq. (3.42) reduces to:

$$A_v = \frac{1}{\beta} \tag{3.43}$$

Here we disregarded the negative sign, because it refers to the direction and not the magnitude of the relationship between the input and output. Consequently, the overall amplifier gain A_v is no longer dependent on the amplifier but on the feedback component, making the amplifier stable. The sensitivity of an amplifier expresses the portion of variation in the overall gain, A_v, of the feedback amplifier as a result of the variation in the open-loop gain, A; in other words, how the internal dynamics of the amplifier, such as thermal noise and drift, affect its overall gain:

$$\sigma = \frac{dA_v/A_v}{dA/A} \tag{3.44}$$

Differentiating A_v with respect to A and rearranging terms produces the following:

$$\frac{dA_v}{dA} = \frac{(1 - \beta A) - (-\beta)A}{(1 - \beta A)^2} = \frac{1}{(1 - \beta A)^2} \tag{3.45}$$

Rewriting Eq. (3.44) in terms of Eq. (3.45) yields:

$$\sigma = \frac{1}{(1 - \beta A)} \tag{3.46}$$

which confirms our statement that the variation of the open-loop gain does not greatly affect the gain of the feedback amplifier.

Figure 3.27 The signal induced by the power line is coupled into the sensing system (ECG) and is difficult to isolate using a low-pass filter, because its frequency domain overlaps with the frequency domain of the signal that the human body generates.

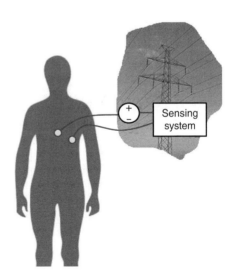

3.5.2 Difference Amplifier

Filters and amplifiers are useful to condition the output of a sensor, but they are unable to suppress noise having a spectrum overlapping with the useful signal. Consider Figure 3.27, where the spectrum of the desired signal that can be generated by the ECG overlaps with the spectrum of the signal coming from a nearby power line. Thermal noise, vibration, and surrounding pressure produce similar effects. Hence, the desired signal and the noise will be captured by the sensor and the output is a superposition of the two signals (see Figure 3.28). For our example, the magnitude of the desired signal

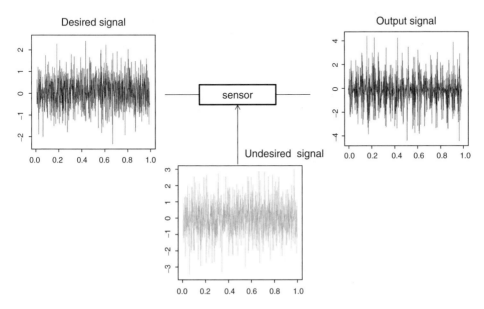

Figure 3.28 Where a sensor picks up two signals with the same frequency spectrum, it is difficult to isolate the desired signal using a filter.

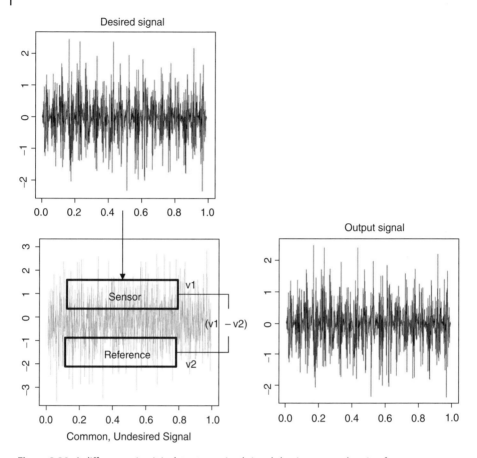

Figure 3.29 A difference circuit isolates two mixed signals having an overlapping frequency spectrum. The idea is to have a reference signal producing a comparable effect on differentially connected components, but the mixed signal is supplied only to one of the components, which is a sensor for our case.

(approximately 5 mV is not sufficiently large compared to the noise signal, so that it is not helpful to use a limiter circuit to suppress the noise.

Almost all practical sensor implementations employ some form of differential circuits to deal with this situation. The idea is as follows: the sensor is placed together with a reference element inside the process being monitored. The reference element does not respond to the physical process, but the noise induces in both the sensor and the reference element approximately the same amount of voltage. Then the voltages of the two elements are added differentially (subtractively), thereby effectively suppressing the voltage common to both. This is depicted in Figure 3.29.

The Wheatstone bridge (Figure 3.30) is a typical example of a differential circuit widely used in the design of sensors (Boisen et al. 2000; Lim et al. 2005; Pallas-Areny and Webster 2001; Rife et al. 2003). Three fixed and one variable (the sensor) resistors are connected to form a bridge, which has two wings: the left wing and the right wing.

Figure 3.30 The Wheatstone bridge as a differential circuit.

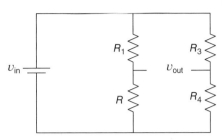

The output voltage is taken from the middle of the two wings and can be expressed as:

$$v_{\text{out}} = v_{\text{in}} \left(\frac{R_4}{R_4 + R_3} - \frac{R}{R + R_1} \right) \tag{3.47}$$

Under normal conditions – when the physical process does not yield anything to sense – the bridge should be balanced and V_{out} should be zero, even if there is noise from the surroundings. For this to happen, the resistance of all the resistors should be the same. If noise affects the sensor (R), it should also affect all the other resistors, making the net noise voltage at the output zero. When the physical process releases energy, signifying an interesting phenomenon, then the value of R changes and the bridge becomes unbalanced, the degree of unbalance being proportional to the signal generated by the physical process.

The problem with the Wheatstone bridge is that it is made up of resistive components only. Passive electrical components, most significantly resistors, produce a voltage drop across them and a power dissipation. This means, the amplitude of the desirable signal will be attenuated when it passes through them. Since the signal produced by the sensor is inherently weak, forcing it to pass through the Wheatstone bridge makes it even weaker. Therefore, the alternative solution is to combine difference circuits with amplification.

However, care must be taken when amplifiers are used in combination with a difference circuit. Due to the presence of a feedback path (as we have already seen above), differential amplifiers produce not only the desired differential voltage, but also a common mode voltage, which introduces unwanted effects on the output of the amplifier. To understand this, consider Figure 3.31a. As can be seen, we wish to mix at the amplifier (A) the output of the sensor (which now has an impedance of Z_1) and the reference element (with impedance Z_2) differentially and pass on the output to the next stage via Z_3. To complete the circuit, the output is connected to both elements. In doing so, however, we have created two feedback paths for currents to flow. The first path (the middle figure) establishes the differential path because the current that flows in this circuit is the difference of I_1 and I_2. The second path (the bottom figure) permits the current to flow through Z_3 to the next stage and it is a common mode circuit, because the current is the addition of I_1 and I_2. A good differential amplifier is one that rejects the common mode current (it should be noted that no differential current flows through Z_3).

The differential mode and the common mode currents depend on the differential and the common mode input impedances, respectively. To compute these impedances, we shall define (simply for convenience) two types of voltages and currents. Let the voltage that produces the differential mode current be called the differential voltage (V_d) and the voltage that produces the common mode current be termed the common mode voltage (V_c).

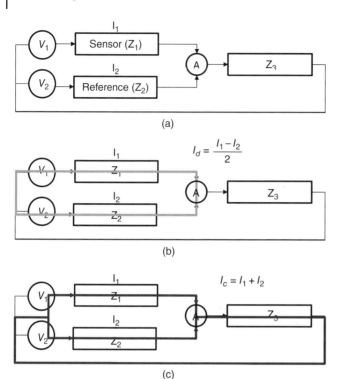

Figure 3.31 Common and differential mode voltages in a differential amplifier: (a) abstraction of the common and the differential modes; (b) the current flowing in the loop is the difference of I_1 and I_2 – it is a differential current; (c) the current flowing in the loop (through Z_3) is the confluence of I_1 and I_2 – it is a common mode current.

$$V_c = \frac{V_1 + V_2}{2} \tag{3.48}$$

And,

$$V_d = V_1 - V_2 \tag{3.49}$$

Similarly, let the differential and the common mode currents be defined as follows:

$$I_c = I_1 + I_2 \tag{3.50}$$

and,

$$I_d = \frac{I_1 - I_2}{2} \tag{3.51}$$

With these definitions, it is possible to replace Figure 3.31 with Figure 3.32 (because it simplifies the analysis). By applying the superposition theorem, the differential and common mode input impedances can be determined. For the differential mode, let $V_c = 0$ (see Figure 3.33). Hence,

$$I_c = I_1 + I_2 = \frac{1}{Z_1}\left(\frac{V_d}{2} - V_x\right) + \frac{1}{Z_2}\left(-\frac{V_d}{2} - V_x\right) = \frac{-2V_x}{Z} \tag{3.52}$$

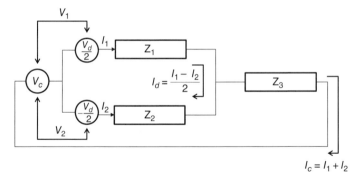

Figure 3.32 A modified version of the common and differential mode voltages in a differential amplifier.

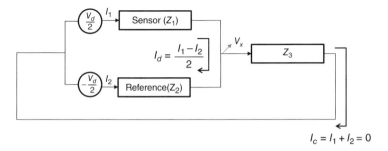

Figure 3.33 The differential amplifier with the common mode voltage set to zero.

Where we have assumed $Z_1 = Z_2 = Z$. If $V_x = 0$, then $I_c = 0$. Similarly, the differential current can be expressed as:

$$I_d = I_1 - I_2 = \frac{V_d}{2Z_1} - \left(\frac{V_d}{2Z_2}\right) = V_d\left(\frac{1}{Z_1 + Z_2}\right) = \left(\frac{V_d}{2Z}\right) \tag{3.53}$$

The input impedance as seen by the differential mode current is defined as the ratio of the differential mode voltage to the differential mode current:

$$Z_{\text{in}-\text{dm}} = \frac{V_d}{I_d} = 2Z \tag{3.54}$$

For the common mode, let $V_d = 0$ (see Figure 3.34). Thus,

$$I_c = \frac{V_c}{Z_1 \| Z_2 + Z_3} = \frac{V_c}{\frac{Z}{2} + Z_3} \tag{3.55}$$

The differential current is zero because:

$$2I_d = I_1 - I_2 = \frac{I_c}{2} - \frac{I_c}{2} = 0 \tag{3.56}$$

The input impedance as seen by the common mode current is defined as the ratio of the common mode voltage to the common mode current:

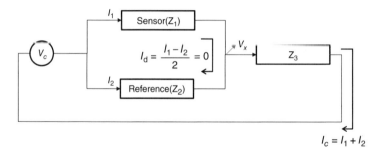

Figure 3.34 The differential amplifier with the differential mode voltage set to zero.

$$Z_{in-cm} = \frac{V_c}{I_c} = \frac{Z}{2} + Z_3 \tag{3.57}$$

From Eq. (3.57) it is apparent that to prevent the common mode current from flowing into the next stage, Z_3 should be large; we cannot change Z without affecting the differential mode input impedance.

References

Becciolini B 1974 Impedance matching networks applied to RF power transistors. *Motorola Appl, Note AN-721.*

Boisen A, Thaysen J, Jensenius H and Hansen O 2000 Environmental sensors based on micromachined cantilevers with integrated read-out. *Ultramicroscopy*, **82** (1), 11–16.

Bosco BA, Emrick RM and Franson SJ 2003 Integrated impedance matching and stability network. US Patent 6,531,740.

Haykin SS, Moher M and Song T 1989 *An Introduction to Analog and Digital Communications*, vol. 1. John Wiley.

Jackson LB 2013 *Digital Filters and Signal Processing: With MATLAB® Exercises.* Springer Science & Business Media.

Kimura J 2013 *Electrodiagnosis in Diseases of Nerve and Muscle: Principles and Practice.* Oxford University Press.

Kong N, Ha DS, Erturk A and Inman DJ 2010 Resistive impedance matching circuit for piezoelectric energy harvesting. *Journal of Intelligent Material Systems and Structures*, **21** (13), 1293–1302.

Li M, Tang HX and Roukes ML 2007 Ultra-sensitive NEMS-based cantilevers for sensing, scanned probe and very high-frequency applications. *Nature Nanotechnology*, **2** (2), 114–120.

Lim HC, Schulkin B, Pulickal M, Liu S, Petrova R, Thomas G, Wagner S, Sidhu K and Federici JF 2005 Flexible membrane pressure sensor. *Sensors and Actuators A: Physical*, **119** (2), 332–335.

Liu N, Mesch M, Weiss T, Hentschel M and Giessen H 2010 Infrared perfect absorber and its application as plasmonic sensor. *Nano Letters* **10** (7), 2342–2348.

Nemati E, Deen MJ and Mondal T 2012 A wireless wearable ECG sensor for long-term applications. *Communications Magazine, IEEE*, **50** (1), 36–43.

Oppenheim AV, Schafer RW, Buck JR *et al.* 1989 *Discrete-time Signal Processing*, vol. 2. Prentice Hall.

Pallas-Areny R and Webster JG 2001 *Sensors and Signal Conditioning*, vol. 1. Wiley-Interscience.

Parks TW and Burrus CS 1987 *Digital Filter Design*. Wiley-Interscience.

Pennisi S 2005 High-performance and simple CMoS interface circuit for differential capacitive sensors. *Circuits and Systems II: Express Briefs, IEEE Transactions on*, **52** (6), 327–330.

Rife J, Miller M, Sheehan P, Tamanaha C, Tondra M and Whitman L 2003 Design and performance of GMR sensors for the detection of magnetic microbeads in biosensors. *Sensors and Actuators A: Physical*, **107** (3), 209–218.

Su KL 2012 *Analog Filters*. Springer Science & Business Media.

Ueno A, Akabane Y, Kato T, Hoshino H, Kataoka S and Ishiyama Y 2007 Capacitive sensing of electrocardiographic potential through cloth from the dorsal surface of the body in a supine position: a preliminary study. *Biomedical Engineering, IEEE Transactions on*, **54** (4), 759–766.

Webster JG 2009 *Medical Instrumentation Application and Design*, 4th edn. Wiley Publishing.

Yang GZ and Yacoub M 2006 *Body Sensor Networks*. Springer.

4

Electrical Sensing

Traditionally, a transducer is the basic element that transforms one form of energy to another form of energy. When this transducer integrates additional elements such as a conditioning circuit, the entire assembly is called a sensor. The term sensing refers to the spectrum of interaction between the measurand and the transducer or the sensing element. Regardless of the nature of the measurand:

- if its initial effect upon the sensing element is modifying the electrical characteristic of the sensing element, we call this interaction an electrical sensing
- if it is modifying the quantum characteristics of the sensing element (that is, if the measurand in some way interacts with quantum level particles such as photons and individual electrons), we call this interaction collectively optical sensing
- if the interaction modifies the magnetic properties of the sensing element, then the interaction is referred to as magnetic sensing.

In most cases, it is possible to determine the type of sensing by referring to the energy the measurand releases. For example, if the measurand releases ultrasonic energy, then this energy can be picked up by an ultrasonic sensor. Therefore, we can refer to the sensing type as an ultrasonic or ultrasound sensing. But there are situations where the nature of the measurand may not correspond with the sensing type. For example, a mechanical force or pressure may modify the electrical, optical, or magnetic characteristics of a sensing element, but we rarely refer to it as a mechanical sensing.

In this chapter, we shall be dealing with how a measurand modifies the electrical properties of different sensing elements. Specifically, we shall be dealing with resistive, capacitive, and inductive sensing as well as the thermoelectric effect. In resistive sensing, we shall investigate how the resistance or the resistivity of a sensing element changes as a result of a change in the measurand; similarly, in capacitive and inductive sensing, we shall investigate how the capacitance (inductance) of a capacitive (inductive) sensing element changes in response to a change in some aspect of the measurand. Finally, in the last section, we shall deal with thermocouples, investigating how they transform heat energy to electrical energy.

Regardless of the initial interaction between the measurand and the sensing element, the final output of most existing sensors is an electrical quantity such as a voltage, a current, or power, because the sensor is an integral part of a more complex electrical system, as considered in Chapter 1.

Principles and Applications of Ubiquitous Sensing, First Edition. Waltenegus Dargie.
© 2017 John Wiley & Sons, Ltd. Published 2017 by John Wiley & Sons, Ltd.
Companion Website: www.wiley.com/go/dargie2017

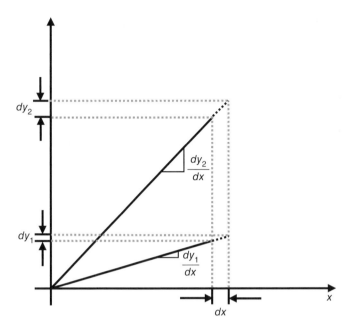

Figure 4.1 The proportion of change in a dependent variable is determined by both the amount of change in the independent variable and the slope of the linear equation.

4.1 Resistive Sensing

Resistive sensors change their resistance in response to a change in the physical process in which they are placed or embedded. The perceived change in resistance is typically due to an applied pressure or force. Recall that the electrical resistance of a resistor depends on

- its resistivity (ρ) – the material from which it is made
- its length (l)
- the cross-sectional area (A).

That is:

$$R = \rho \frac{l}{A} \tag{4.1}$$

The resistance changes if any of the parameters in Eq. (4.1) changes. The amount of change each parameter introduces depends on its relationship with R. To highlight this point, consider Figure 4.1. Both y_1 and y_2 change in response to the change in x, dx, but the amount of change they experience depends on the slopes of the graphs:

$$\frac{dy_1}{dx}$$

and

$$\frac{dy_2}{dx}$$

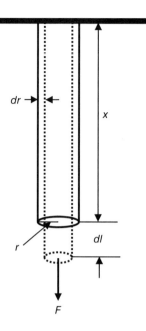

Figure 4.2 A change in the resistance of a resistor in response to an applied force and a change in the dimension of the resistor.

For the resistance in Eq. (4.1), the overall change can be expressed as:

$$dR = \frac{\partial R}{\partial \rho} d\rho + \frac{\partial R}{\partial l} dl + \frac{\partial R}{\partial A} dA \tag{4.2}$$

Inserting the right-hand side of Eq. (4.1) into Eq. (4.2) yields,

$$dR = \frac{l}{A} d\rho + \frac{\rho}{A} dl - \frac{\rho l}{A^2} dA \tag{4.3}$$

Often, the relative change in resistance to a nominal value (the original R) is of interest. Taking into account the fact that for a cylindrical structure (as shown in Figure 4.2), $A = \pi r^2$ and $dA = 2\pi r dr$, the relative change of resistance due to an applied force or pressure can be expressed as:

$$\frac{dR}{R} = \frac{d\rho}{\rho} + \frac{dl}{l} - \frac{2dr}{r} \tag{4.4}$$

Two of the parameters in Eq. (4.1) are mutually dependent. For instance, if the length of the resistor changes as a result of a force that pulls the resistor downwards (see Figure 4.2), then the change in length makes the radius of the resistor smaller, forcing the cross-sectional area to shrink. For most materials the relative change in radius as a result of a relative change in length can be experimentally determined, and is described by the Poisson number, v:

$$v = -\frac{dr/r}{dl/l} \tag{4.5}$$

The relative change in length (or, in general, the relative change in the structure along the direction of force) is known as the mechanical strain, $\epsilon = \frac{dx}{x}$. With this in mind,

$$-v\epsilon = \frac{dr}{r} \tag{4.6}$$

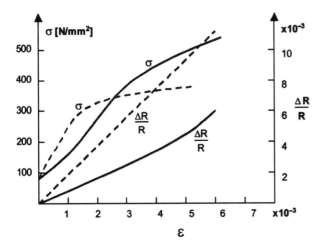

Figure 4.3 The relationship between strain, (ϵ), stress (σ), and relative change in resistance for manganin (solid line) and constantan (dashed line).

In summary, the relative change in resistance can be expressed as:

$$\frac{dR}{R} = \epsilon(1 + 2v) + \frac{d\rho}{\rho} \tag{4.7}$$

The right-hand side of Eq. (4.7) has two components: the first quantifies the effect of dimension on the relative change in resistance while the second quantifies the effect of resistivity. Figure 4.3 displays the relationship between strain (ϵ), stress $\left(\sigma = \frac{F}{A}\right)$, and the relative change in resistance $\left(\frac{\Delta R}{R}\right)$ for two different metal alloys. In practice, the gauge factor (GF), which is the ratio of the relative change in resistance to the mechanical strain (which is a measure of structural deformity), is used to measure the change in resistance as a result of force or pressure:

$$GF = \frac{\frac{dR}{R}}{\epsilon} = (1 + 2v) + \frac{1}{\epsilon}\frac{d\rho}{\rho} \tag{4.8}$$

For most metals, GF is between 1.6 and 2, whereas for platinum it is approximately 6. Moreover, the change in resistance is below 2%.

Figure 4.4 (reproduced here from Chapter 2) shows a piezoresistive sensor – a sensor whose resistance changes as a result of applied pressure – attached to a supporting cable to measure the horizontal oscillation of a building to a forced or ambient excitation (impact hammer, shaker, wind, or the movement of cars). As the building moves back and forth, it elongates and compresses the piezoresistive sensor, thereby changing its length and cross-sectional area and changing its resistance. The magnitude of the building's oscillation can be determined by directly measuring the current that flows through a Wheatstone bridge in which the piezoresistive sensor makes up a part of one of the wings.

Similarly, Figure 4.5 shows a simple water-quality sensor consisting of two electrodes. When the water is clean, it has a resistance of R, which is taken as a reference, but as the quality of the water changes, its resistivity changes as well. The relative change in

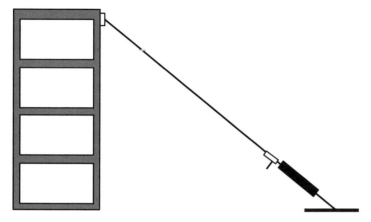

Figure 4.4 A piezoresistive sensor attached to a supporting cable can be used to measure the natural and forced responses of a building.

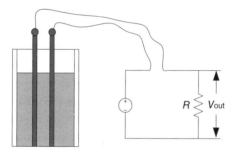

Figure 4.5 As the quality of water changes, its resistivity changes as well. The relative change in resistance can be sensed by measuring the output voltage across the resistance R.

resistance can be accounted for by measuring the output voltage across the resistance R. The same principle is used in medicine to measure skin conductance, skin moisture, and sweat.

One of the most widely used resistive sensors is the strain gauge. It is essentially a thin wire (approximately 25 μm in length) that can be stretched within its elastic limit (so that the wire can relax back to its original size). As the wire is stretched, its dimensions change and with them its resistance. The change in resistance can be mapped to pressure, force, or displacement. As can be seen in Figure 4.6, the strain gauge is fixed between a fixed frame and a moveable armature (which can be interfaced with a diaphragm). Initially, the wire is stretched to its maximum limit in order to ensure that the measurand has a contraction effect. A single strain gauge, however, is rarely used. The most widely used configuration is a combination of four strain gauges, as shown in Figure 4.7. This configuration measures both stretching and contraction at the same time thereby improving the sensing accuracy. Figure 4.8 shows the equivalent electrical circuit, which is a Wheatstone bridge with four active elements.

Example 4.1 We wish to employ a metal foil having a GF of 4 and a resistance of 150 Ω to measure the linear displacement of a suspension cable. If the maximum strain the cable applies to the foil is 10 microstrain, what should be the change in resistance in the metal foil that should be measured?

Figure 4.6 The basic principle of a strain gauge.

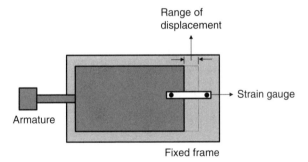

Figure 4.7 A combination of four strain gauges. When two of the strain gauges (the ones connected to the armatures on the left) are stretched, the other two (the ones connected to the armatures on the right) are contracted. Since the change in stretching and contraction should be proportional, the sensing sensitivity is better than when a single strain gauge is used.

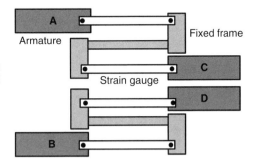

Figure 4.8 The electrical equivalent circuit of the four strain gauges forming a Wheatstone bridge. The potentiometer with a resistance R_p and the tuning component with a resistance R_t are used to calibrate (balance) the Wheatstone bridge. The four active elements are mapped to Figure 4.7 as follows: $R_1 = A, R_2 = B, R_3 = C, R_4 = D$.

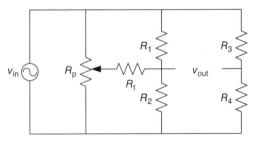

From Eq. (4.8), we know the relationship between GF and the change in resistance:

$$GF \times \epsilon = \frac{\Delta R}{R}$$

Hence,

$$\Delta R = GF \times \epsilon \times R = 4 \times 150\,\Omega \times 10 \times 10^{-6} \quad = 6\,\text{m}\Omega$$

Example 4.2 We wish to set up a Wheatstone bridge consisting of a strain gauge as shown in Figure 4.9. If the nominal value (the value in the absence of a strain) of the strain gauge (R_G) equals $R = R_1 = R_2 = R_3$, derive an expression for the transfer function of the Wheatstone bridge in terms of GF and ϵ of the strain gauge when it is active (that is, when the Wheatstone bridge is no longer balanced).

In the absence of a strain, the Wheatstone bridge is said to be balanced and $V_{out} = 0$. Moreover,

$$V_{out} = \left(\frac{R_3}{R_3 + R_G} - \frac{R_2}{R_2 + R_1} \right) V_{in} \tag{4.9}$$

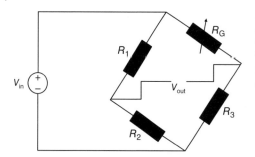

Figure 4.9 A strain gauge sensor making up one of the elements of a Wheatstone bridge. In the absence of strain, the nominal resistance of the strain gauge is equal to $R_1 = R_2 = R_3 = R$ But when the strain gauge is active, then $R_G = R + \Delta R$.

From which we have

$$H = \frac{V_{out}}{V_{in}} = \frac{R_3}{R_3 + R_G} - \frac{R_2}{R_2 + R_1} = \frac{R}{R + (R + \Delta R)} - \frac{R}{R + R} = \frac{1}{2 + \frac{\Delta R}{R}} - \frac{1}{2} \quad (4.10)$$

Taking into account that $\Delta R/R = \epsilon GF$ and rearranging terms will yield:

$$H = \frac{-\epsilon GF}{2(2 + \epsilon GF)} \quad (4.11)$$

4.2 Capacitive Sensing

Due to the flexibility and relative ease with which they can be designed and produced, capacitive sensors are the most pervasive and extensively used sensor types. They are used to measure, among other things:

- touch (as in the design of touch screens in smart phones; (Ningrat et al. 2009; Schulz et al. 2004; Snyder and Seguine 2006)
- displacement (Kim et al. 2006)
- force, velocity, acceleration, pressure (Shaeffer 2013)
- flow (Dagamseh et al. 2013; Xie et al. 1992)
- proximity (Lee et al. 2009)
- thickness (Ueno et al. 2007)
- humidity (Rivadeneyra et al. 2014)
- vibration (Chiu and Tseng 2008; Kuehne et al. 2008)
- chemical and biological phenomena (Beyeler et al. 2007; Satyanarayana et al. 2006; Waggoner and Craighead 2007)

Whenever two conductive plates are separated by a dielectric material (essentially an electrical insulator that can be polarized by an electric field) as in Figure 4.10, they create a capacitor. The capacitance of the capacitor (namely, its ability to store an electric charge) depends on the cross-sectional area, A, of the two plates, the distance of separation between the plates, l, and the dielectric material:

$$C = \epsilon_0 \epsilon_r \frac{A}{l} \quad (4.12)$$

where ϵ_0 is the permittivity of free space and ϵ_r is the relative static permittivity of the dielectric material. The permittivity of a material is the measure of how well it stores (or leaks) charged particles.

Figure 4.10 The basic structure of a capacitor.

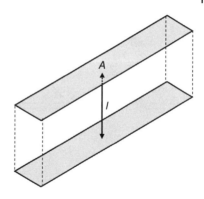

Figure 4.11 A moving armature changes the dielectric of two capacitors that are connected in series.

Apparently, the capacitance changes if any of the capacitor's parameters changes. This can be done in a number of ways. One way is to change the relative permittivity. For example, a humidity sensor measures the change in capacitance (or the voltage induced as a result of this) as water vapour changes the composition of the dielectric material. Alternatively, a dielectric can be inserted into and removed from two parallel plates, thereby changing the permittivity of the capacitor they form. Figure 4.11 shows how a moving armature can change the dielectric of a capacitor, so that the change in the capacitance of the capacitor can be mapped to a displacement. Similarly, most touch-sensitive sensors are capacitive sensors and react to a change in the dielectric composition of a capacitor.

As can be seen in Figure 4.12, when a potential difference exists between the two plates of a capacitor, a magnetic field is set up, emanating from the positive plate and entering into the negative plate. The density and distribution of this field partly depends on the relative permittivity of the dielectric material. When a finger rests or slides on a pad which is placed on top of the two plates, it disturbs the electric field distribution and changes the capacitance of the capacitor. In a simple sensor, the desired event is the detection of the presence or absence of a touch, which can easily be determined by

Figure 4.12 A simple touch can influence the distribution of the electric field between the two plates of a capacitor. The change in the electric field can be attributed to the presence or absence of a touch.

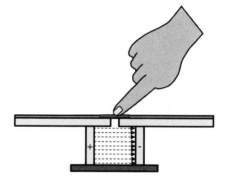

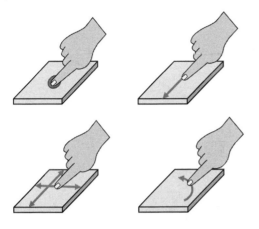

Figure 4.13 The basic design principle of single and multidimensional Atmel capacitive buttons, sliders, and wheel sensors. As the finger rests or slides on the pad, it disturbs the electric field distribution between the two plates of the capacitor, thereby modifying the relative permittivity of the dielectric material. This in turn changes the capacitance of the capacitor.

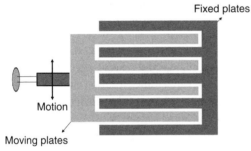

Fixed plates

Motion

Moving plates

Figure 4.14 A capacitor consisting of stationary and rotary plates. The change in the capacitance depends on the degree of alignment between the two plates.

quantifying the magnitude of the disturbance of the field. Figure 4.13 shows the essential design approach for different Atmel capacitive touch sensors (buttons, sliders, and wheels sensors).

The distribution of electric field can also be changed by changing the alignment of the two plates. This is useful to measure quantities such as angular displacement and rotation. Figure 4.14 displays multiple stationary and rotary plates constituting a capacitor. When the rotary plates change position, the alignment between them and the static plates changes too, thereby changing the amount of electric field that can be coupled between the two sets of plates. As a result, the overall capacitance changes. The change in capacitance is a measure of the degree of alignment between the rotary and the stationary plates. Alternatively, the measurand can change the distance of separation, as in Figure 4.15. The change in the distance of separation, Δl, can correspond with displacement or pressure, in which case the change in capacitance can be expressed as:

$$\frac{dC}{dl} = -\epsilon_0 \epsilon_r \frac{A}{l^2} = \frac{C}{l} \tag{4.13}$$

Hence,

$$\frac{dC}{C} = -\frac{dl}{l} \tag{4.14}$$

which is independent of ϵ_0 and ϵ_r. For a small change in l, the relationship is linear and dC/dl can be approximated by $\Delta C/\Delta l$. The linearity of Figure 4.15 can be improved by

Figure 4.15 The distance separating the two plates of a capacitor can be varied as a function of some aspects of the measurand in order to establish a quantitative relationship between the measurand and the capacitance of a capacitive sensor.

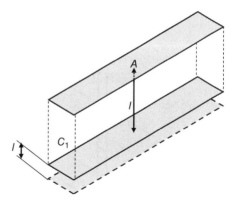

Figure 4.16 A differential capacitive sensor which increases the linear sensing range.

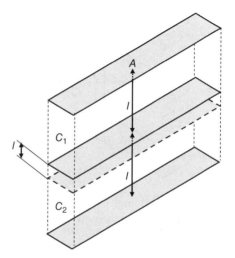

designing a differential capacitance, as in Figure 4.16. The change in capacitance (ΔC) can be expressed as:

$$\Delta C = C_2 - C_1 = \epsilon_0 \epsilon_r A \left(\frac{1}{l - \Delta l} - \frac{1}{l + \Delta l} \right) = \epsilon_0 \epsilon_r A \frac{2\Delta l}{l^2 - (\Delta l)^2} \tag{4.15}$$

From which we can approximate:

$$\frac{\Delta C}{C} \approx \frac{2\Delta l}{l} \tag{4.16}$$

because:

$$1 - \left(\frac{\Delta l}{l} \right)^2 \approx 1 \tag{4.17}$$

Example 4.3 We wish to increase the linearity of a differential capacitive sensor to measure applied pressure or force by carefully inserting three movable capacitive plates as shown in Figure 4.17 which undergo displacement when a force or a pressure of known magnitude is applied to the capacitive sensor. The two outer plates of the sensor are unmoving. Derive an expression for relating the change in the overall capacitance to the change in the overall displacement of the capacitive sensor.

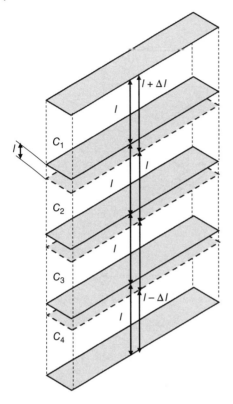

Figure 4.17 A differential capacitive sensor with three movable plates enclosed inside two fixed parallel plates for measuring force or pressure as a function of the overall displacement of the movable plates.

The overall change in the capacitance of the capacitive sensor can be expressed as the difference between the new capacitance and the initial or nominal capacitance:

$$\Delta C = C_{new} - C_{initial} = \epsilon_0 \epsilon_r A \left(\frac{1}{l + \Delta l} + \frac{1}{l} + \frac{1}{l} + \frac{1}{l - \Delta l} - \frac{1}{4l} \right)$$

Notice that the two central capacitors do not have a net change in displacement. After simplifying, we have:

$$\Delta C = \epsilon_0 \epsilon_r A \left(\frac{2l}{(l + \Delta l)(l - \Delta l)} - \frac{2}{l} \right)$$

From which we have,

$$\frac{\Delta C}{C} = \frac{2(\Delta l/l)^2}{1 - (\Delta l/l)^2} \approx \left(\frac{2\Delta l}{l} \right)^2$$

Example 4.4 A differential capacitive accelerometer is modelled as a spring–mass system, as shown in Figure 4.18. When a force is applied to the outer frame of the accelerometer (for example, when the frame is shaken), the mass of the accelerometer experiences acceleration, and oscillates up and down. If the mass is attached to a differential capacitive sensor (CS), derive an expression relating the acceleration with the displacement of the capacitive sensor, assuming that both the external frame and the mass of the accelerometer undergo a sinusoidal oscillation with the same radial frequency (but with different magnitudes).

Figure 4.18 A differential capacitive accelerometer modelled as spring–mass system.

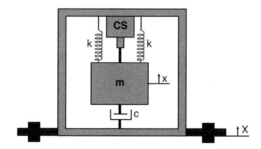

When the accelerometer is shaken, three types of forces will act and react upon it:

- the first is proportional to its mass and the acceleration of the oscillating mass.
- the second is proportional to the spring constant k and the displacement of the oscillating mass (following Hook's law)
- the third is proportional to the damping constant c and the velocity of the oscillating mass (following Newton's third law).

Assuming that $X(t)$ is the position of the outer frame of the acceleration and $x(t)$ is the displacement of the oscillating mass, then we have:

$$k(X(t) - x(t)) + c\frac{d(X(t) - x(t))}{dt} = m\frac{d^2x(t)}{dt^2} \tag{4.18}$$

Letting:

$$Z(t) = X(t) - x(t)$$

we can rewrite Eq. (4.18) as follows:

$$kZ(t) + c\frac{dZ(t)}{dt} + m\frac{d^2Z(t)}{dt^2} = m\frac{d^2X(t)}{dt^2} \tag{4.19}$$

Since we have assumed that the oscillation of both the outer reference frame and the mass are sinusoidal, we can express $X(t) = X_0 e^{j\omega t}$ and $Z(t) = Z_0 e^{j\omega t}$. Substituting these expressions in Eq. (4.19) yields:

$$-m\omega^2 Z_0 e^{j\omega t} + kZ_0 e^{j\omega t} + j\omega c e^{j\omega t} = -m\omega^2 X_0 e^{j\omega t} \tag{4.20}$$

From which we have:

$$Z_0 = \frac{\omega^2 X_0}{\omega^2 - \frac{k}{m} - j\frac{c\omega}{m}} \tag{4.21}$$

If we let:

$$\omega_0 = \sqrt{\frac{k}{m}}$$

and

$$\zeta = \frac{b}{2\sqrt{km}} = \frac{b}{2\omega m}$$

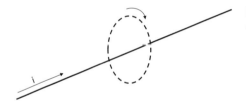

Figure 4.19 A current induces a magnetic field around a conductor through which it flows.

then

$$Z_0 = \frac{\omega^2 X_0}{\omega^2 - \omega_0^2 - j^2\omega^2\zeta} \tag{4.22}$$

ζ is called the damping ratio.

4.3 Inductive Sensing

Whenever an electric current flows through a conductor, a magnetic field will be set up around the conductor (Figure 4.19). The direction of this field depends on the direction of the electric current but its strength depends on both the magnitude of the current and the "reluctance" of the conductor. The reluctance is a measure of the ease with which magnetic moments can be aligned with one another inside the conductor and depends on the material from which the conductor is made (permeability) as well as the cross-sectional area and the length of the material:

$$R = \frac{l}{\mu_0\mu_r A} \tag{4.23}$$

where μ_0 is the permeability in vacuum ($\mu_0 = 4\pi \times 10^{-7}$ N/A^2) and μ_r is the relative permeability. One way to increase the magnetic field strength is by winding the conductor around a ferromagnetic material (core) whose permeability is relatively high, as shown in Figure 4.20. In this case, a significant portion of the magnetic field will be contained within the core. The magnetic flux, Φ, is a measure of the portion of the magnetic field that is aligned with (or normal to) the surface area of the core. The reluctance and the magnetic flux are related to each other as follows:

$$\mathcal{F} = iN = \Phi \times R \tag{4.24}$$

where $\mathcal{F}$ is the magnetomotive and i is the current that passes through the coil. From Eq. (4.24), it is clear that:

$$\Phi = \frac{Ni}{R} \tag{4.25}$$

If the current is an alternating current, then the tendency of the magnetic field induced by the current is to oppose the change in current. This opposition is expressed in terms of the inductance of the whole set up:

$$L = N\frac{d\phi}{di} = \frac{N^2}{R} \tag{4.26}$$

Similar to Eq. (4.2), the reluctance of an inductive sensor can be varied by varying any of the parameters in Eq. (4.26). The most frequently used approach, however, is creating

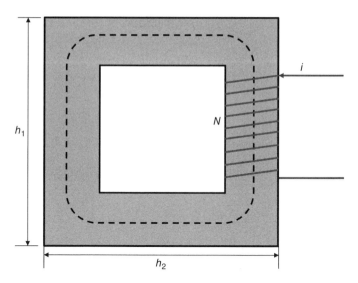

Figure 4.20 A current flowing through a coil induces a magnetic flux proportional to the number of coils inside the core.

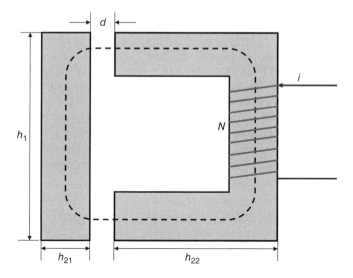

Figure 4.21 The basic principle of an inductive sensor.

an air gap in the core (see Figure 4.21) and varying the width of the air gap as a function of some physical phenomenon, for instance, as a function of physical displacement.

Using Figure 4.21, the reluctance of the inductive sensor can be computed by summing up the reluctance of the four different sections of the core and the reluctance of the air gap:

$$R = \sum_{i=1}^{4} \frac{l_i}{\mu A} + \sum_{j=1}^{2} \frac{d_j}{\mu_0 A} = \frac{2h_1}{\mu A} + \frac{2h_{22}}{\mu A} + \frac{2d}{\mu_0 A} \qquad (4.27)$$

where $\mu = \mu_0 \mu_r$. If we define R_0 as the displacement-free reluctance of the inductor:

$$R_0 = \frac{2h_1 + 2h_{22}}{\mu A}, \tag{4.28}$$

and define k as a constant that describes the static aspect of the air gap:

$$k = \frac{2}{\mu_0 A}, \tag{4.29}$$

then the reluctance can be simplified as follows:

$$R = R_0 + kd \tag{4.30}$$

Alternatively, R can be described as:

$$R = R_0(1 + x), x = \frac{kd}{R_0} \tag{4.31}$$

The inductance can now be expressed in terms of the simplified reluctance that separates the flux path through the core from the air gap:

$$L = \frac{N^2}{R_0(1 + x)} = \frac{L_0}{1 + x} \tag{4.32}$$

where

$$L_0 = \frac{N^2}{R_0} \tag{4.33}$$

In Figure 4.16, we demonstrated that the linear region of a capacitive sensor can be improved by using a differential set up. The same approach can be used to increase the linear region of an inductive sensor. As can be seen in Figure 4.22, the size of the air gap between the two cores can be varied differentially, so that when the air gap of one of the cores increases, the air gap of the other core decreases (the core in the middle is used by both inductors). For the first inductor, the width of the air gap after the displacement of the middle core to the right by a distance of dx is $y = x + dx$. Therefore, its reluctance becomes:

$$R_1 = R_0(1 + y) \tag{4.34}$$

and its inductance becomes:

$$L_1 = \frac{L_0}{R_0(1 + y)} \tag{4.35}$$

In the same way, the inductance of the second inductor can be expressed as:

$$L_2 = \frac{L_0}{R_0(1 - y)} \tag{4.36}$$

One of the widely used differential inductive sensors is the linear variable differential transformer (LVDT). It has one primary and two secondary coils and a movable core in the middle. The location of the movable core is a function of the desired physical parameter which we wish to measure. Figure 4.23 displays the internal arrangement of a LVDT and Figure 4.24 illustrates the basic principle. When an AC excitation voltage is supplied to the primary coil, it sets up a magnetic field across all the coils. The amount of magnetic flux that is coupled to the two secondary coils depends on the mutual inductance,

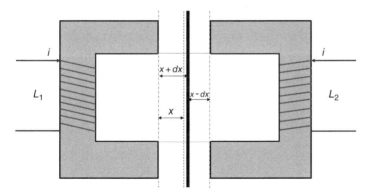

Figure 4.22 A differential inductive setup.

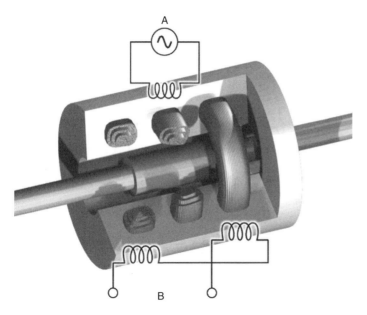

Figure 4.23 The internal setup of a linear variable differential transformer. The transformer consists of three coils: one primary coil in the middle and two secondary coils placed on the left- and right-hand sides of the primary coil. A moveable core in the middle moves in and out of the structure as a result of a physical process that displaces it. The voltage induced at the secondary coils is a function of the displacement of the moveable core. A, the primary coil; B, the differential coil. Source: Wikipedia.

M, which in turn depends on the relative position of the moveable core. The variable magnetic flux coupled to the secondary coils induces voltages e_1 and e_2 across the two coils, which are proportional to the magnitude of the magnetic flux. When the moveable core is exactly in the middle, e_1 and e_2 are in phase with each other and equal in magnitude. When, however, the moveable core is not in the middle, the two voltages are no longer equal (as shown in Figure 4.25). We are usually interested in the difference between these two voltages to determine the relative displacement of the core and the direction of its movement.

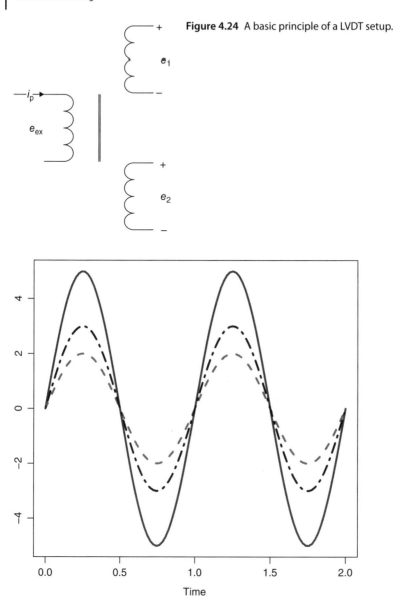

Figure 4.24 A basic principle of a LVDT setup.

Figure 4.25 When the movable core is not located in the middle, the voltages induced on the secondary coils by the magnetic flux of the primary coil will be different. The magnitude of these two secondary voltages (measured in volts) depends on the relative position of the moveable core.

Figure 4.26 shows a differential configuration and how the output voltage can be obtained. Since the negative potential of e_1 is connected to the negative potential of e_2, the output voltage is the difference of the two secondary voltages. In this case, the two voltages are out of phase by $180°$ and their magnitude depends on the relative position of the moveable core. Figure 4.27 displays the the relationship between the input excitation voltage and the output voltage.

Figure 4.26 A differential configuration of the LVDT.

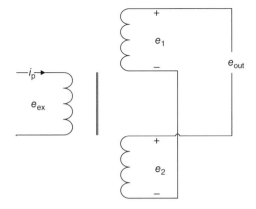

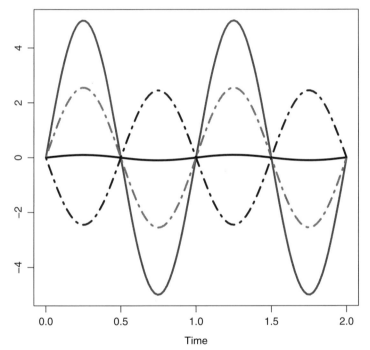

Figure 4.27 The relationship between the excitation voltage (the one with the largest magnitude, measured in volts) and the output voltage (the one with the smallest magnitude, measured in volts) of a LVDT. The dashed lines indicate the differential secondary voltages which are out of phase with each other by 180°.

To quantitatively describe the relationship between the excitation and the output voltages, consider Figure 4.28. The mutual inductance due to the magnetic flux between the primary coil and the first secondary coil is M_1 and produces an inductive reactance that equals sM_1, where $s = j\omega$. M_2 and M_3 are established likewise, as shown in the figure, producing inductive reactances sM_2 and sM_3. Applying Kirchhoff's voltage law results in:

$$e_{ex} = i_p(R_p + sL_p) + i_s(sM_2 - sM_1) \tag{4.37}$$

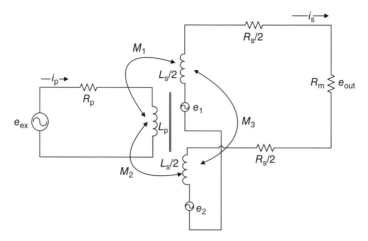

Figure 4.28 The circuit set up of a LVDT.

$$e_{out} = i_s R_m \tag{4.38}$$

The net voltage drop inside the secondary loop can be expressed as:

$$0 = \left(\frac{R_s}{2} + R_m + \frac{R_s}{2}\right) i_s + \left(s\frac{L_s}{2} + s\frac{L_s}{2} - sM_3\right) i_s + (sM_2 - sM_1)i_p \tag{4.39}$$

Which is the same as:

$$0 = (R_s + R_m)i_s + s(L_s - M_3)i_s + s(M_2 - M_1)i_p \tag{4.40}$$

Ideally, the fixed coils are independent of the position of the moveable core, in which case, $L_s - M_3$ should be independent of the moveable core. If we designate $L_s - M_3 = L_2$, then we have:

$$\frac{e_{out}}{e_{ex}} = \frac{s\left(M_2 - M_1\right) R_m}{s^2\left[(M_2 - M1)^2 - L_2L_p\right] - s\left(L_2R_p + L_pR_s + L_pR_m\right) - R_p(R_s + R_m)} \tag{4.41}$$

Usually L_2L_p is very much greater than $(M_2 - M1)^2$, in which case:

$$\frac{e_{out}}{e_{ex}} = \frac{s\left(M_2 - M_1\right) R_m}{-s^2L_2L_p - s\left(L_2R_p + L_pR_s + L_pR_m\right) - R_p(R_s + R_m)} \tag{4.42}$$

If $R_m \to \infty$, then

$$\frac{e_{out}}{e_{ex}} = \frac{s(M_1 - M_2)}{sL_p + R_p} \tag{4.43}$$

As can be seen, the numerator of Eq. (4.43) is a function of the position of the moveable core whereas the denominator is not. If the excitation voltage has a fixed frequency, then s will be a fixed quantity as well and the output voltage can be expressed as:

$$e_{out} = e_{ex}kf(x) \tag{4.44}$$

where

$$k = \frac{1}{sL_p + R_p} \tag{4.45}$$

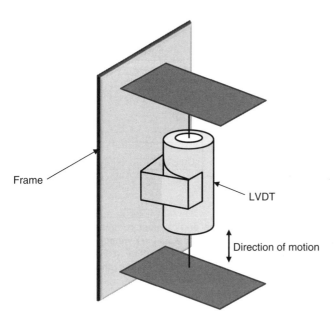

Figure 4.29 A one-dimensional accelerometer consisting of an oscillating LVDT core. The AC voltage generated by the LVDT is a function of the acceleration of the seismic mass.

and,

$$f(x) = s \left(M_1 - M_2 \right) \tag{4.46}$$

Example 4.5 Similar to the differential capacitive accelerometer we considered in Example 4.4, we wish to develop a LVDT accelerometer. Show a potential architecture for realising such a device.

The accelerometer can be developed by taking advantage of the natural linear displacement of a linear variable differential transformer (LVDT), the core of which serves as a seismic mass. The acceleration induced on the LVDT can be transformed into a linear up and down displacement of the LVDT core and the displacement of the core in turn can be translated directly into a proportional AC voltage. This architecture is illustrated by Figure 4.29.

4.4 Thermoelectric Effect

If two conductors of dissimilar materials are connected to each other and heat energy is applied at the junction, the conductors respond to the heat differently. As a result, a voltage difference is produced across the two conductors (see Figure 4.30a). This phenomenon is called the thermoelectric effect and the conductors are said to be a thermocouple. The voltage produced as a result is known as the Seebeck voltage, in honour of the scientist who first observed the phenomenon. For a small change in the junction

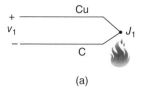

Figure 4.30 A thermocouple: (a) free electrons in the copper and constantan wires react differently to a temperature at the copper–constantan junction giving raise to a Seebeck voltage; (b) additional temperature-sensitive junctions are created when the thermocouple is connected with a thermoelectric circuit; (c) the net voltage difference between a copper–copper junction is zero.

temperature, the Seebeck voltage is linearly proportional to the temperature and can be expressed as:

$$\Delta v_1 = \alpha \Delta T \tag{4.47}$$

where α is the Seebeck coefficient and is considered to be a constant as long as the change in temperature is small.

Since a thermocouple is usually a part of an electrical circuit (as all sensors are), the way it is connected with the rest of the circuit can complicate the reading of the Seebeck voltage, which should be mapped to a temperature. This can be seen in Figure 4.30b and c where a thermocouple consisting of copper (Cu) and constantan (C) is connected to the next stage with copper wires. Due to a potential temperature difference at the two interconnection points (J_2 and J_3), there will be additional Seebeck voltages here as well. As J_3 is a copper-to-copper junction V_3 is zero, but J_2 is a copper-to-constantan junction and, as a result, V_2 is not zero. Moreover, v_2 is evidently in opposition to v_1. Therefore, the net Seebeck voltage is proportional to the temperature difference between J_1 and J_2. In other words, it is not possible to determine the temperature of J_1 without first determining the temperature of J_2.

One way to condition (control) the temperature of J_2 is to physically force its temperature to a reference temperature, for example, by putting the junction inside an ice box. Hence, the Seebeck voltage that can be coupled to the next stage can be computed as follows:

$$v = (v_1 - v_2) = \alpha(T_1 - T_2) \tag{4.48}$$

If the temperature of J_2 is specified in degrees Celsius (t_2), then its absolute temperature equals: $T_2 = t_2 + 273.15$. Thus,

$$v = (v_1 - v_2) = \alpha[(t_1 + 273.15) - (t_2 + 273.15)] = \alpha[(t_1 - t_2)] \tag{4.49}$$

Since our reference temperature is 0 °C, the above equation reduces to:

$$v = (v_1 - v_2) = \alpha t_1 \tag{4.50}$$

Table 4.1 Reference temperature as defined by the International Temperature Scale of 1990

Equilibrium point	K	°C	°F
Triple point of hydrogen	13.8033	259.3467	434.8241
Triple point of neon	24.5561	248.5939	415.4690
Triple point of oxygen	54.3584	218.7916	361.8249
Triple point of argon	83.8058	189.3442	308.8196
Triple point of mercury	234.3156	38.8344	37.9019
Triple point of water	273.16	0.01	32.02
Melting point of gallium[a]	302.9146	29.7646	85.5763
Freezing point of indium	429.7485	156.5985	313.8773
Freezing point of tin	505.078	231.928	449.470
Freezing point of zinc	692.677	419.527	787.149
Freezing point of aluminium	933.473	660.323	1,220.581
Freezing point of silver	1,234.93	961.78	1,763.20
Freezing point of gold	1,337.33	1,064.18	1,947.52
Freezing point of copper	1,357.77	1,084.62	1,984.32

a) Due to its unique molecular structure, the melting point of gallium can be uniquely determined. Hence, this point rather than the freezing point, is taken as a reference temperature.

Note that, depending on the specific application of the thermocouple, different reference points may be chosen. Normally these reference points are selected from the table of International Temperature Scale (ITS; see Table 4.1).

The strength of Eq. (4.50) is that the reference temperature can be accurately controlled. The configuration of Figure 4.31 is relatively simple because one of the terminals of the thermocouple is the same as the terminals of the next stage. This may not always be the case; there are other junctions (for example, chromel–constantan, iron–constantan, chromel–alumel) that have better thermal responses than copper–constantan, and if used the thermocouple configuration becomes more complicated. For example, when the thermocouple consists of an iron–constantan junction, as shown in Figure 4.32, which has to be coupled to the next stage by copper terminals, then in addition to J_2 a third Cu–Fe junction, J_3, is added to the configuration. In order to deal with this concern, both J_2 and J_3 can be put together in an isothermal block with a fixed reference temperature, so that Eq. (4.50) can be rewritten as:

$$v = \alpha \left(T_1 - T_{\text{ref}} \right) \tag{4.51}$$

A thermistor can be attached to the isothermal block to monitor and control a variation in the reference temperature. One may ask at this point, if one is supposed to employ a thermistor anyway – the resistance of which is a function of temperature – what is the use of also employing a thermocouple to sense temperature? The answer is that the thermocouple responds to a wide range of temperature variations whereas the thermistor does not. But the thermistor is useful because the variation in the reference temperature inside the isothermal block is not appreciable.

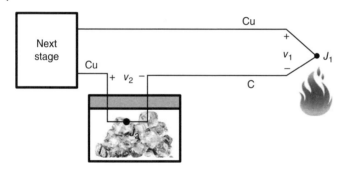

Figure 4.31 Setting one of the junctions of the interconnecting terminals at a reference temperature simplifies the determination of the Seebeck voltage and thereby the temperature at J_1.

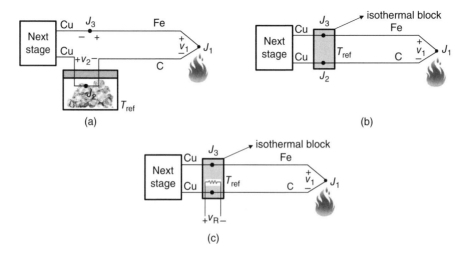

Figure 4.32 Putting all the junctions save the desired junction (J_1) into an isothermal block puts the temperature of all the junctions at a fixed reference temperature. The variation in this temperature can be managed by employing a thermistor.

References

Beyeler F, Neild A, Oberti S, Bell DJ, Sun Y, Dual J and Nelson BJ 2007 Monolithically fabricated microgripper with integrated force sensor for manipulating microobjects and biological cells aligned in an ultrasonic field. *Journal of Microelectromechanical Systems*, **16** (1), 7–15.

Chiu Y and Tseng VF 2008 A capacitive vibration-to-electricity energy converter with integrated mechanical switches. *Journal of Micromechanics and Microengineering*, **18** (10), 104004.

Dagamseh A, Bruinink C, Wiegerink R, Lammerink T, Droogendijk H and Krijnen G 2013 Interfacing of differential-capacitive biomimetic hair flow-sensors for optimal sensitivity. *Journal of Micromechanics and Microengineering* **23** (3), 035010.

Kim M, Moon W, Yoon E and Lee KR 2006 A new capacitive displacement sensor with high accuracy and long-range. *Sensors and Actuators A: Physical*, **130**, 135–141.

Kuehne I, Frey A, Marinkovic D, Eckstein G and Seidel H 2008 Power MEMS – a capacitive vibration-to-electrical energy converter with built-in voltage. *Sensors and Actuators A: Physical*, **142** (1), 263–269.

Lee HK, Chang SI and Yoon E 2009 Dual-mode capacitive proximity sensor for robot application: Implementation of tactile and proximity sensing capability on a single polymer platform using shared electrodes. *Sensors Journal, IEEE*, **9** (12), 1748–1755.

Ningrat KA, Noviello G and Italia F 2009 Capacitive-inductive touch screen. US Patent App. 12/491,990.

Rivadeneyra A, Fernández-Salmerón J, Agudo M, López-Villanueva JA, Capitan-Vallvey LF and Palma AJ 2014 Design and characterization of a low thermal drift capacitive humidity sensor by inkjet-printing. *Sensors and Actuators B: Chemical*, **195**, 123–131.

Satyanarayana S, McCormick DT and Majumdar A 2006 Parylene micro membrane capacitive sensor array for chemical and biological sensing. *Sensors and Actuators B: Chemical*, **115** (1), 494–502.

Schulz SC, Chernefsky AF and Geaghan B 2004 Flexible capacitive touch sensor. US Patent 6,819,316.

Shaeffer DK 2013 MEMS inertial sensors: A tutorial overview. *Communications Magazine, IEEE*, **51** (4), 100–109.

Snyder W and Seguine R 2006 Capacitive touch sense device having polygonal shaped sensor elements. US Patent App. 11/605,819.

Ueno A, Akabane Y, Kato T, Hoshino H, Kataoka S and Ishiyama Y 2007 Capacitive sensing of electrocardiographic potential through cloth from the dorsal surface of the body in a supine position: a preliminary study. *Biomedical Engineering, IEEE Transactions on*, **54** (4), 759–766.

Waggoner PS and Craighead HG 2007 Micro-and nanomechanical sensors for environmental, chemical, and biological detection. *Lab on a Chip*, **7** (10), 1238–1255.

Xie C, Huang S, Hoyle B, Thorn R, Lenn C, Snowden D and Beck M 1992 Electrical capacitance tomography for flow imaging: system model for development of image reconstruction algorithms and design of primary sensors *Circuits, Devices and Systems, IEE Proceedings G*, **139** (1), 89–98.

5

Ultrasonic Sensing

Ultrasonic sensing is extensively used to detect the presence, proximity, or distance of an object of interest, in fields such as:

- underwater monitoring (Fairfield et al. 2007; Kinsey and Whitcomb 2007; Marani et al. 2009; Vasilescu et al. 2005)
- structural health monitoring (Gao and Rose 1687; Terzic et al. 2010)
- robotics and indoor localisation (Choi et al. 2011; Ladd et al. 2005; Park and Yoon 2006; Roberts et al. 2007)
- environment monitoring (Turner et al. 2008)
- manufacturing (Carullo and Parvis 2001; Henning and Rautenberg 2006; Lynnworth 2013)
- parking lot regulation (Han et al. 2008; Lee et al. 2008; Park et al. 2008)
- medicine (Hata et al. 2007; Khuri-Yakub and Oralkan 2011; Krill 2006; Zhang et al. 2008).

Its popularity stems mainly from three factors: its safety, its non-invasive nature, and the relatively simple construction of ultrasonic sensors. However, ultrasonic sensing requires a careful modelling of the medium through which ultrasonic waves propagate (Bossy et al. 2005), (Hughes et al. 2007), (Protopappas et al. 2007), (Treeby et al. 2012), (Laugier and Haïat 2011), (Schmerr Jr 2013).

Ultrasonic sensing essentially consists of highly sensitive transmitter (speaker) and receiver (microphone) or a single device which serves both as a transmitter and a receiver. The transmitter generates and sends ultrasonic signal in short bursts towards a target and the target reflects the signal, which is then received by the receiver. When a single device is used both as a transmitter and a receiver, the reflected signal returns back to its source. Often, the time of arrival, angle of arrival, amplitude, wavelength, and speed of the reflected signal (echo) are examined – some of which undergo modification – to determine the nature, proximity, size, or distance of the target. Figure 5.1 displays a simple layout of an ultrasonic device and Figure 5.2 illustrates the set up of the device for measuring the distance of a target from the ultrasonic sensors. The transmitter sends pulses of 250 μs duration in burst which then reflect back by the target to the ultrasonic receiver. The distance of the target is then computed as:

$$d = \frac{1}{2}c\tau \tag{5.1}$$

Figure 5.1 An example layout of an ultrasonic sensor having separate transmitter and receiver. Courtesy of Innovation One Inc. (2005).

Figure 5.2 A simple set up and wave propagation model for an ultrasonic sensing in air. Courtesy of Innovation One Inc. (2005).

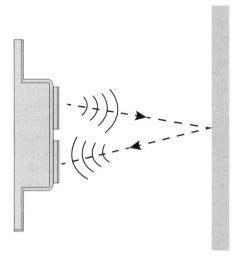

where c is the speed of the ultrasonic pulse (wave) in air in metres per second and τ is the round trip time in seconds. As we shall see shortly, c depends on many factors and they should be taken into account to determine d.

Equation (5.1) is interesting to re-examine, not only because of its simplicity, but also because it reveals the different modelling assignments one can consider to solve different problems. For example, if d and τ are known, one can compute c and reason about the particular medium through which the wave propagated, in which case, the assignment is to determine the nature of the medium. In medicine, this is the case when one wishes to determine the composition of tissues or the presence or absence of foreign artefacts (such as tumours). Alternatively, d and c might be known and the assignment is to determine τ. This is the case, for example, when one wants to determine the depth of penetration of an ultrasonic signal in a known medium. Consequently, the modelling of ultrasonic sensing should address some of the following critical aspects (Massa 1999):

- how the wavelength of an ultrasonic signal varies as the speed and frequency undergo changes, and how this variation in turn affects the resolution, accuracy, the minimum target size, and the minimum and maximum distances of the target from the ultrasonic sensing device;
- how the transmission medium attenuates ultrasonic signals as the frequency of the signal and the humidity (if air is used as the transmission medium) of the medium vary, and how the variation in attenuation in turn affects the maximum target distance;
- how background noise affects the maximum target distance and the minimum target size as the frequency of the ultrasonic signal varies;

- how the ultrasonic radiation pattern (beam angle) affects the maximum target distance and the exclusion of spurious targets;
- how the amplitude of a returning echo varies as the target distance, geometry, surface, and size vary, and how in turn this variation affects the maximum target distance that can be measured by an ultrasonic sensing.

Example 5.1   We wish to diagnose an organ of the body by transmitting ultrasonic pulses and by examining the nature of the reflected wave: the echo. If we regard the transmission of a single pulse only, what will be the expected shape of the reflected ultrasonic wave? Explain the nature of the echo.

When an ultrasonic wave propagates in a homogeneous medium, it does not undergo any reflection or refraction. As soon as it encounters a change in a medium, however, it can reflect, refract, or scatter, depending on the boundary of interface. Figure 5.3 illustrates the expected wave shape of the reflected ultrasonic wave. There will be a minimum delay (d seconds) between the transmission of the ultrasonic pulse and the detection of the echo, which depends on the distance between the transmitter and the skin. When the ultrasonic pulse interacts with the skin, some of its energy will be reflected back while the remaining energy will penetrate the skin and propagate further forward. Because of the change in the medium, the speed of propagation changes. Since the region between the skin and the destination organ consists of different tissues, the propagation entails further reflections and refractions. Assuming that the boundary between the skin and the organ is fluid, then we should expect weak reflection until the ultrasonic wave interacts with the front boundary of the organ, where we expect a significant amount of reflection. Then a further penetration and propagation is

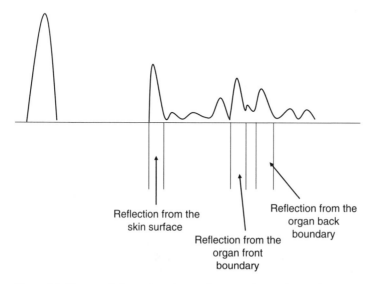

Figure 5.3 The use of ultrasonic pulses to diagnose the condition of an organ and the expected wave form of the echo as the propagating pulses encounter different media on their way and experience different magnitudes of reflections.

expected until finally the pulse interacts with the back-end boundary. And so on. From this, it is apparent that knowledge of the composition of the medium as well as the organ is required to meaningfully interpret the reflected ultrasonic wave.

Example 5.2   One of the applications of ultrasonic sensing is indoor localisation. In this scenario, ultrasonic transmitters are carefully placed in strategic locations in an indoor environment and regularly transmit ultrasonic pulses in which identity, coordinates, and timestamp are embedded. A robot with an ultrasonic receiver can receive these signals and use them to determine its position, assuming that it has the geographic coordinates of the site. Show how the two-dimensional position of both the transmitters and the receiver can affect the accuracy with which the robot can determine its position.

If the robot has a map of the indoor area, it can take advantage of the propagation time to determine its location. Assuming that the transmitter is omnidirectional, localisation with a single transmitter yields the poorest accuracy. The robot may determine its proximity by applying Eq. (5.1), but it may not be able to determine its direction. As a result, it can be on the circumference of the circle which is found inside the area of interest, as shown in Figure 5.4.

With two transmitters, the localisation accuracy increases, but still there can be a considerable overlap area between the two transmitters, as a result of which the two likely locations of the robot can be far from each other. One of the reasons for the difficulty in identifying the disappearance spot of Malaysia Airlines Flight 370 (which disappeared on 8 March 2014 while flying from Kuala Lumpur International Airport, Malaysia, to Beijing Capital International Airport) was the extensive size of the overlapping areas identified by different satellites. Figures 5.5 and 5.6 compare the likely positions of the robot for two different scenarios.

With three transmitters, the localisation accuracy considerably increases, as can be seen in Figure 5.7. In all the figures we have considered, it is assumed that the relationship between distance and propagation time is accurate. This is, however, not the case in practical scenarios, which means that the robot's location may potentially be different from the ones described by the circumferences of the circles. Figure 5.8 shows the likely locations of the robot, assuming that the propagation time computed by the robot is the maximum propagation time. In order to compensate for the uncertainty stemming

Figure 5.4 Indoor localisation with a single ultrasonic transmitter.

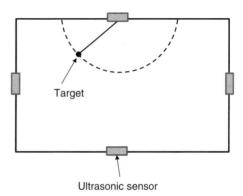

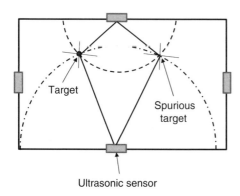

Figure 5.5 Indoor localisation with two parallel ultrasonic transmitters.

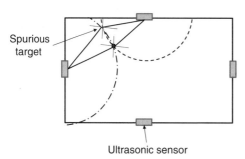

Figure 5.6 The accuracy with which the localisation can be achieved with two transmitters depends on the relative positions of the transmitters as well as the robot itself.

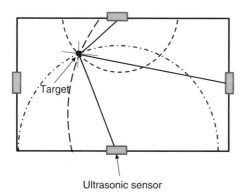

Figure 5.7 The localisation of the two-dimensional position of the robot can be considerably improved with three ultrasonic transmitters.

from the error in modelling the propagation characteristics of the ultrasonic pulses, more than three transmitters should be used in practice.

5.1 Ultrasonic Wave Propagation

An ultrasonic signal is generated by the vibration of the diaphragm of the transmitting device (which is why it is also referred to as a speaker). The ultrasonic wave, coming into existence in consequence of the vibration, propagates by longitudinal motion; in other words, by compressing and expanding the particles of the medium through which it propagates. The amplitude and the wavelength of the wave as well as the speed at

Figure 5.8 Due to the error in the propagation model for ultrasonic pulses, the uncertainty of localising the robot increases. As a result, the robot can be found in any of the points described by the shaded region.

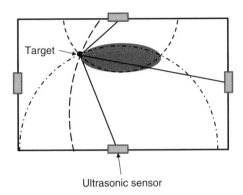

Target

Ultrasonic sensor

which it propagates depend on the compressibility (k) and the density (ρ) of the medium, in accordance with the laws of conservation of mass and momentum. Thus ultrasonic waves travel at the speed of sound (c) and, for a particular medium, this speed is given as:

$$c = \sqrt{\frac{1}{\rho k}} \qquad (5.2)$$

As the particles of the medium compress and expand, they apply pressure on their neighbouring particles and pressure is applied on them by their neighbouring particles too. For a longitudinal wave, the ultrasonic pressure and the particle velocity are related to one another as follows:

$$p = vZ \qquad (5.3)$$

where Z is the characteristic impedance of the medium. The characteristic impedance, in turn, is a function of the speed of the ultrasonic wave (which is different from the particle velocity) and the density of the medium:

$$Z = \rho c \qquad (5.4)$$

If we insert Eq. (5.2) into Eq. (5.4), then we obtain:

$$Z = \rho c = \sqrt{\frac{\rho}{k}} \qquad (5.5)$$

Tables 5.1 and 5.2 summarise the relationship between the density and speed of sound for different materials and body parts. Table 5.1 summarises some of the materials of interest in structural health monitoring and Table 5.2 summarises the media of interest in medicine.

As the ultrasonic wave propagates, its field distribution in space and its amplitude (intensity) experience change. As far as its amplitude is concerned, one can observe two phenomena:

- axially – along the centre of the transmitter – the magnitude of the wave changes gradually as a function of distance
- laterally – at the sides – at any plane perpendicular to the direction of propagation, the magnitude of the wave changes rapidly as a function of distance.

Table 5.1 A summary of the density, wave propagation speed, and characteristic impedance of different materials

Material	Density $\rho(kg/m^3)$	Speed $c(m/s)$	Characteristic impedance Z (MΩ)
Aluminium	2700	6400	17
Brass	8500	4490	38
Polyethylene	920	2000	1.8
Polymethylmethacrylate	1190	2680	1.5

Table 5.2 A summary of the density, wave propagation speed, and characteristic impedance of different body parts (Goss et al. 1978 1980)

Material	Density $\rho(kg/m^3)$	Speed $c(m/s)$	Characteristic impedance Z (MΩ)
Blood	1060	1566	1.66
Bone	1380–1810	2070–5350	3.75–7.38
Brain	1030	1505–1612	1.55–1.66
Fat	920	1446	1.33
Kidney	1040	1567	1.62
Liver	1060	1566	1.66
Lung	400	650	0.26
Muscle	1070	1542–1626	1.65–1.74

As far as the spatial field distribution is concerned, one can likewise observe two phenomena:

- Initially, the wave shape is narrow up to a distance D, having a beam width which is approximately equal to the diameter of the transmitter. The region up to the distance D is referred to as the near field or the Fresnel zone
- As the distance from the transmitter increases, the wave begins to spread sideways and the assumption of longitudinal propagation no longer holds. The region beyond D is referred to as the far field or the Fraunhofer zone.

Figure 5.9 summarises the propagation aspects of an ultrasonic wave. The length of the Fresnel zone (D), the beam width d in the Fresnel zone, and the angle of divergence (θ) beyond the Fresnel zone are important parameters in ultrasonic sensing and depend mainly on two factors:

- the size and shape of the ultrasonic transmitter and,
- the wave frequency.

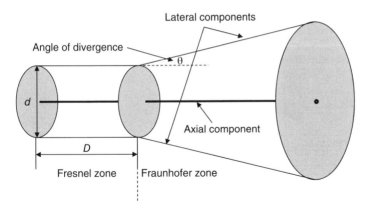

Figure 5.9 The propagation characteristics of an ultrasonic wave. The field or beam width in the Fresnel zone is narrow and the wave can be considered as a longitudinal wave (with no lateral components). The field in the Fraunhofer zone, on the other hand, contains lateral components and, therefore, cannot be considered longitudinal.

For a transmitter with a circular transmitting surface, the length of the Fresnel zone can be approximated by:

$$D = \frac{r^2}{\lambda} \tag{5.6}$$

where r is the radius of the surface area and λ is the wavelength of the wave. Hence, as the radius of the transmitter increases, the length of the Fresnel zone increases exponentially. Moreover, the angle of divergence beyond the Fresnel zone decreases proportionally. The advantage of a longer Fresnel zone is that the ultrasonic wave can penetration deeper into the target of interest when this is desired. This is, for example, the case, in the diagnosis of tissues and organs with ultrasound sensing. However, the beam width also increases proportionally with the increment in the transmitter's radius. One of the side effects of a large beam width is poor resolution. Figure 5.10 summarises the relationship between the length of the Fresnel zone, D, and the angle of divergence, θ.

We can rewrite Eq. (5.6) in terms of the frequency f of the ultrasonic wave:

$$D = \frac{r^2 f}{c} \tag{5.7}$$

From Eq. (5.7) it is apparent that the length of the Fresnel zone increases as the wave frequency increases. Conversely, the length decreases as the frequency decreases. The angle of divergence beyond the Fresnel zone increases as the frequency decreases, which means the resolution and sensing quality suffer with a decrease in wave frequency. A high-frequency wave is robust to divergence but the wave attenuation becomes high as the wave frequency increases.

Example 5.3 Analysing the time-domain features of ultrasonic echoes is often not enough to determine the nature of objects with which the ultrasonic signal interacted. Demonstrate how frequency-domain features can provide additional and useful information.

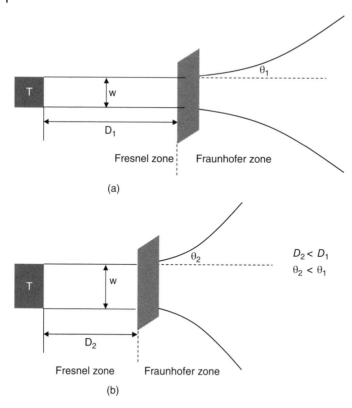

Figure 5.10 A summary of the length of the Fresnel zone (*D*) and the angle of divergence (*θ*).

Any function of time (a time series; $f(t)$), can be expressed in terms of frequency. When the function is described in terms of time, its description is called a time-domain description. Similarly, when a function is described in terms of frequency, it is description is called a frequency-domain description. The basis for the frequency-domain description is the Fourier series. There are several advantages in transforming the time-domain description of a signal to a frequency-domain description. One of these advantages is that it is possible to estimate at what frequency the power of the signal is concentrated. While the mathematical expression of transforming from the time domain to the frequency domain may appear complex, the concept itself is rather straightforward and intuitive. We shall illustrate this step with the help of Figure 5.11.

Suppose the top plot in Figure 5.11 describes an ultrasonic echo in the time domain. As can be seen, this echo is an under-damped sinusoidal signal. To transform this signal to its frequency-domain representation, we produce infinite sinusoidal signals, each having one specific frequency (remember that frequency is the number of cycles a function makes in a second). In the figure, three such sinusoidal signals are shown. For each sinusoidal signal, at each infinitesimal (dt) interval (from $-\infty$ to $+\infty$), we multiply the magnitude of the ultrasonic signal by the magnitude of the sinusoidal signal at that point and then we add all the results. The sum is proportional to the power density of the ultrasonic signal at that frequency. We repeat the process for the second sinusoidal

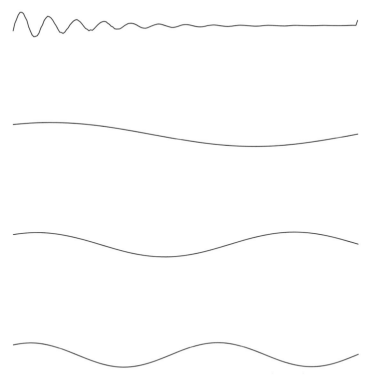

Figure 5.11 Transforming a time-domain description of an ultrasonic signal into a frequency domain description: top, time-domain description of the echo; other traces, the basic orthogonal sinusoidal functions that are used to compute the power spectral density of the echo.

signal to obtain the power density of the ultrasonic signal for the second frequency, and so on for subsequent sinusoidal signals up to infinity. One of the desirable aspects of using infinite sinusoidal functions is that they all are single-frequency, orthogonal functions; in other words, each function encodes a unique feature of the time-domain function; in our case, the echo.

Figure 5.12 shows the spectral power density of the ultrasonic signal. As can be seen, the power of the signal is concentrated in a specific region. In some situations, analysing this region makes it much easier to determine the nature of the organ with which the ultrasonic signal interacted than analysing the time-domain signal. Mathematically, the transformation from time to frequency domain and vice versa can be described by either the Fourier transform or the Laplace transform. For the case of Fourier transform, this is given as:

$$\mathcal{F}(j\omega) = \int_{-\infty}^{\infty} f(t)e^{-j\omega t}\,dt \tag{5.8}$$

where $\omega = 2\pi f$. Likewise, the inverse Fourier transform returns the time-domain description of the signal:

$$f(t) = \int_{\infty}^{\infty} \mathcal{F}(j\omega)e^{j\omega t}\,dt \tag{5.9}$$

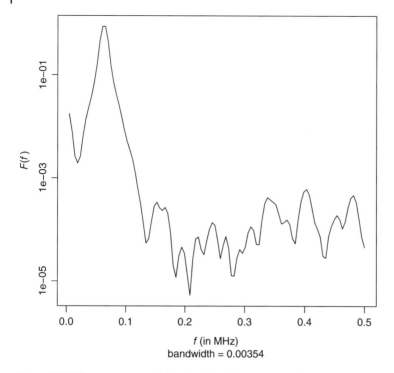

Figure 5.12 The power spectral density of an ultrasonic signal.

5.2 Wave Equation

The frequency spectrum of interest in ultrasonic sensing ranges from just above the threshold of hearing for humans (20 kHz) to several megahertz. Most existing proximity sensors which use air as their propagation medium use frequencies between 40 and 250 kHz, whereas in medicine, where the propagation media are the skin, tissues, organs, blood, and blood vessels, the frequencies of interest are typically between 2 and 20 MHz. The biggest challenge in using ultrasonic sensing is identifying the major factors that modify the characteristics of an ultrasonic wave and relating them to it, both before and after its interaction with the target. Fortunately, the extent to which they influence the propagation characteristics of the wave can be determined by examining the spatial (in three-dimensional coordinates) and temporal behaviour of the wave, $\psi(x, y, z, t)$. Alternatively, one can examine particle displacements, $d(x, y, z, t)$, particle velocities, $v(x, y, z, t)$, or particle pressures, $p(x, y, z, t)$, which are associated with the compression and expansion of the ultrasonic wave.

Within the Fresnel zone, we can assume that the ultrasonic wave is longitudinal, and hence, the spatial dimension of interest is only one; say z, $\psi(z, t)$. Furthermore, for many practical reasons, the short bursts of signals generated and transmitted by the ultrasonic transmitter are sinusoidal in nature. Assuming that the corresponding wave is both longitudinal and sinusoidal, it can be described as follows:

$$\psi(z, t) = e^{-j(zk+\omega t)} \tag{5.10}$$

where $k = 2\pi/\lambda$ is the wave number and quantifies the spatial variation (frequency) of the wave in radians per unit distance whereas ω quantifies the temporal variation (frequency) of the wave in radians per unit time. In other words, the kz term describes the spatial characteristic of the wave (because it is a travelling wave) and the ωt describes the temporal characteristic of the wave (because it is a time-varying wave). We have used Euler's formula, $e^{jx} = \cos x + j\sin x$, to describe the sinusoidal nature of the wave. The rate of change of the travelling wave in space and time can be determined by differentiating $\psi(z, t)$ with respect to z and t. Hence,

$$\frac{\partial \psi(z, t)}{\partial z} = (-jk)e^{-j(zk+\omega t)} = (-jk)\psi(z, t) \tag{5.11}$$

Recall that:

$$\frac{d(e^x)}{dx} = e^x$$

$$\frac{d(e^{ax})}{dx} = ae^x$$

and

$$\frac{d(e^{-ax})}{dx} = -ae^x$$

In order to get rid of the complex number, let us differentiate Eq. (5.11) one more time with respect to z,

$$\frac{\partial^2 \psi(z, t)}{\partial z^2} = (-jk)^2 e^{-j(zk+\omega t)} = -k^2 \psi(z, t) \tag{5.12}$$

Similarly, differentiating $\psi(z, t)$ with respect to t yields,

$$\frac{\partial^2 \psi(z, t)}{\partial t^2} = -\omega^2 e^{-j(zk+\omega t)} = -\omega^2 \psi(z, t) \tag{5.13}$$

Combining Eqs. (5.12) and (5.13) yields an important expression:

$$\left(\frac{k}{\omega}\right)^2 \frac{\partial^2 \psi(z, t)}{\partial t^2} = \frac{\partial^2 \psi(z, t)}{\partial z^2} \tag{5.14}$$

This is because, in both equations, $\psi(z, t)$ can be expressed in terms of the partial differential terms. Since $k = 2\pi/\lambda$, $\omega = 2\pi f$, and $c = \lambda f$ we can rewrite Eq. (5.14) as follows:

$$\frac{\partial^2 \psi(z, t)}{\partial z^2} = \frac{1}{c^2} \frac{\partial^2 \psi(z, t)}{\partial t^2} \tag{5.15}$$

Equation (5.14) is the ultrasonic wave equation and provides sufficient information about the temporal and spatial characteristics of the propagating wave. For example, if the wave is a purely cosine wave travelling at a speed c in a specific medium (which determines c) – that is, $\psi(z, t) = \cos(kz - 2\pi ft)$ – the frequency of the wave can be determined as follows:

$$\frac{\partial \psi(z, t)}{\partial z} = (-jk)\sin(kz - 2\pi ft) \tag{5.16}$$

and,

$$\frac{\partial^2 \psi(z, t)}{\partial z^2} = -(jk)^2 \cos\left(kz - 2\pi ft\right) = k^2 \cos\left(kz - 2\pi ft\right) \tag{5.17}$$

Similarly,

$$\frac{\partial^2 \psi(z,t)}{\partial t^2} = (2\pi f)^2 \cos\left(kz - 2\pi ft\right) \tag{5.18}$$

Using the equality of Eq. (5.15), we have:

$$k^2 \cos(kz - 2\pi ft) = \frac{1}{c^2}(2\pi f)^2 \cos(kz - 2\pi ft) \tag{5.19}$$

from which it follows that

$$f = \frac{kc}{2\pi} \tag{5.20}$$

which, of course, leads to the important relation (taking into consideration that $k = 2\pi/\lambda$):

$$c = f\lambda \tag{5.21}$$

5.3 Factors Affecting Ultrasonic Wave Propagation

When an ultrasonic wave propagates through different media, apart from the reflection and refractions it experiences, its speed undergoes a change, since different media have different compressibility and density. The wave's frequency depends mainly on the source generating it, and hence can be regarded as constant (assuming that it is not changed at the source); as a result of the change in c, the wavelength also changes, as shown in Figure 5.13. Even within the same medium, the speed of ultrasonic wave undergoes change as a function of the temperature of the medium, because temperature affects the density of the medium. Table 5.3 summarises how the speed of sound in air changes as the temperature of air changes. Different models exist for different media to approximate the relationship between temperature and wave propagation speed. For example, when the medium is air, the relation can be approximated by Eq. (5.22)

$$c(\text{m/s}) = 331 + 0.6 \times T(\text{in } °C) \tag{5.22}$$

Similarly, as the ultrasonic wave propagates through a medium, it interacts with the particles of the medium, which inevitably absorb its energy and scatter and reflect it. The collective effect of these phenomena is a reduction in the intensity (amplitude) of the wave. The term "attenuation" is used for the loss in the wave's amplitude. The amount of attenuation a wave undergoes depends on the nature of the medium, the frequency of the wave, and the distance it travels. A model for attenuation should therefore consist of these parameters. Thus the intensity of a wave as a function of the pressure it exerts on the particles of the medium through which it propagates and the characteristic impedance of the medium can be expressed as the time average transfer of energy in a unit area (see Figure 5.14):

$$A = \frac{1}{T} \int_0^T p(t)u(t)dt \tag{5.23}$$

Equation (5.23) simply relates the amplitude of a wave at a particular location to the pressure the wave exerts on the particles of the medium and the resulting particle speed at that particular location. Since an ultrasonic wave is a mechanical wave (created as a

Figure 5.13 As an ultrasonic wave propagates through different media, its amplitude, speed, and wavelength change in accordance with the difference in the attenuation constants, compressibilities, and densities of the media.

Table 5.3 A summary of the relationship between temperature, density, and wave propagation speed in air

Temperature (°C)	Density ρ (kg/m³)	Propagation speed c (m/s)
0	1.2922	331.30
5	1.2690	334.32
10	1.2466	337.31
15	1.2250	340.27
20	1.2041	343.21
25	1.1839	346.13
30	1.1644	349.02
35	1.1455	351.88

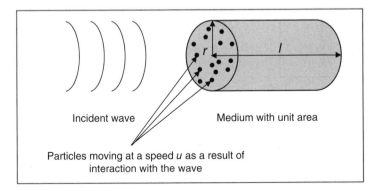

Incident wave   Medium with unit area

Particles moving at a speed *u* as a result of interaction with the wave

Figure 5.14 The intensity of an incident wave at a particular location can be expressed by the pressure the wave exerts on local particles and the resulting speed of the particles.

result of the mechanical oscillation of particles), the pressure can be expressed as (recall Eq. (5.2)):

$$p(z, t) = p_0 e^{-(kz+\omega t)} \tag{5.24}$$

With this, the intensity of the wave at a particular location can be expressed as:

$$A = \frac{1}{T}\frac{p_0^2}{Z} \int_0^T e^{-(kz+\omega t)} \, e^{-(kz+\omega t)} dt = \frac{P_0^2}{2Z} \tag{5.25}$$

where we make use of Eq. (5.3): $u(t) = p(t)/Z$. Likewise, the change in the amplitude of a propagating wave can be described by

$$A(z) = A_0 e^{-\mu_A z} \tag{5.26}$$

where A_0 is the amplitude of the wave at a reference point and μ_A is related to the frequency of the wave and the medium through which the wave propagates. As can be seen, the amplitude diminishes exponentially as a function of distance. The attenuation coefficient is a ratio relating the reference amplitude to the amplitude of the changing wave:

$$\frac{A_0}{A(z)} = e^{\mu_A z} \tag{5.27}$$

or, in decibels (dB):

$$\alpha(\text{dB/m}) = 20 \log \left(\frac{A_0}{A(z)} \right) = 20 \log (e^{\mu_A z}) = \mu_A z \, 20 \log (e) = 8.7 \mu_A z \tag{5.28}$$

As can be seen, Eq. (5.28) consists of two terms. One of them – $20 \log (e)\mu_A$ – is independent of distance but depends on μ_A, which in turn depends on the nature of the wave as well as the medium (frequency, humidity, and temperature) whereas the other

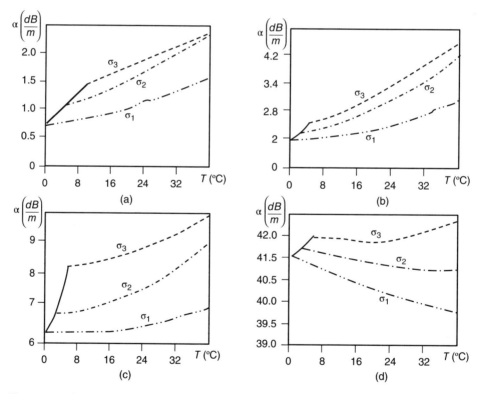

Figure 5.15 The attenuation coefficient of air in an isochronous pressure for different ultrasonic waves as a function of temperature and relative humidity: (a) 50 kHz; (b) 100 kHz; (c) 200 kHz; (d) 500 kHz. $\sigma_1 = 20\%$, $\sigma_2 = 40\%$, $\sigma_3 = 60\%$. The solid line is the plot of attenuation versus temperature when the relative humidity is 100%.

Table 5.4 Attenuation coefficients for different human tissues

Tissue	Attenuation ($dB/MHz.cm$)
Bone	16.0–23.0
Blood	0.17–0.24
Fat	1.0–2.0
Kidney	0.8–1.0
Liver	0.6–0.9

Source: Haney and O'Brien, (1986).

term is a function of distance. Figure 5.15 displays the attenuation of different ultrasonic waves propagating through air as a function of temperature and relative humidity. Table 5.4 lists the attenuation coefficients of different human tissues.

References

Bossy E, Padilla F, Peyrin F and Laugier P 2005 Three-dimensional simulation of ultrasound propagation through trabecular bone structures measured by synchrotron microtomography. *Physics in Medicine and Biology,* **50** (23), 5545.

Carullo A and Parvis M 2001 An ultrasonic sensor for distance measurement in automotive applications. *IEEE Sensors Journal,* **1** (2), 143–147.

Choi BS, Lee JW, Lee JJ and Park KT 2011 A hierarchical algorithm for indoor mobile robot localization using RFID sensor fusion. *Industrial Electronics, IEEE Transactions on,* **58** (6), 2226–2235.

Fairfield N, Kantor G and Wettergreen D 2007 Real-time slam with octree evidence grids for exploration in underwater tunnels. *Journal of Field Robotics,* **24** (1–2), 3–21.

Gao H and Rose J 2006 Ultrasonic sensor placement optimization in structural health monitoring using evolutionary strategy *AIP Conference Proceedings,* vol. 820, p. 1687.

Goss S, Johnston R and Dunn F 1978 Comprehensive compilation of empirical ultrasonic properties of mammalian tissues. *The Journal of the Acoustical Society of America,* **64** (2), 423–457.

Goss S, Johnston R and Dunn F 1980 Compilation of empirical ultrasonic properties of mammalian tissues. II. *The Journal of the Acoustical Society of America,* **68** (1), 93–108.

Han D, Kim S and Park S 2008 Two-dimensional ultrasonic anemometer using the directivity angle of an ultrasonic sensor. *Microelectronics Journal,* **39** (10), 1195–1199.

Haney M and O'Brien W 1986 Temperature dependency of ultrasonic propagation properties in biological materials. In: Greenleaf J (ed), *Tissue Characterization with Ultrasound.* CRC Press.

Hata Y, Kamozaki Y, Sawayama T, Taniguchi K and Nakajima H 2007 A heart pulse monitoring system by air pressure and ultrasonic sensor systems. *System of Systems Engineering, 2007. SoSE'07. IEEE International Conference on,* pp. 1–5.

Henning B and Rautenberg J 2006 Process monitoring using ultrasonic sensor systems. *Ultrasonics,* **44**, e1395–e1399.

Hughes ER, Leighton TG, White PR and Petley GW 2007 Investigation of an anisotropic tortuosity in a biot model of ultrasonic propagation in cancellous bone. *The Journal of the Acoustical Society of America*, **121** (1), 568–574.

Khuri-Yakub BT and Oralkan Ö 2011 Capacitive micromachined ultrasonic transducers for medical imaging and therapy. *Journal of Micromechanics and Microengineering*, **21** (5), 054004.

Kinsey JC and Whitcomb LL 2007 In situ alignment calibration of attitude and Doppler sensors for precision underwater vehicle navigation: Theory and experiment. *Oceanic Engineering, IEEE Journal of*, **32** (2), 286–299.

Krill JA 2006 Ingestible medical payload carrying capsule with wireless communication. US Patent 7,118,531.

Ladd AM, Bekris KE, Rudys A, Kavraki LE and Wallach DS 2005 Robotics-based location sensing using wireless ethernet. *Wireless Networks* **11** (1–2), 189–204.

Laugier P and Haïat G 2011 *Bone Quantitative Ultrasound*, vol. 576. Springer.

Lee S, Yoon D and Ghosh A 2008 Intelligent parking lot application using wireless sensor networks. *Collaborative Technologies and Systems, 2008. CTS 2008. International Symposium on*, pp. 48–57.

Lynnworth LC 2013 *Ultrasonic Measurements for Process Control: Theory, Techniques, Applications*. Academic Press.

Marani G, Choi SK and Yuh J 2009 Underwater autonomous manipulation for intervention missions AUVs. *Ocean Engineering*, **36** (1), 15–23.

Massa DP 1999 Features – choosing an ultrasonic sensor for proximity or distance measurement Part 1: Acoustic considerations – the first step toward identifying the right proximity sensor for your application. *Sensors – the Journal of Applied Sensing Technology*, **16** (2), 34–37.

Park WC and Yoon MH 2006 The implementation of indoor location system to control Zigbee home network. *SICE-ICASE, 2006. International Joint Conference*, pp. 2158–2161.

Park WJ, Kim BS, Seo DE, Kim DS and Lee KH 2008 Parking space detection using ultrasonic sensor in parking assistance system. *Intelligent Vehicles Symposium, 2008 IEEE*, pp. 1039–1044.

Protopappas VC, Kourtis IC, Kourtis LC, Malizos KN, Massalas CV and Fotiadis DI 2007 Three-dimensional finite element modeling of guided ultrasound wave propagation in intact and healing long bones. *The Journal of the Acoustical Society of America*, **121** (6), 3907–3921.

Roberts JF, Stirling T, Zufferey JC and Floreano D 2007 Quadrotor using minimal sensing for autonomous indoor flight. LIS-CONF-2007-006 in *European Micro Air Vehicle Conference and Flight Competition (EMAV2007)*.

Schmerr Jr, LW 2013 *Fundamentals of Ultrasonic Nondestructive Evaluation: A Modeling Approach*. Springer Science & Business Media.

Terzic J, Nagarajah C and Alamgir M 2010 Fluid level measurement in dynamic environments using a single ultrasonic sensor and support vector machine (SVM). *Sensors and Actuators A: Physical*, **161** (1), 278–287.

Treeby BE, Jaros J, Rendell AP and Cox B 2012 Modeling nonlinear ultrasound propagation in heterogeneous media with power law absorption using a k-space pseudospectral method. *The Journal of the Acoustical Society of America*, **131** (6), 4324–4336.

Turner IL, Russell PE and Butt T 2008 Measurement of wave-by-wave bed-levels in the swash zone. *Coastal Engineering*, **55** (12), 1237–1242.

Vasilescu I, Kotay K, Rus D, Dunbabin M and Corke P 2005 Data collection, storage, and retrieval with an underwater sensor network .*Proceedings of the 3rd International Conference on Embedded Networked Sensor Systems*, pp. 154–165.

Zhang E, Laufer J and Beard P 2008 Backward-mode multiwavelength photoacoustic scanner using a planar Fabry–Perot polymer film ultrasound sensor for high-resolution three-dimensional imaging of biological tissues. *Applied Optics*, **47** (4), 561–577.

6

Optical Sensing

Optical sensors have applications in many areas. In structural health monitoring, they (X-ray sensors) are used to detect microscopic fractures in structures (Diamanti and Soutis 2010; Jang et al., 2010; López-Higuera et al., 2011; Majumder et al., 2008); in medicine, they have both simple and complex applications ranging from sensing blood flow and oxygen concentration in the blood (Schurman and Shakespeare 2009; Yao et al., 2010) to the diagnosis and treatment of cancer (Baldini et al., 2006; Barone et al., 2005; Jain et al., 2008); in biochemistry, optical sensors and optical spectroscopy are routinely used to determine the composition of biological and chemical components (Gordon et al., 2008; Kruss et al., 2013; White et al., 2006). In this chapter, we shall consider the most important principles of optical sensor design.

An optical sensing essentially consists of a light source, an object or a process we wish to monitor, and an optical receiver, all of which directly determine the scope of the sensing assignment. When light interacts with matter, it can be absorbed or scattered by it; sometimes both absorption and scattering take place at the same time. During the interaction, the light imparts energy to the matter and, as a result, some of the properties of the matter as well as of the light undergo changes. Some of the most important properties of interest are the wavelength, the momentum, the intensity, and the angle of deflection of the light. At the quantum level, the properties of an electron, an atom, or a molecule of the matter can be of interest. The aim of an optical sensing is to establish a quantitative relationship between states of both the light and the matter of interest before and after an interaction.

The nature of interaction between light and matter primarily depends on the energy carried by the photons of the light, the atomic or molecular structure of the matter, and the relative size of the matter with respect to the wavelength of the light. As far as optical sensing is concerned, the most important optical phenomena are the photoelectric effect, the Compton effect, pair production, Rayleigh scattering, and Raman scattering. The description of these phenomena has two important aspects: the kinematics aspect and the cross-sectional aspect. The kinematics aspect deals with the energy and momentum transfer between particles during an interaction while the cross-sectional aspect deals with the probability of observing the desired phenomena. The latter depends on the electron density of the matter and the intensity of the incident light. In this book we shall not be concerned with the cross-sectional aspect. Readers interested in the full spectrum of optical sensing, particularly in the context of radiology,

Principles and Applications of Ubiquitous Sensing, First Edition. Waltenegus Dargie.
© 2017 John Wiley & Sons, Ltd. Published 2017 by John Wiley & Sons, Ltd.
Companion Website: www.wiley.com/go/dargie2017

Type	Gamma rays		Hard x-rays	Soft x-rays	Extreme ultraviolet		Near ultraviolet	Near infrared	Mid infrared		Far infrared
Frequency	300 EHz	30 EHz	3 EHz	300 PHz	30 PHz	3 PHz	300 THz	30 THz	3 THz		300 GHz
Wavelength	1 pm	10 pm	100 pm	1 nm	10 nm	100 nm	1 µm	10 µm	100 µm		1 mm
Energy	1.24 MeV	124 keV	12.4 keV	1.24 keV	124 eV	12.4 eV	1.24 eV	124 meV	12.4 meV		1.24 meV

Figure 6.1 The spectrum of light.

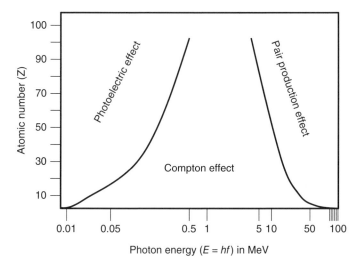

Figure 6.2 The dominant effects observed during the interaction of gamma rays with electrons and atoms.

are recommended to refer to Alpen (1997) for a more exhaustive treatment of the subject.

Figure 6.1 displays the spectrum of light (the relationship between the frequency of light and its energy). As one goes from the infrared region towards gamma rays, the frequency of photons increases and their wavelength decreases. As a result, their energy increases, since it is directly proportional to the frequency of photons ($E = hf$, where h is Planck's constant). At very low wavelengths, photons can directly interact with electrons but as the wavelength increases, interaction is more likely to take place with atoms and molecules than with electrons. Interaction with electrons produces either one of the following effects: pair production, the Compton effect, or the photoelectric effect. All of these occur with gamma rays. Figure 6.2 shows the three regions where gamma rays produce dominant effects. As the wavelength of light increases, interaction with light mainly affects entire atoms, or even molecules, and the collective response of electrons in these particles becomes more interesting. The two essential effects in these sense are the Rayleigh and Raman scatterings. In the subsequent sections of this chapter, we shall discuss the kinematics aspects of these phenomena more closely.

6.1 Photoelectric Effect

When a light of frequency f interacts with matter, it may emit electrons from the matter if f is greater than a certain threshold: $(f : f > f_{th})$. The emitted electrons are most likely valence electrons—those found in the most outer orbit of an atom—because they are the least tightly held. It must be emphasised that both the emission probability and the kinetic energy with which emitted electrons leave the matter depend on the frequency of light instead of its intensity. Interestingly, the kinetic energy of emitted electrons cannot be arbitrary; instead it is a discrete quantity. It was by observing this essential characteristic that Einstein suggested that light must be made up of discrete wave packets (photons) each of which has an energy that equals the multiplication of its frequency and Planck's constant, $E = fh$. The emission process is known as photoelectric effect and it is an absorption process, because for each electron that is emitted, a photon from the incident light is absorbed by the atom.

Figure 6.3 describes the photoelectric effect. As can be seen in part (a), electron emission does not take place until the frequency of the incident light exceeds f_{th}, which means a minimum energy of $E_{min} = hf_{th} = E_b$ is required to liberate an electron from the atom. E_b is called the work function and corresponds to the energy (coulomb force) that binds the electron to the positively charged nucleus, shown in part (b). Once this energy is exceeded, electrons begin to be ejected from the matter at an angle of θ relative to the

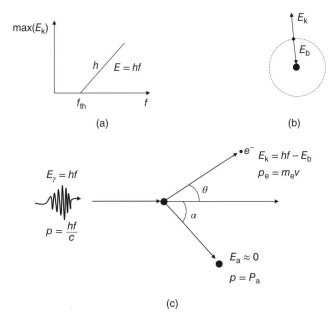

(a)

(b)

(c)

Figure 6.3 The basic principle of the photoelectric effect: (a) the incident light should have a minimum frequency f_{th} in order to set an electron free from the energy that binds it to the nucleus; (b) the minimum energy required to eject an electron from its atom is called a work function ; (c) the maximum kinetic energy with which an electron leaves its atom is directly proportional to the frequency of the incident photon. The energy absorbed by the atom as a result of its interaction with light is negligible; however, to conserve momentum, the atom will have a momentum P_a, so that $p_\gamma = P_a + p_e$ (assuming that the electron was initially at rest).

photon's direction of incidence, carrying a momentum of p as shown in part (c). The maximum kinetic energy of the electron is constrained by the following expression:

$$E = hf - E_b - E_a \tag{6.1}$$

where E_a is the kinetic energy of the recoiling atom. This is illustrated in part (c). Compared to E, E_a is significantly small and the maximum kinetic energy of the emitted electron can be approximated by:

$$E = hf - E_b \tag{6.2}$$

The direction taken by the recoiling atom is likewise of little consequence, since it carries a negligible amount of kinetic energy.

The photoelectric event will not be an isolated event. When an electron from a lower-energy shell (say, from a K-shell) is emitted, it leaves a vacancy with an attraction force (like a black hole), which will immediately be filled by an electron from a higher energy level. Suppose a k-shell electron is emitted as a result of a photoelectric effect with a kinetic energy that equals:

$$E_k = hf - (E_b)_k \tag{6.3}$$

The k-shell will be immediately filled by an electron from the L, M or N or other higher energy shells, but most likely from the L-shell. Let us suppose it is filled by an electron from the L-shell. Upon taking a lower energy level, the L-shell electron releases energy in the form of an X-ray photon the magnitude of which is equal to:

$$hf_k = (E_b)_k - (E_b)_l \tag{6.4}$$

where $(E_b)_k$ is the attraction energy of the K-shell and $(E_b)_l$ is the energy by which the electron was bound by the L-shell. Interestingly, this photon can be absorbed by another electron in one of the higher energy levels, which is then emitted from the atom (recall that as the distance of an electron from the nucleus increases, the binding energy of the nucleus decreases). This is called the Auger effect. If we assume that the newly emitted electron was an M-shell electron (as shown in Figure 6.4), its kinetic energy is given as:

$$E_m = (E_b)_l - (E_b)_m \tag{6.5}$$

where $(E_b)_m$ is the energy by which the newly emitted electron was bound in the M-shell and $(E_b)_l$ is the attraction energy of the L-shell. Now two electrons are emitted from the atom, leaving vacant spaces in two shells: one in the L and one in the M-shells. Suppose these vacancies are filled by two N-shell electrons which, as a result, emit energy because they are occupying shells that have lower energy levels. For the electron descending into the L-shell, the emitted energy corresponds to:

$$E'_{n1} = (E_b)_l - (E_b)_n \tag{6.6}$$

For the other electron:

$$E'_{n2} = (E_b)_m - (E_b)_n \tag{6.7}$$

If E'_{n1} and E'_{n2} are absorbed by higher energy level electrons (suppose by N-shell electrons) which are then emitted from the atom, the kinetic energy of these two electrons will be:

$$E_{n1} = E'_{n1} - (E_b)_n = (E_b)_l - 2(E_b)_n \tag{6.8}$$

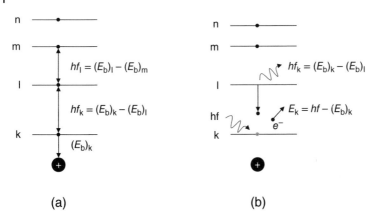

(a)   (b)

Figure 6.4 The Auger effect : (a) the energy absorbed or emitted by an electron to ascend or descend a single energy level; (b) a vacant lower level shell can pull down an electron residing in a higher shell. In the process the electron emits energy which may be sufficient enough to set another electron from a higher energy level free. The chain reaction thus sets multiple electrons free before an equilibrium is achieved.

and,

$$E_{n2} = E'_{n2} - (E_b)_n = (E_b)_m - 2(E_b)_n \tag{6.9}$$

Thus the total kinetic energy of the emitted electrons is given as:

$$E_{tot} = E_k + E_m + E_{n1} + E_{n2} \tag{6.10}$$

Example 6.1 One of the initial materials in which the photoelectric effect was observed was the tungsten wire, which was used to produce electric bulbs. Calculate the longest wavelength that can produce a photoelectric current in a tungsten wire.

The work function (E_b) of a tungsten wire equals 4.52 eV. Hence,

$$E_b = hf_{th} = h\frac{c}{\lambda_{th}}$$

From which we have,

$$\lambda_{th} = \frac{hc}{E_b} = \frac{1240\,\text{eV nm}}{4.52\,\text{eV}} = 274.3\,\text{nm}.$$

Example 6.2 What will be the maximum kinematic energy that can be measured from emitted electrons if we shine an infrared light of 10 μm on a tungsten wire?

Even though the answer for this question is readily available from the above example, we can address it in a different way. The energy of the incident light with a wavelength of $\lambda = 10$ μm equals 124 meV (see Figure 6.1). This energy is far less than the energy required to liberate an electron from a tungsten atom (which has a work function of 4.52 eV). Therefore, no electron will be emitted.

Example 6.3 An electron is emitted from a tungsten electrode using light of 200 nm wavelength and accelerated towards another tungsten electrode. If we wish to stop the

Figure 6.5 The kinetic energy of the emitted electron should be balanced by a corresponding potential energy for the electron to stay in its orbit.

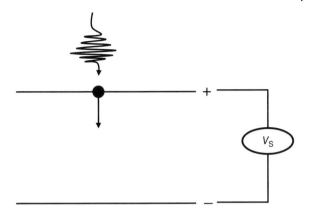

electron emission by a counter effect, as shown in Figure 6.5, what is the magnitude of the voltage required?

The maximum kinetic energy with which the electron is emitted as a result of the photoelectric effect equals:

$$E_k = hf - E_b = h\frac{c}{\lambda} - 4.52\,\text{eV} = \frac{1240\,\text{eV}\,\text{nm}}{200\,\text{nm}} - 4.52\,\text{eV} = 1.68\,\text{nm}$$

This energy should be counterbalanced by a potential energy in order to keep the electron in its orbit. The potential energy is the product of the charge of an electron (1 eV) and the applied voltage:

$$E_p = eV$$

Since e is already a known quantity, we have:

$$V = \frac{E_k}{e} = \frac{1.68\text{eV}}{1\,\text{eV}} = 1.68\text{V}$$

Example 6.4   We wish to develop a photo diode for inspecting the structural integrity of buildings and bridges (see Figure 6.6). The photo diode requires at least 3.28×10^{16} photons/s to generate an appreciable current. If the corresponding laser diode emitting the photons uses light having a wavelength of 1310 nm, what is the magnitude of the received power required to produce the desired current?

The energy of a single photon at the specified wavelength should be:

$$E = hf = h\frac{c}{\lambda} = 0.95\,\text{eV}$$

The collective energy produced by the incident photons will be:

$$E_{\text{incident}} = (3.28 \times 10^{16})\ \text{photons}\ \times (0.95\,\text{eV/photon}) = 3.116 \times 10^{16}\ \text{eV}$$

Considering that a 1 W electric power equals to 1J s^{-1} and $1\text{J} = 6.25 \times 10^{18}$ eV, the power of the incident light emitted by the laser diode should be:

$$P = \frac{3.116 \times 10^{16}\,\text{eV}}{6.25 \times 10^{18}\,\text{eV}} \frac{1}{s} = 5\,\text{mW}$$

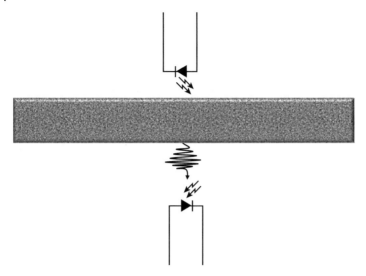

Figure 6.6 A simple optical sensing arrangement for sensing the integrity of a structure. The laser diode emits a light having a wavelength of 1310 nm. This passes through the structure (if there is a microscopic fracture) and is collected by the photo-diode, which transforms the incident photon energy into an electric current.

6.2 Compton Effect

When a photon of sufficient energy (approximately 1 MeV; see also Figure 6.2) collides with an electron, it imparts part of its momentum to the electron, thereby ejecting it from its shell. The electron departs at an angle θ with respect to the angle of incidence. Unlike the photoelectric effect, the photon is not absorbed by the electron; instead, it is scattered at an angle ϕ with respect to the incident angle (see Figure 6.7) (from the

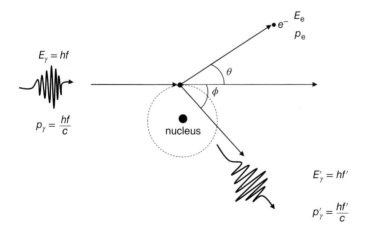

Figure 6.7 The Compton effect or Compton scattering. An incident photon gives an electron sufficient momentum to leave its orbit. The photon is scattered with a momentum lower than its original momentum. The photon's scattering angle and its initial and final energies are related with one another.

required energy, it is apparent how short the wavelength of the incident photon should be for a Compton interaction to take place). As a result of the interaction, the photon's frequency changes (reduces) and the electron acquires momentum and kinetic energy. This phenomenon is relativistic, due to the high speed at which the electron leaves its orbit after the collision. The photon's new frequency and momentum and its scattering angle, the electron's angle of departure, kinetic energy and momentum are all related to one another by theory of special relativity. By measuring some of these quantities, it is possible to determine the others.

Suppose the incident photon has a zero rest mass and $\mathbf{p}_\gamma$ momentum vector. Similarly, the electron, having a rest mass of m_e, is initially at rest and unbound. [1] After the collision, the photon's momentum vector changes to $\mathbf{p}'_\gamma$ and the electron achieves a momentum of $\mathbf{p}'_e$. By the conservation of momentum, the initial and final momenta are related to each other as follows:

$$\mathbf{p}_\gamma = \mathbf{p}'_\gamma + \mathbf{p}'_e \tag{6.11}$$

Rearranging and squaring both sides of Eq. (6.11) yields:

$$p_\gamma^2 + p_\gamma'^2 - 2\mathbf{p}_\gamma \mathbf{p}'_\gamma = \mathbf{p}_e'^2 \tag{6.12}$$

Notice that some boldface expressions are replaced by normal fonts. This is not accidental. The dot product of a vector with itself is a scalar quantity:

$$\left(p \cos \theta \; p \sin \theta\right) \cdot \begin{pmatrix} p \cos \theta \\ p \sin \theta \end{pmatrix} = p^2(\cos^2\theta + \sin^2\theta) = p^2 \tag{6.13}$$

From the special relativity theory, we know that when a particle having a rest mass of m_e moves at a speed that approaches the speed of light, its kinetic energy is equal to:

$$E_e = \sqrt{(p'_e c)^2 + (m_e c^2)^2} \tag{6.14}$$

Hence the total relativistic energy E_e and momentum $\mathbf{p}'_e$ of the electron are related to its rest mass m_e by the invariant relation: $(\mathbf{p}'_e \cdot \mathbf{p}'_e)c^2 - E_e^2 = -m_e^2 c^4$; this is obtained simply by squaring both sides of Eq. (6.14) and rearranging terms. Since the energy of the system before collision should be equal to the energy after collision (due to conservation of energy), we have the expression:

$$hf + m_e c^2 = hf' + \sqrt{(p'_e c)^2 + m_e^2 c^4} \tag{6.15}$$

where $E_\gamma = hf$ is the initial energy of the photon, and $m_e c^2$ is the initial energy of the electron. Dividing both sides of Eq. (6.15) by c results in:

$$p_\gamma + m_e c = p'_\gamma + \sqrt{p_e'^2 + m_e^2 c^2} \tag{6.16}$$

Note that for a massless photon, the momentum is:

$$p_\gamma = \frac{E_\gamma}{c} = \frac{hf}{c}$$

---

1 In reality, these assumptions are not rigorous, as long as electrons occupy various atomic energy levels. But the assumptions simplify the analysis and the resulting error is inconsequential (Alpen 1997; Attix 2008).

This realisation is a consequence of one of Einstein's most remarkable insights. If we rearrange Eq. (6.16) and square both sides of the equation, we obtain:

$$p_\gamma^2 + p_\gamma'^2 - 2p_\gamma p_\gamma' + 2m_e c (p_\gamma - p_\gamma') + m_e^2 c^2 = m_e^2 c^2 + p_e'^2 \tag{6.17}$$

If we subtract Eq. (6.12) from Eq. (6.17) and rearrange terms, we obtain:

$$m_e c (p_\gamma - p_\gamma') = p_\gamma p_\gamma' - \mathbf{p}_\gamma \mathbf{p}_\gamma' \tag{6.18}$$

Notice that the vector dot product in Eq. (6.18) takes into account the fact that $\mathbf{p}_\gamma$ has only an x-component (its y-component is zero), whereas $\mathbf{p}_\gamma'$ has both x- and y-components; furthermore, the x-component of $\mathbf{p}_\gamma'$ equal $p_\gamma' \cos \phi$ and its y-component equals $p_\gamma' \sin \phi$ (see Figure 6.7). Hence, the dot product of the vectors yields the following outcome:

$$\mathbf{p}_\gamma \cdot \mathbf{p}_\gamma' = \begin{pmatrix} p_\gamma & 0 \end{pmatrix} \cdot \begin{pmatrix} p_\gamma' \cos \phi \\ p_\gamma' \sin \phi \end{pmatrix} = p_\gamma p_\gamma' \cos \phi \tag{6.19}$$

Finally, by dividing both sides of Eq. (6.18) by $m_e c p_\gamma p_\gamma'$ we can establish a relationship between the energies of the incident and the scattered photon:

$$\frac{1}{E_\gamma'} - \frac{1}{E_\gamma} = \frac{1}{m_e c^2} (1 - \cos \phi) \tag{6.20}$$

Since $1/(m_e c^2)$ is a constant, the incident and scattered energies are related to one another by the scattered angle ϕ. The angle of departure of the electron can also be derived in the same way and it is related to ϕ as follows:

$$\cot \theta = \left(1 + \frac{hf}{m_0 c^2} \right) \tan \left(\frac{\phi}{2} \right) \tag{6.21}$$

Example 6.5 A photon having a wavelength of 6000 nm collides with an electron which is at rest. After the interaction, the wavelength of the scattered photon changes by exactly one Compton wavelength of the electron. Calculate:

1) the scattering angle of the photon
2) the scattering angle of the electron
3) the change in the energy of the electron due to scattering.

In order to understand the meaning and significance of the Compton wavelength of the electron, we can rewrite Eq. (6.20) in terms of the initial and final wavelengths of the photon. Since $E = hf$ and $f = c/\lambda$,

$$\frac{\lambda_\gamma'}{hc} - \frac{\lambda_\gamma}{hc} = \frac{1}{m_e c^2} (1 - \cos \phi) = \lambda_\gamma' - \lambda_\gamma = \frac{h}{m_e c} (1 - \cos \phi) \tag{6.22}$$

The value of

$$\frac{h}{m_e c} = \lambda_c$$

is a constant and it is known as the standard Compton wavelength of the electron. According to the 2010 position of the Committee on Data for Science and Technology, it has a value of $2.4263102389(16) \times 10^{12}$ m. With this in mind and using Figure 6.8, we can calculate all the unknown quantities.

Figure 6.8 A photon of 6000 nm wavelength collides with an electron and scatters at an angle ϕ.

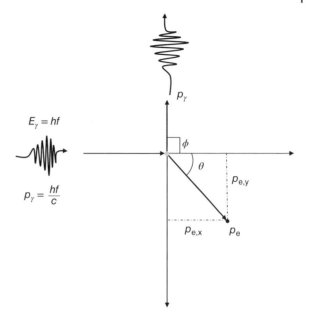

$E_\gamma = hf$

$p_\gamma = \dfrac{hf}{c}$

1) *The scattering angle of the photon.*
   From Eq. (6.22), the scattering angle of the photon can be expressed as:

   $$\phi = \cos^{-1}\left(1 - \frac{1}{\lambda_c}(\lambda'_\gamma - \lambda_\gamma)\right) = \cos^{-1}\left(1 - \frac{1}{2426\,\text{nm}}(8426\,\text{nm} - 6000\,\text{nm})\right)$$

   From the above question, it is clear that the photon is scattered at 90 °.
2) *The scattering angle of the electron.*
   Because of the conservation of momentum (Eq. (6.11)), we can solve this problem in a simple way. The initial momentum of the photon is:

   $$p_\gamma = \frac{hf}{c} = \frac{h}{\lambda_\gamma} = \frac{6.63 \times 10^{-34}}{6 \times 10^{-12}} = 1.105 \times 10^{-22}\ \text{kg.m/s}$$

   This momentum is in the x-direction (or in vector representation, $\hat{\imath}$). Likewise, the final momentum of the photon is:

   $$p'_\gamma = \frac{hf'}{c} = \frac{h}{\lambda'_\gamma} = \frac{6.63 \times 10^{-34}}{8.4 \times 10^{-12}} = 7.9 \times 10^{-23}\ \text{kg.m/s}$$

   This momentum is in the y-direction (or in vector representation, $\hat{\jmath}$). For the momentum to be conserved the electron should have the following final momentum (see also Figure 6.8):

   $$\left(1.105 \times 10^{-22}\right)\hat{\imath} - \left(7.9 \times 10^{-23}\right)\hat{\jmath}$$

   Consequently,

   $$\theta = \tan^{-1}\left(\frac{-7.9}{11.05}\right) = 35.6\text{o}$$

3) *The change in the energy of the electron due to scattering.*
Once again, due to the conservation of energy, the change in the energy of the electron should be the negative of the change in the energy of the photon, which is:

$$\Delta_E = hc \left(\frac{1}{\lambda'_\gamma} - \frac{1}{\lambda_\gamma} \right) = 9.47 \times 10^{-15} \text{J} = 59.2 \text{ eV}$$

Example 6.6 We wish to determine the nature of a sample by exposing it to a beam of X-rays having a wavelength of 0.2400 nm and an incident angle of 0 °. The electrons in the sample are assumed to be initially at rest. After the X-rays interact with and impart energy to the electrons in the sample, they scatter in different angles and with different momentums. The electrons likewise scatter with different angles and momentums. We surround the sample by a detector, which detects the scattered photons at a particular angle and determines their final wavelength, as shown in Figure 6.9.

1) What is the longest wavelength that can be detected?
2) What will be the kinetic energy of an electron that is scattered by a photon, the final wavelength of which is the longest?
3) Determine the final wavelength of a photon that is scattered at an angle of $\pi/3$
4) What will be the scattering angle of an electron which interacts with a photon that is scattered at an angle of $\pi/3$?

The answers are as follows:

1) *The longest wavelength that can be detected.*
From the relationship between the initial and final wavelength of a photon in a Compton scattering (Eq. (6.22)), it is clear that the longest final wavelength occurs when $\cos \theta = -1$ because:

$$\lambda'_\gamma = \lambda_\gamma + \lambda_c(1 - \cos \theta) = 0.2400 + 0.0049 = 0.2449 \text{ nm}$$

2) *The kinetic energy of the electron scattered by a photon the final wavelength of which is 0.2449 nm.*

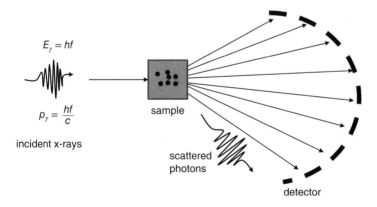

$E_\gamma = hf$

$p_\gamma = \dfrac{hf}{c}$

incident x-rays

sample

scattered photons

detector

Figure 6.9 A sample exposed to X-rays scatter the incident photons at different angles and with different momentums. By determining the scattering angle and momentum of the photons and the electrons in the sample, it is possible to determine the nature of the sample.

From the conservation of energy (Eq. (6.15)), we know that:

$$E + m_e c^2 = E' + \sqrt{(p'_e c)^2 + m_e^2 c^4}$$

Hence,

$$E - E' = \sqrt{(p'_e c)^2 + m_e^2 c^4} - m_e c^2$$

is the kinetic energy of the electron (because the difference between the initial and the final energy of a photon is the energy imparted to an electron as a result of which the electron recoils, assuming that the electron was at rest before the interaction, of course):

$$E = \frac{hc}{\lambda_\gamma} = \frac{1240}{0.24} = 5167\,\text{eV}$$

and

$$E' = \frac{hc}{\lambda'_\gamma} = \frac{1240}{0.2449} = 5063\,\text{eV}$$

Finally,

$$E - E' = 5167\,\text{eV} - 5063\,\text{eV} = 511\,\text{eV}$$

3) *The final wavelength of the photon that is scattered at an angle of $\pi/3$.*
From Eq. (6.16) we have:

$$\lambda'_\gamma - \lambda_\gamma = \lambda_c (1 - \cos\theta)$$

Since $\cos \pi/3 = 0.5$, we have

$$\lambda'_\gamma = \lambda_\gamma + 0.5\lambda_c = 0.2400 + (0.5)(0.0024) = 0.2412\,\text{nm}$$

4) *The scattering angle of an electron which interacted with a photon scattered at angle of $\pi/3$.*
Once again we can exploit the conservation of momentum to determine this angle: With the final wavelength determined above, the final momentum of the photon should be:

$$p'_\gamma = \frac{hf'}{c} = \frac{h}{\lambda'_\gamma} = \frac{6.63 \times 10^{-34}}{241.2 \times 10^{-12}} = 2.748756 \times 10^{-24}\,\text{kg.m/s}$$

Apparently, the photon travels with this momentum at an angle of $\phi = \pi/3$, from which it is possible to decompose the final momentum into its x- and y-components.

$$p'_{\gamma,x} = 2.748756 \times 10^{-24} \quad \cos \pi/3 = 1.374378 \times 10^{-24}\,\text{kg.m/s}$$

$$p'_{\gamma,y} = 2.748756 \times 10^{-24} \quad \sin \pi/3 = 2.391418 \times 10^{-24}\,\text{kg.m/s}$$

Once again, equating the x-component of the initial momentum with the x-components of the final momentum (the x-component of the final momentum of the photon plus the x-component of the final component of the momentum of the electron) yields:

$$p_\gamma = p'_{\gamma,x} + p'_{e,x}$$

from which we have:

$$p'_{e,x} = p_\gamma - p'_{\gamma,x}$$

Likewise, we have:

$$p'_{e,y} = -p'_{\gamma,y}$$

Recall that the initial momentum of the incident photon does not have a y-component (we assume that it was directed along the x-axis). Furthermore, the y-component of the final momentum of the electron should be equal but opposite in direction to the y-component of the momentum of the photon. Thus,

$$p'_{e,x} = \frac{h}{\lambda_\gamma} - p'_{\gamma,x} = 2.7625 \times 10^{-24} - 1.374378 \times 10^{-24} = 1.3881 \times 10^{-24} \text{ kg.m/s}$$

Finally,

$$\theta = \tan^{-1}\left(-\frac{2.391418}{1.374378}\right) = -59.8°$$

6.3 Pair Production

In order to understand pair production, it is useful to briefly summarise quantum field theory first. At a subatomic level, there are certain physical phenomena that cannot be adequately explained by classical physics. For example, when two protons travelling at a speeds approaching the speed of light collide inside the Large Hadron Collider, new particles emerge, as it were, out of nothing. These particles are not, as some may imagine, the building blocks from which the two protons are made. In fact, a proton is made up of two up quarks and one down quark, all of which are basic particles; the newly emerged particles are something else. Similarly, when an electron and a positron collide, both particles completely disappear and a gamma particle (photon) is created instead. It is also known that inside the sun, in a process called beta decay, a single proton is transformed into a neutron (which is made up of two down quarks and an up quark), a positron, and an electron neutrino.

Quantum field theory is a theory that attempts to explain these phenomena. According to the theory, reality is but a web of elementary fields extending through out space. Each elementary particle (quarks, bosons, leptons) has its own field. When these fields are excited by a sufficient amount of energy, the excitation manifests itself in the form of particles coming into existence (in most cases, in the creation of a mass). Likewise, when the energy is removed from the excitation, the particles vanish and the fields collapse into a resting state. Furthermore, an excitation in or a collapse of one field can excite another field, causing a particle of that field to emerge.

Pair production is an absorption process. When a photon of very high energy (of the order of 10s MeV) penetrates into a coulomb field, it will be absorbed by the field giving rise to the emergence of an electron and a positron. The most probable condition for the occurrence of pair production is in the vicinity of a nucleus, where the coulomb field is the strongest. However, it can also take place, with a lower probability, in the field of an electron. If it happens in an electron's field, the host electron providing the coulomb field also obtains a significant kinetic energy to be ejected from the atom. Thus, two

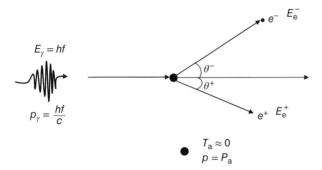

Figure 6.10 A pair production.

electrons and a positron are ejected from the site. As a result, the effect is called "triplet production", even though one of the electrons is not actually created but ejected.

A minimum photon energy of $2m_ec^2 = 1.022\,\text{MeV}$(cf. Eq. (6.15)) is required for pair production to take place in a nuclear field, but $4m_ec^2$ is the threshold for triplet production because two electrons should be unbound and ejected. Figure 6.10 illustrates a pair production process in a nuclear field. The incident photon with a quantum energy $E_\gamma = hf$ is absorbed by a nuclear field and as a result an electron–positron pair is created with kinetic energies E_e^+ and E_e^-. Ignoring the small amount of kinetic energy imparted to the nucleus, the energy-conservation equation can be stated as:

$$E_\gamma = hf = 2m_ec^2 + E_e^+ + E_e^- \tag{6.23}$$

The electron and positron do not necessarily have equal kinetic energies, but their average is given by:

$$\overline{E}_e = \frac{hf - 1.022\,\text{MeV}}{2} \tag{6.24}$$

where we replace $2m_ec^2$ with 1.022 MeV. For hf values well above the threshold energy $2m_ec^2$, the electron and positron are strongly forward directed. Their average angle of departure with respect to the incident angle can be approximated by:

$$\overline{\theta} \cong \frac{m_0c^2}{\overline{E}_e} \tag{6.25}$$

6.4 Raman Scattering

For light having a long wavelength, direct interaction with individual subatomic particles is improbable. Instead, interaction is more likely to take place at atomic and molecular levels; besides the size of the wavelength, the photon energy also becomes smaller as the wavelength increases.

When two or more atoms form a bond to become a molecule, some interesting phenomena can be observed. Firstly, even though the net electric charge of the molecule is zero, the atoms may not share the free electrons equally at all times. As a result, some of the atoms become slightly negatively charged (those possessing the shared electrons much of the time) and some of the atoms become slightly positively charged (those which do not posses the shared electrons much of the time). The result is the set up of a permanent dipole moment and the molecule is said to be polar. Not all molecules are polar or permanently polar. Molecules that have symmetric structures such as oxygen, nitrogen, and carbon dioxide do not have permanent dipole moments. However, even if a molecule does not have a permanent dipole moment, it is possible to induce one using an external electric field. This process is called polarisation and the magnitude of the dipole moment induced by this process depends on the polarisability of the molecular species.

Secondly, the inter-nuclear distances within the molecule are never fixed due to the different types of motions (rotational, translational, and vibrational [2]) the molecule undergoes. Of these motions, the vibrational motion—stretching of the bonds between the atoms—is the most interesting phenomenon. Figure 6.11 shows the different types of vibrational motions a water molecule undergoes. Each mode of vibration has its own fundamental frequency and can be considered as a simple harmonic motion that can be expressed by Hook's law. The polarisability of the molecule, which is a measure of the ease with which the electron cloud of the molecule (see Figure 6.12) can be distorted by an external electric field, depends on the vibrational motion.

If a monochromatic (having only a single frequency) light interacts with a polarisable molecule, its electric field induces a dipole moment (remember that light has both electric and magnetic field which are orthogonal to each other and to the direction of propagation). Due to the oscillation of the external electric field and the internal

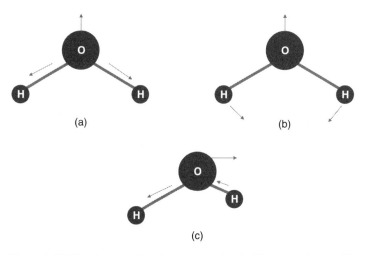

Figure 6.11 Vibrational motions in a water molecule: (a) symmetric stretching; (b) bending ; (c) asymmetric stretching.

---

2 Molecular motion, heat, and temperature are interrelated. Temperature is the average kinetic energy of molecules in a substance. Heat is a measure of the energy transfer between two substances having different temperatures. The translational motion of molecules is a motion resulting from their temperature. The higher the temperature the higher the translational motion. Rotational motion typically involves the rotation of an entire molecule or the rotation of parts of the molecule relative to another molecule in a torsional motion. Such motion can only occur in gas and liquid phases, as molecules are bonded in fixed positions in solids.

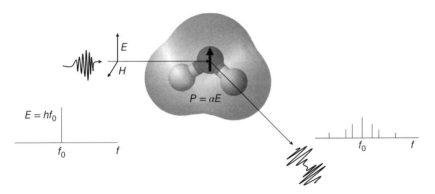

Figure 6.12 An electron cloud in a water molecule and Raman scattering.

vibrations of the molecule, the dipole moment, too, is a time-varying quantity. Depending on its magnitude and frequency, the dipole moment emits radiation, which can be expressed as the amplitude modulation of the incident light. In order to highlight the significance of this statement, we shall closely investigate the mathematical expressions of the dipole moment (**p**) as a function of the incident electric field and the polarisability tensor α of a molecule:

$$\mathbf{p} = \tilde{\alpha}\mathbf{E} \tag{6.26}$$

where **E** is the oscillating electric field of the incident light. The dipole moment is a three-dimensional quantity relating the ease with which the incident electric field modifies the dipole moment in a specific direction. Hence, the polarisability tensor is a 3×3 matrix:

$$\begin{pmatrix} p_x \\ p_y \\ p_z \end{pmatrix} = \begin{pmatrix} \alpha_{xx} & \alpha_{xy} & \alpha_{xz} \\ \alpha_{yx} & \alpha_{yy} & \alpha_{yz} \\ \alpha_{zx} & \alpha_{zy} & \alpha_{zz} \end{pmatrix} \begin{pmatrix} E_x \\ E_y \\ E_z \end{pmatrix} \tag{6.27}$$

Suppose the oscillating electric field of the incident light can be expressed as: $\mathbf{E} = \mathbf{E}_0 \cos(2\pi f_0 t)$, where f_0 is the oscillating frequency. Thus the induced dipole moment is given as:

$$\mathbf{p} = \tilde{\alpha}\mathbf{E}_0 \cos(2\pi f_0 t) \tag{6.28}$$

Since the polarisability of a molecule is directly related to the fundamental frequencies of its molecular vibrations, we can express $\tilde{\alpha}$ as:

$$\tilde{\alpha} = \alpha_0 + \sum_{k=1}^{M} \alpha_k \cos(2\pi f_k t) \tag{6.29}$$

where $k = 1, 2,...,M$ are the fundamental frequencies of the molecular vibrations. Then the induced dipole moment can be expressed as:

$$\mathbf{p} = \alpha_0 \mathbf{E}_0 \cos(2\pi f_0 t) + \sum_{k=1}^{M} \alpha_k \mathbf{E}_0 \cos(2\pi f_0 t) \cos(2\pi f_k t) \tag{6.30}$$

Using the trigonometric equality:

$$\cos x \cos y = \frac{1}{2}(\cos(x + y) + \cos(x - y))$$

Eq. (6.30) can be rewritten as:

$$\mathbf{p} = \alpha_0 \mathbf{F}_0 \cos(2\pi f_0 t) + \frac{\mathbf{E}_0}{2} \sum_{k=1}^{M} \alpha_k \left[\cos(2\pi (f_0 + f_k)t) + \cos(2\pi (f_0 - f_k)t) \right] \qquad (6.31)$$

Equation (6.31) is an important realisation. As can be seen, the dipole moment has three components. The first component has the same frequency as the incident light, even though its magnitude is attenuated by the factor α_0. It is called a Rayleigh or elastic scattering. The components inside the summation, however, shift the frequency of the incident light by $\pm f_k$. The $+f_k$ components emit radiation with higher energies that equal: $E = h(f_0 + f_k)$; the $-f_k$ components emit radiation with lower energies that equal: $E = h(f_0 - f_k)$. The lower-energy components are called Stokes Raman scattering and the higher-energy components are called anti-Stokes Raman scattering.

Another way of looking at the three phenomena of Eq. (6.31) is to consider the energy transitions some of the electrons of the molecule make during interaction with the incident light. These electrons are sufficiently excited to leave their shell, but after reaching a virtual state (an evanescent energy state that will be abandoned quickly), the electrons bounce back to a lower energy level. Some of the electrons return where they have been before, thereby emitting the same amount of energy they absorbed when they were ejected from their shell. Some of the electrons fall to some energy level higher than the original level, hence emitting a smaller amount of energy than they have absorbed. Some others return to an energy level lower than the original one, thereby emitting a larger amount of energy than they have absorbed. These three phenomena are illustrated in Figure 6.13.

Example 6.7 A given molecule has a vibrational mode at 5714.3 nm. If this molecule is excited by a beam of light at 623.8 nm, at what frequencies and wavelengths is the Raman (Stokes) scattering expected to be observed?

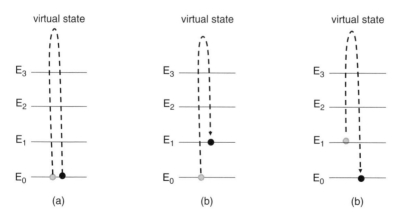

Figure 6.13 Raman scattering: (a) Rayleigh scattering occurs when bouncing electrons emit the exact same amount of energy they absorbed when they were ejected from their shell and they return to their original energy level; (b) Stokes Raman scattering occurs when bouncing electrons return to a higher energy level than the original, thereby emitting less energy than they absorbed when they were ejected from their shell; (c) anti-Stokes Raman scattering occurs when bouncing electrons return to an energy level lower than the original, thereby emitting more energy than they absorbed when they were ejected from their shell.

Since the incident light is travelling at a speed of light, its frequency will be:

$$f_0 = \frac{3 \times 10^8 \text{ m/s}}{623 \times 10^{-9} \text{ nm}} = 481 \text{ THz}$$

The vibrational frequency translated into the Raman spectroscopy will be:

$$f_R = \frac{3 \times 10^8 \text{ m/s}}{5714.3 \times 10^{-9} \text{ nm}} = 52.5 \text{ THz}$$

Hence, a Raman (Stokes) scattering is expected to be observed at the following frequencies

1)

$$f_0 - f_R = 481 \text{ THz} - 52.5 \text{ THz} = 428.5 \text{ THz}$$

2)

$$f_0 + f_R = 481 \text{ THz} + 52.5 \text{ THz} = 533.5 \text{ THz}$$

The corresponding wavelengths are $\lambda_{f_0 - f_R} = 700.1 \text{nm}$ and $\lambda_{f_0 + f_R} = 562.3 \text{nm}$, respectively.

6.5 Surface Plasmon Resonance

From the energy spectrum of the scattered light during a Raman interaction, it is possible to infer the molecular vibrations and, thereby, the composition of a molecule that gave rise to the Raman scattering. This is because the fundamental frequencies of the molecular vibrations of different molecules are known or can be determined. The process is reciprocal; once the fundamental frequencies of the molecular vibrations are known, it is possible to determine a priori the energy spectrum of a Raman scattering. One of the biggest challenges, however, is that the probability of experiencing Raman scattering is very small. In water molecules, for example, one in a million photons produces a Raman effect. Consequently, for Raman scattering to take place, either a high concentration of incident light or a long scan time is required. As shown in Figure 6.14, the interaction probability can be increased by confining light within an enclosure, so that multiple reflections can take place (Mayer and Hafner 2011; Willets and Van Duyne 2007; Zijlstra et al., 2012).

If this enclosure is a metal, such as a waveguide, then the incident light can collectively excite the electron cloud of the metal, which oscillates at a specific frequency. This phenomenon is called "plasmon". The most frequently used approach to generate surface plasmon is to interface two different types of materials having opposite dielectric constants; for example, a metal and a dielectric material. Figure 6.15 displays the interface of a prism with a metal. As the incident light passes through the prism, it reflects multiple times within the prism, finally leaving the prism at an angle θ. The intensity of the scattered light is a function of the scattered angle θ. If the light is scattered at a certain critical angle θ_{sers}, one can observe a sharp dip in the intensity of the scattered light. This is because a significant portion of the incident light is absorbed by the electric cloud at the surface of the metal to give rise to a plasmon (hence, the name "surface plasmon resonance").

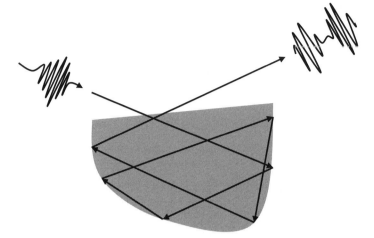

Figure 6.14 The probability of observing a Raman scattering increases by confining light within an enclosure, so that through multiple reflections, incident photons have ample opportunity to interact with the molecular electron cloud which absorbs them and gives rise to surface plasmon resonance.

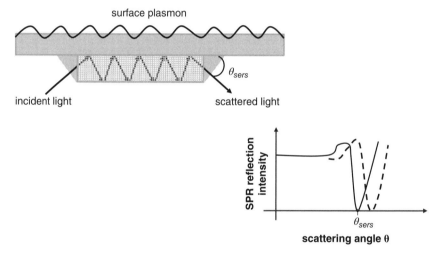

Figure 6.15 The resonance of plasmon (the collective excitation of the electron cloud) depends on the angle of scattering (θ) of the incident light. At a critical angle, θ_{sers}, the intensity of the reflected light is very low because all the photons will be absorbed by the electric cloud which gives rise to the electric dipole moment which in turn gives rise to surface plasmon resonance.

If artefacts having molecular affinity to the metal are adsorbed by the metal surface, they disturb the surface plasmon in two different ways. Firstly, they shift the critical angle θ_{sers} to the right, as shown in Figure 6.15 (Luk'yanchuk et al., 2010; Yu et al., 2011). Secondly, it will take some time for the molecules of the artefacts to bond with the metal surface. This time can be determined by observing the surface plasmon resonance response characteristics (see Figure 6.16). Depending on the affinity and quantity of the artefacts, the time to bond can be slow or fast and by observing the resonance reflection angle and the characteristic resonance curve, it is possible to

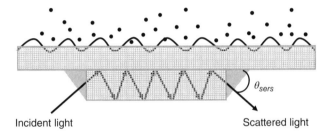

Incident light   Scattered light

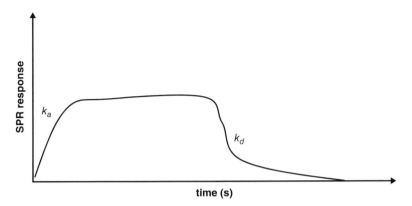

Figure 6.16 The surface plasmon resonance binding characteristic as a function of time is influenced by the association and dissociation properties of the molecules interacting with the metal surface. When molecules of high binding affinity interact with the surface of the metal (optical sensor) molecular binding and unbinding takes place. Initially, there are enough molecules to which incoming artefacts can bind and, hence, the binding rate in the time-resolved surface plasmon resonance response of kinetic behaviour increases, depending on the binding constant of the artefacts; then the binding rate achieves its maximum after which the binding and unbinding rates are balanced and no net binding rate is observed (the system is known to be in an equilibrium state). After a while, no free molecules are available to bind to and a rapid decline in the binding rate is observed. The declining binding rate depends on the dissociation constant of the artefacts. The system is unresponsive for artefacts which have no affinity to metal surface.

determine the type of artefact that is adsorbed on the surface of the metal (Ringe et al., 2010; Scarano et al., 2010; Verellen et al., 2011). This approach is extensively used in biomedical, electrochemical, and biochemical analysis, among others. For example, different types of nanoparticles interfacing metals with dielectric materials are used to diagnose cancer and microscopic particles in the body (El-Sayed et al., 2005; Eustis and El-Sayed 2006; Ghosh and Pal 2007). Moreover, chemical laboratories around the world use surface plasmon resonance to study chemical adsorption, contamination, and emerging behaviour in biochemical organisms (El-Sayed et al., 2005; Homola 2008).

References

Alpen EL 1997 *Radiation Biophysics*. Academic Press.
Attix FH 2008 *Introduction to Radiological Physics and Radiation Dosimetry*. John Wiley & Sons.

Baldini F, Chester AN, Homola J and Martellucci S 2006 *Optical Chemical Sensors.* Springer Science & Business Media.

Barone PW, Baik S, Heller DA and Strano MS 2005 Near-infrared optical sensors based on single-walled carbon nanotubes. *Nature Materials*, **4**(1), 86–92.

Diamanti K and Soutis C 2010 Structural health monitoring techniques for aircraft composite structures. *Progress in Aerospace Sciences* **46**(8), 342–352.

El-Sayed IH, Huang X and El-Sayed MA 2005 Surface plasmon resonance scattering and absorption of anti-EGFR antibody conjugated gold nanoparticles in cancer diagnostics: applications in oral cancer. *Nano Letters*, **5**(5), 829–834.

Eustis S and El-Sayed MA 2006 Why gold nanoparticles are more precious than pretty gold: noble metal surface plasmon resonance and its enhancement of the radiative and nonradiative properties of nanocrystals of different shapes. *Chemical Society Reviews*, **35**(3), 209–217.

Ghosh SK and Pal T 2007 Interparticle coupling effect on the surface plasmon resonance of gold nanoparticles: from theory to applications. *Chemical Reviews*, **107**(11), 4797–4862.

Gordon R, Sinton D, Kavanagh KL and Brolo AG 2008 A new generation of sensors based on extraordinary optical transmission. *Accounts of Chemical Research*, **41**(8), 1049–1057.

Homola J 2008 Surface plasmon resonance sensors for detection of chemical and biological species. *Chemical Reviews*, **108**(2), 462–493.

Jain PK, Huang X, El-Sayed IH and El-Sayed MA 2008 Noble metals on the nanoscale: optical and photothermal properties and some applications in imaging, sensing, biology, and medicine. *Accounts of Chemical Research*, **41**(12), 1578–1586.

Jang S, Jo H, Cho S, Mechitov K, Rice JA, Sim SH, Jung HJ, Yun CB, Spencer Jr BF and Agha G 2010 Structural health monitoring of a cable-stayed bridge using smart sensor technology: deployment and evaluation. *Smart Structures and Systems*, **6**(5-6), 439–459.

Kruss S, Hilmer AJ, Zhang J, Reuel NF, Mu B and Strano MS 2013 Carbon nanotubes as optical biomedical sensors. *Advanced Drug Delivery Reviews*, **65**(15), 1933–1950.

López-Higuera JM, Rodriguez Cobo L, Incera AQ and Cobo LR 2011 Fiber optic sensors in structural health monitoring. *Lightwave Technology, Journal of*, **29**(4), 587–608.

Luk'yanchuk B, Zheludev NI, Maier SA, Halas NJ, Nordlander P, Giessen H and Chong CT 2010 The fano resonance in plasmonic nanostructures and metamaterials. *Nature Materials*, **9**(9), 707–715.

Majumder M, Gangopadhyay TK, Chakraborty AK, Dasgupta K and Bhattacharya DK 2008 Fibre Bragg gratings in structural health monitoring—present status and applications. *Sensors and Actuators A: Physical*, **147**(1), 150–164.

Mayer KM and Hafner JH 2011 Localized surface plasmon resonance sensors. *Chemical Reviews*, **111**(6), 3828–3857.

Ringe E, McMahon JM, Sohn K, Cobley C, Xia Y, Huang J, Schatz GC, Marks LD and Van Duyne RP 2010 Unraveling the effects of size, composition, and substrate on the localized surface plasmon resonance frequencies of gold and silver nanocubes: a systematic single-particle approach. *The Journal of Physical Chemistry C*, **114**(29), 12511–12516.

Scarano S, Mascini M, Turner AP and Minunni M 2010 Surface plasmon resonance imaging for affinity-based biosensors. *Biosensors and Bioelectronics*, **25**(5), 957–966.

Schurman MJ and Shakespeare WJ 2009 Using optical coherence tomography (OCT) for use in monitoring blood glucose and diagnosing diabetes. US Patent 7,510,849.

Verellen N, Van Dorpe P, Huang C, Lodewijks K, Vandenbosch GA, Lagae L and
    Moshchalkov VV 2011 Plasmon line shaping using nanocrosses for high sensitivity
    localized surface plasmon resonance sensing. *Nano Letters*, **11**(2), 391–397.
White IM, Oveys H and Fan X 2006 Liquid-core optical ring-resonator sensors. *Optics
    Letters*, **31**(9), 1319–1321.
Willets KA and Van Duyne RP 2007 Localized surface plasmon resonance spectroscopy
    and sensing. *Annual Review of Physical Chemistry*, **58**, 267–297.
Yao J, Maslov KI, Shi Y, Taber LA and Wang LV 2010 In vivo photoacoustic imaging of
    transverse blood flow by using Doppler broadening of bandwidth. *Optics Letters*, **35**(9),
    1419–1421.
Yu N, Genevet P, Kats MA, Aieta F, Tetienne JP, Capasso F and Gaburro Z 2011 Light
    propagation with phase discontinuities: generalized laws of reflection and refraction.
    Science, **334**(6054), 333–337.
Zijlstra P, Paulo PM and Orrit M 2012 Optical detection of single non-absorbing molecules
    using the surface plasmon resonance of a gold nanorod. *Nature Nanotechnology*, **7**(6),
    379–382.

7

Magnetic Sensing

Broadly speaking, magnetic sensors can be classified into three major categories. In the first category we find sensors that can sense magnetic fields weaker than the earth's magnetic field (less than 1 μG). They are called low-field sensors. In the second category we find sensors capable of sensing magnetic fields in the range of 1 μG to 10 G. This is the range in which the earth's magnetic fields can be contained. The sensors are called medium-field sensors. In the third category we find sensors that can detect fields above 10 G. They are called high-field sensors. A magnetic field is a vector quantity, but some sensors are capable of sensing only the magnitude, some only the direction, and others both the direction and the magnitude of a magnetic field.

Low-field sensors are often large and power-hungry, but they have indispensable places in many areas. In medicine, for example, a magnetoencephalogram is used to measure magnetic fields generated in the brain (Sander et al. 2012; Uchiyama et al. 2012; Yasui 2009). Low-field sensors are also fundamental building blocks of nuclear magnetic resonance and nuclear quadrupole resonance (Andrew 2009; Emsley et al. 2013; Fukazawa et al. 2009; Hore 2015; Lee et al. 2006). In the semiconductor industry, they are employed to detect structural flows and impurities (Inoue and Nishiyama 2007; Sze and Ng 2006; Tonouchi 2007). In fundamental physics, they are used to measure electromagnetic radiation originating from weakly interacting massive particles (Campbell and Wynne 2011; Fattori et al. 2008; Knoll 2010).

Medium- and high-field sensors are small and can easily be embedded into everyday objects. Mostly they are fabricated as parts of microelectromechanical systems and can be employed to detect the type, proximity, direction, speed, angular momentum, and size of moving objects including cars (for monitoring congestion in cities) (Cheung et al. 2005; Dutta et al. 2005; Yun et al. 2007).

In the computer industry, magnetic sensors are presently revolutionising data storage technology due to their capacity to store a large amount of data (terabytes and exabytes of data) in a small space, consuming a very small amount of power (Chappert et al. 2007; Parkin et al. 2008). In the following sections, the fundamental principles of magnetic sensors will be explained.

7.1 Superconducting Quantum Interference Devices

When certain materials, such as niobium, are cooled below a critical temperature (77 K), some interesting phenomena occur. Firstly, they exhibit zero electrical resistance, as a

Principles and Applications of Ubiquitous Sensing, First Edition. Waltenegus Dargie.
© 2017 John Wiley & Sons, Ltd. Published 2017 by John Wiley & Sons, Ltd.
Companion Website: www.wiley.com/go/dargie2017

Figure 7.1 The Meissen effect. When superconductors are cooled to a temperature below a certain critical point and placed in a magnetic field, they cancel nearly all interior magnetic fields.

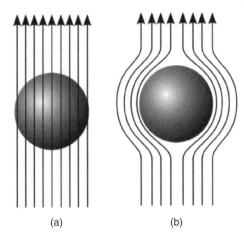

(a)   (b)

result of which Ohm's law does not apply to them. The materials are said to be super-conductors and the state superconductivity (Fagaly 2006; Koch et al. 1987; Koelle et al. 1999). Superconductivity, however, is observed as long as the applied current is below a critical value. If the current exceeds the critical value, it experiences resistance, similar to a current flowing in a conductor experiences resistance. Secondly, if the superconductors are placed in a magnetic field, they cancel nearly all interior magnetic fields (they become perfectly demagnetised) and an area of strong external magnetic field is observed around them, as shown in Figure 7.1. This phenomenon is called the Meissen effect. The advantage of this phenomenon is that if a superconductor forms a loop, the magnetic flux that is enclosed in the loop can be quantised in units of the flux quantum, which equals:

$$\Phi_0 = \frac{h}{(2e)} \approx 2.067833758 \times 10^{-15} \text{ Wb} \tag{7.1}$$

where h is Planck's constant and e is the electron charge. The implication of this phenomenon is profound; for instance, superconductors can be used to detect extremely weak magnetic fields such as the magnetic fields produced by the brain (10^{-13} T) (Huettel et al. 2004; Ogawa et al. 1990) or the heart (10^{-10} T) (Geselowitz 1970). Thirdly, the superconductor electrons in the conduction band behave like bosons instead of fermions (electrons), which means that they can no longer be described by the Pauli exclusion principle which states that no two identical fermions can occupy the same quantum state simultaneously. Precisely speaking, these electrons, instead of repelling each other as they normally do, establish a bond to become a pair—a Cooper pair—and start to vibrate in unison at a single fundamental frequency. If sufficient voltage is applied to these electron pairs, they flow together to generate a supercurrent.[1]

---

1 From Chapter 6 we know that an electron residing at a given energy level can give up a quantum of energy ($E = hf$) in the form of a single photon, allowing it to fall to a lower energy level. Similarly, it can absorb a quantum of energy to ascend to the next energy level up. When the free electrons of a conductor (valence electrons) each absorb a quantum of energy, then they transit into a conduction band to generate an electric current. Interestingly, the Cooper pair of electrons ascend or descend between energy levels in unison and the energy required to make a Cooper pair of electrons transit into a conduction band equals the flux quantum, Φ_0.

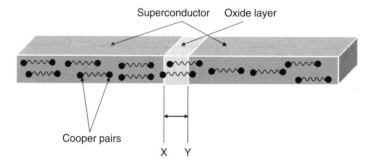

Figure 7.2 The Josephson junction measures a change in a magnetic field at a quantum level.

When two superconductors are loosely connected by a thin oxidized barrier as shown in Figure 7.2, they form a Josephson junction. If this junction is placed in a superconductivity state as described above and a biasing voltage is applied across the two ends, assuming that there is no thermal fluctuation in the system, it is possible to tunnel the Cooper pairs through the barrier without introducing a voltage drop across the barrier (effortlessly, so to speak). This can be done, however, only as long as the current flowing through the superconductors is below the critical value; otherwise, the Cooper-pair phenomenon is no longer observed and the flow of electrons takes place in the normal sense and a voltage drop across the barrier can be measured.

The Josephson junction has two interesting properties:

1) The electron pairs (Cooper pairs) can be described by a macroscopic wave function, Ψ, which has a well-defined phase ϕ.
2) The supercurrent that flows through the junction can be expressed in terms of the biasing current I, the critical current I_0, and the phase difference in the order parameters of the two superconductors, δ:

$$I = I_C \sin \delta \tag{7.2}$$

An order parameter is an aspect of a system undergoing a state transition. For the Josephson junction, due to the variation of the medium through which the supercurrent flows, the two superconductors will not have identical Cooper-pair densities. The order parameter is a measure of these differences (Emery and Kivelson 1995; Maki 1968; Mazin et al. 2008). Recall that in an oscillating current, the transfer of electric energy takes place in the form of a propagating electromagnetic wave rather than in the form of an actual drift of free electrons. When the electromagnetic wave propagates in a homogeneous or quasi-homogeneous medium, it often oscillates without experiencing an appreciable deformation, but when the medium is not homogeneous, there can be a phase shift between wavefronts, as a result of which two wavefronts may not propagate in a constructive way. The degree of destructiveness between local wavefronts depends on the phase difference. For the Josephson junction shown in Figure 7.2, the phase difference can be affected by both the current density within the superconductors and an external magnetic field (which we wish to sense). The phase difference due to the current density is expressed as:

$$[\delta]_i = \frac{4\pi m}{h n_s e} \int_x^y \mathbf{J} \cdot dl \tag{7.3}$$

where m is the mass and e is the charge of an electron, respectively; n_s is the local Cooper-pair density, $\mathbf{J}$ is the current density, and $\overline{xy} < 2$nm. Similarly, the phase difference due to an external magnetic field is:

$$[\delta]_B = \frac{4\pi e}{h} \int_x^y \mathbf{A} \cdot dl \tag{7.4}$$

where $\mathbf{A}$ is the magnetic potential. Moreover,

$$[\delta]_B + 2[\delta]_i = 2\pi n_s \tag{7.5}$$

In other words, the overall change in the phase difference due to an external magnetic field and a current flowing through the Josephson junction must be quantised.

A device built on the principle of superconductivity and the Josephson junction to detect very weak magnetic fields is the superconducting quantum interference device (SQUID). Existing technologies can be classified into DC-SQUID and RF-SQUID. In the next two subsections we shall discuss these devices in some detail.

7.1.1 DC-SQUID

The DC-SQUID consists of two Josephson junctions forming a superconducting loop, as shown in Figure 7.3. Assuming the two junctions are identical, an equal amount of DC current flows in each branch when a DC biasing voltage V is applied to the system. In the absence of an external magnetic field inside the loop, the two superconductors are coupled by an energy amounting to $I_C\Phi_0/2\pi$. If an external magnetic field is set up in the loop, as shown in Figure 7.4, it will induce its own current, I_B, the tendency of which is to disturb the balance of the net current flowing between the two branches. At the same time, however, the current that circulates inside the loop sets up its own opposing magnetic field surrounding the outer part of the loop. The tendency of this magnetic field is to oppose the external magnetic field according to Lenz's law. If the magnetic flux in the loop increases, the voltage across the loop oscillates with a period Φ_0, and by detecting a small change in this voltage, it is possible to sense a change in the magnetic flux as low as $10^{-6}\Phi_0$ (Kleiner et al. 2004).

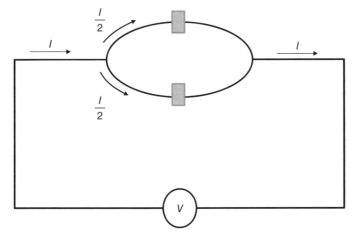

Figure 7.3 The basic set up of a DC superconducting quantum interference device (DC-SQUID).

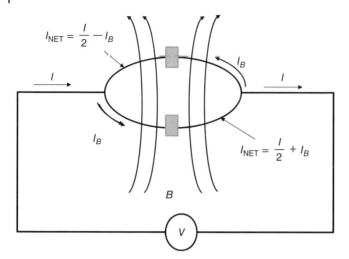

Figure 7.4 An external magnetic field induces a current in a DC-SQUID, which in turn creates a current imbalance between the two branches. The net imbalance is a function of the magnetic field strength.

For a low cooling temperature (77 K), the DC-SQUID can be approximated by two RC shunt circuits connected to a superconducting loop with an inductance of L, as shown in Figure 7.5. In order to observe a barrier voltage that is proportional to the supercurrent, the loop should be non-hysteretic. This precondition is fulfilled when:

$$\frac{2\pi I_C R^2 C}{\Phi_0} \equiv \beta_c \leq 1 \tag{7.6}$$

If $\beta_c \ll 1$, then the RC model reduces to a simple shunt-resistive circuit and the V−I characteristic in the absence of thermal noise can be approximated by:

$$V = R\sqrt{\left(I^2 - I_C^2\right)} \tag{7.7}$$

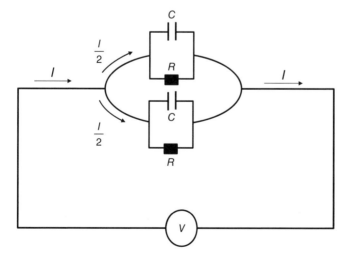

Figure 7.5 The electric equivalent circuit of a DC-SQUID.

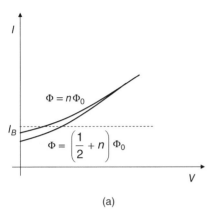

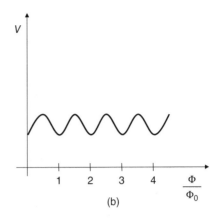

(a)
(b)

Figure 7.6 The V–I characteristic of a DC-SQUID: (a) the influence of the flux in the loop on the V–I characteristic; (b) V versus $\frac{\Phi}{\Phi_0}$ at a constant bias current, I_b. Figure adapted from (Kleiner et al. 2004).

For $I_B > 2I_C$, the voltage V oscillates with a period Φ_0 as the external magnetic flux Φ changes. To measure small changes in $\Phi << \Phi_0$, the bias current I_B should be chosen such that the flux-to-voltage transfer coefficient ($H_{V\Phi} = \partial V/\partial \Phi$) is maximum and the external magnetic flux is set at:

$$\Phi = (2n + 1)\frac{\Phi_0}{4}, \quad n = 0, 1, 2,... \tag{7.8}$$

Figure 7.6 summarises the V–I characteristics of the DC-SQUID.

7.1.2 RF-SQUID

The RF-SQUID consists of a superconducting loop with a single Josephson junction and an external LC-resonant circuit which is inductively coupled to the loop. The mutual inductance between the superconducting loop and the LC circuit is M (see Figure 7.7). The resonant circuit is driven by an RF current, $i_{RF}(t) = I_0 \sin(\omega_0 t)$, where $\omega_0 = 1/\sqrt{LC}$ is the resonant frequency. The resulting RF voltage across the resonant circuit is a periodic function of the applied flux in the SQUID loop with a period Φ_0.

Unlike the DC-SQUID, the operational region of the RF-SQUID depends on the total magnetic flux of the system (Φ_T), which consists of

- the flux enclosed by the superconducting loop, which we wish to detect (Φ)
- the flux produced by the LC-resonant circuit, which is coupled to the superconducting loop:

$$\Phi_T = \Phi - LI_0 \sin\left(2\pi\frac{\Phi_T}{\Phi_0}\right) \tag{7.9}$$

Depending on the value of $\beta'_L \equiv 2\pi LI_0/\Phi_0$, Eq. (7.9) exhibits two distinct characteristics. When $\beta'_L < 1$, the slope $d\Phi_T/d\Phi$ is everywhere positive and the plot Φ_T versus Φ is non-hysteretic. In other words, it is possible to establish a proportional relationship between the two quantities. When, however, $\beta'_L > 1$, there are regions in which $d\Phi_T/d\Phi$ can be positive, negative, or even divergent, so that the plot Φ_T versus Φ cannot be considered non-hysteretic. Figure 7.8 shows the Φ_T versus Φ plot for two different values

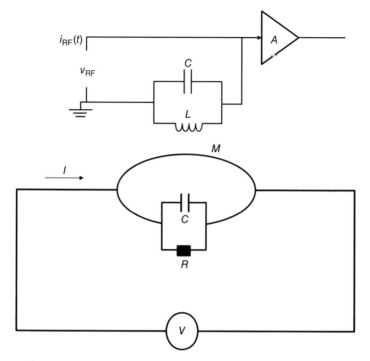

Figure 7.7 The electric equivalent circuit of a RF-SQUID.

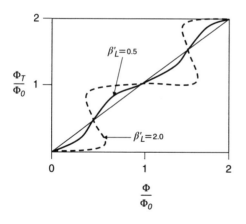

Figure 7.8 A normalised relationship between the total flux of a RF-SQUID and the applied flux for $\beta'_L = 0.5$ and 2. Figure adapted from Kleiner et al. (2004).

of β'_L. Figure 7.9a shows the $V-I$ characteristic of the RF-SQUID when the applied flux is zero and non-zero and Figure 7.9b shows the relationship between the applied flux and the RF voltage.

7.2 Anisotropic Magnetoresistive Sensing

The electron spin in an atom (with an associated angular momentum) is responsible for the creation of magnetic dipoles in ferromagnetic materials (see Figure 7.10). Below

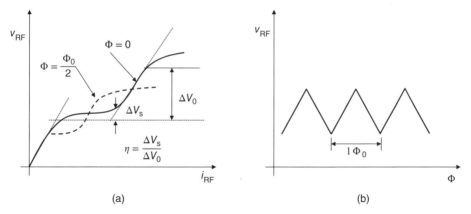

Figure 7.9 (a) The relationship between the peak RF voltage (V_{RF}) across the *LC*-resonant circuit and the peak RF current (I_{RF}) for $\Phi = 0$ (solid line) and $\Phi = \pm\frac{\Phi_0}{2}$. (b) The relationship between the applied flux and the RF voltage across the *LC*-resonant circuit.

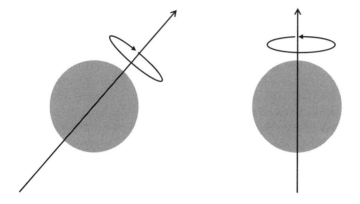

Figure 7.10 The direction and alignment of the spin of electrons in an atom determines the magnetic property of a material.

a certain critical temperature, which is called the Curie temperature, T_C, neighbouring magnetic spins are permanently aligned and therefore set up permanent magnetic dipoles. But when the temperature increases above T_C, the direction of neighbouring magnetic spins becomes random, with a net effect of individual dipoles cancelling the magnetic fields of each other. Under this condition, an external magnetic field is required to realign (or induce) the magnetic spins. The vector field that quantitatively expresses the density of permanent or induced magnetic dipole moments is called magnetisation.

Figure 7.11 shows how the spontaneous magnetisation of an alloy of nickel (99.42%) and cobalt (0.58%) changes as a function of temperature, from absolute zero to the Curie temperature (Freitas et al. 2007). As can be seen, the spontaneous magnetisation decreases approximately as a negative exponential with an increase in temperature. An interesting aspect of this phenomenon is that the electrical resistivity of the material exhibits a directional characteristic. In other words, the resistivity of the material parallel

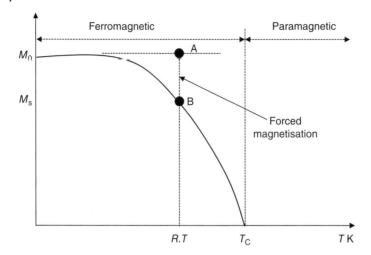

Figure 7.11 The influence of temperature on the saturation magnetisation, M_s. Forced magnetisation takes place between the points A and B. The additional magnetisation as a result of an applied magnetic field produces a corresponding reduction in the resistivity of the ferromagnetic material. Image redrawn from McGuire and Potter (1975).

to the magnetisation vector (or the magnetisation dipole) is different from the resistivity of the material perpendicular to the magnetisation vector. Secondly, the resistivity along the magnetisation vector is greater than the resistivity perpendicular to the magnetisation vector. Furthermore, for a fixed temperature, the spontaneous magnetisation of the ferromagnetic material can be increased to saturation by applying an external magnetic field. For example, in Figure 7.11, the line connecting A and B shows the magnitude of forced magnetisation that can be achieved at room temperature. Another interesting aspect of the forced magnetisation is that the resistivity of the material, the longitudinal as well as the transversal, changes as a function of the applied magnetic field. This change of resistivity is called anisotropic magnetoresistivity (Azevedo et al. 2011; Binasch et al. 1989; Freitas et al. 2007; Hayashi et al. 2006; Moser et al. 2007; Spaldin 2010); anisotropic, because the change in resistivity is directional. Figure 7.12 shows how the resistivity of the nickel–cobalt alloy changes for different temperature values.

The magnetoresistance property is not limited to ferromagnetic materials. Indeed all conductors exhibit this property, but in non-metallic materials the effect is rather weak (it is known as ordinary galvanomagnetic effect) or simply, ordinary magnetoresistance. In ferromagnetic materials, however, the effect is significant enough to establish a quantitative relationship between the change in the resistivity of the ferromagnetic material and the change in the magnitude and direction of the applied magnetic field. When a material exhibits a strong magnetoresistive effect, it is known as anisotropic magnetoresistance (AMR). The resistivity of a thoroughly demagnetised specimen in the absence of an external magnetic field (zero magnetic field), ρ_0, is approximated by the following equation:

$$\rho_0 = \rho_{avg} = \frac{1}{3}\rho_\| + \frac{2}{3}\rho_\perp \tag{7.10}$$

where $\rho_\|$ and $\rho_\perp$ are the longitudinal (parallel to the magnetisation vector) and transversal (perpendicular to the magnetisation vector) resistivity of the material, respectively.

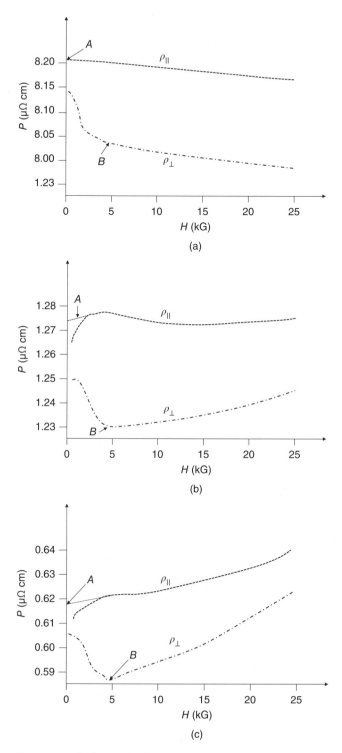

Figure 7.12 The longitudinal ($\rho_\parallel$) and perpendicular ($\rho_\perp$) resistivity of $Ni_{0.9942}Co_{0.0058}$ changes as a function of applied magnetic field, H: (a) at room temperature; (b) at 77 K; (c) at 4.2 K. Image redrawn from McGuire and Potter (1975).

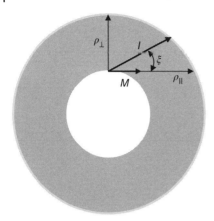

Figure 7.13 The relationship between the magnetisation, current, and resistance of anisotropic magnetoresistive material.

If we define the magnetoresistance as the change in resistivity between a fully magnetised and a fully demagnetised specimen, then the following relationships apply:

$$\frac{\Delta \rho_\perp}{\rho_{avg}} = \frac{\rho_\perp - \rho_{avg}}{\rho_{avg}} \tag{7.11}$$

And

$$\frac{\Delta \rho_\parallel}{\rho_{avg}} = \frac{\rho_\parallel - \rho_{avg}}{\rho_{avg}} \tag{7.12}$$

Suppose a current flows in a circular AMR at an angle ξ with respect to the magnetisation vector as shown in Figure 7.13. The resistivity of the AMR as a function of ξ can be approximated by:

$$\rho(\xi) = \rho_\perp + \Delta \rho \cos^2 \xi \tag{7.13}$$

where $\rho_\perp$ is the minimum resistivity of the ferromagnetic material that can be achieved in the presence of a magnetic field and $\Delta \rho = (\rho_\parallel - \rho_\perp)$ is the anisotropic magnetoresistivity.

From Eq. (7.13), it is clear that maximum sensitivity and linearity can be achieved when the current flows at 45 ° to the magnetisation vector (because of the sinusoidal function in the equation). One way to achieved this is by patterning diagonal stripes of highly conductive metal onto the more resistive AMR material, as shown in Figure 7.14a. The current flows perpendicular to the stripes while the magnetisation vector remains preferentially along the longitudinal direction of the magnetoresistive material. When an external magnetic field is applied to the system, it tends to rotate the magnetization, thereby changing the resistivity, as shown in Figure 7.14b.

Example 7.1 An anisotropic magnetoresistive (AMR) sensor made of a thin ferromagnetic film, as shown in Figure 7.15a, is placed at the centre of a roundabout. In the absence of a moving car, the direction of the current flowing in the film is aligned perpendicular to the earth's magnetic field, so that the resistance the current experiences is a minimum. In the presence of a moving car that disturbs the distribution of the local earth's magnetic field, the magnetisation vector of the sensor varies both in direction and magnitude. Assuming that the car can approach, enter, and leave the roundabout at an angle $\xi : 0 \le \xi \le 360°$, plot the change in the resistance of the AMR sensor as a function of ξ.

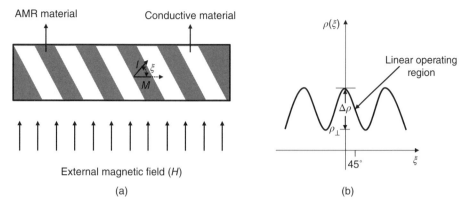

Figure 7.14 (a) Strips of a highly conductive material embedded onto an AMR material constrain the current to flow at 45° to the rest position of the magnetisation. (b) The plot of resistivity versus applied magnetic field.

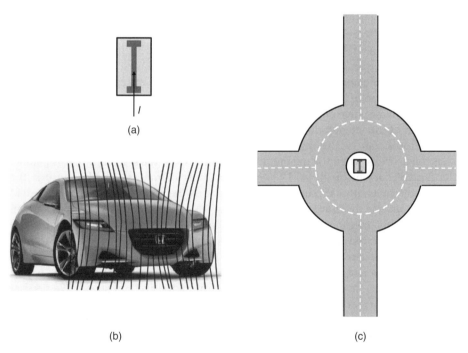

Figure 7.15 An anisotropic magnetoresistive sensor (AMR) employed to determine the speed, direction, and size of a moving car: (a) initially the biasing current is aligned perpendicular to the earth's magnetic field; (b) a moving car disturbs the local distribution of the earth's magnetic field; (c) as the car approaches, enter, and leaves the roundabout, the magnitude and direction of the earth's magnetic field reaching the AMR sensor changes thereby changing the magnetoresistance of the AMR sensor.

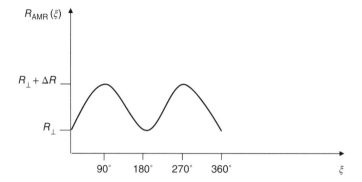

Figure 7.16 The change in the magnitude of the magnetoresistance of the AMR sensor as a function of the angle of arrival of the local earth's magnetic field, which is disturbed by the moving car. The angle of arrival is relative to the original alignment.

The earth's magnetic field emanates from the interior of the earth and extends into the so-called solar wind, a stream of charged particles coming from the sun. The magnitude of this field at the earth's surface ranges from 0.25 to 0.65 G (Budden 2009; Finlay et al. 2010; Merrill and McElhinny 1983). When regarded with respect to the earth's rotational axis, this field deviates approximately by 10 ° towards the east. The local distribution of the earth's magnetic field can be affected by moving metallic objects such as cars, and the degree of disturbance depends on the size, relative direction, and speed of the moving object. For our example, assuming that the car approaches the AMR sensor from the left, the moving magnetic field disturbed by the moving car will be perpendicular to the direction of the current flowing, so there will not be an appreciable change in the magnetoresistance of the sensor. As the car goes around the roundabout, so will the direction of the moving magnetic field change, thereby changing the magnitude of the magnetoresistance. The magnetoresistance reaches its maximum when the car reaches the first exit when the magnetic field coming from the side of the car is parallel to the flow of current in the AMR sensor. For most commercial AMR sensors, the change in the magnetoresistance is proportional to the square of the magnetic field strength H, but the maximum change is below 3% of the nominal value. Figure 7.16 shows how the magnetoresistance changes as the local magnetic field disturbed by the moving car reaches the AMR sensor at different angles.

If the car approaches the AMR sensor from the right, the same pattern will follow. Consequently, it is not possible to determine the direction of the moving car with a single AMR sensor, though it is possible to determine its size and speed. We need at least two AMR sensors to determine the direction (consider the plot in Figure 7.14b). The accuracy with which the speed of the car can be determined also increases when two AMR sensors are employed.

7.3 Giant Magnetoresistance

The magnetoresistance that can be achieved by using a simple AMR construction is typically a few percent (less than 5%). This can be increased to a significant amount (by up to 50–80%) using two or more layers of ferromagnetic materials (such as NiFe, CoFe and

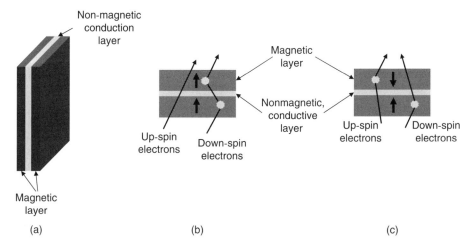

Figure 7.17 Giant magnetoresistance: (a) two magnetic layers separated by a thin non-magnetic conductor; (b) the change in the resistivity for parallel alignment; (c) the change in resistivity for anti-parallel alignment. The circles in (b) and (c) represent the opposition travelling electrons experience as a result of the magnetisation of the specimen. In (b), the down-spinning electrons face opposition in both specimen because in both of them the magnetisation vectors have the same anti-parallel directions. At the same time, the up-spinning electrons do not face a considerable opposition in both specimen. In (c), since the magnetisation is anti-parallel, electrons facing a considerable opposition in one of the specimen will not face a considerable opposition in the other. Therefore, the overall resistivity is averaged.

similar transition metal alloys) separated by a thin non-magnetic metal spacer (as shown in Figure 7.17). The magnetoresistance gained this way is called giant magnetoresistance (GMR) (Baibich et al. 1988; Binasch et al. 1989; Schmaus et al. 2011; Urushibara et al. 1995; Van den Berg et al. 2013). It was independently discovered in the late 1980s by Peter Gruenberg and Albert Fert. The intermediate spacer permits the top and bottom magnetic layers to have different polarisations (or different magnetisation directions). Because of the small dimensions of the spacer, polarised electrons leaving one of the magnetic layers can pass into the next magnetic layer without their polarisation being modified by scattering.

When the magnetisation directions of both metallic layers are aligned (parallel) in the same direction, electrons leaving one of the layers can pass relatively freely into the other layer. If, on the other hand, the magnetisation directions of the layers are opposite (or anti-parallel), then electrons originating in one layer are heavily scattered by the adjacent layer. The disruption of the free movement of the electrons results in an increase in the electrical resistance. When considered in terms of the electron spin, the electrical conductivity can be described by two largely independent conducting channels which correspond to the up-spin and down-spin electrons. Electrical conduction occurs in parallel in these channels. In ferromagnetic materials, the resistance encountered by moving electrons in these channels is different.

It is often assumed that the resistance is stronger for electrons with spin anti-parallel to the magnetisation direction and weaker for electrons with spin parallel to the magnetisation direction. Hence, when the magnetisation of the two magnetic layers in Figure 7.17b are parallel to each other, the up-spin electrons experience a small resistance whereas the down-spin electrons experience a large resistance. The overall resistivity for the parallel magnetisation is proportional to the parallel combination of the two resistivities of the two channels:

$$\rho_\parallel \propto \frac{2\rho_\uparrow \rho_\downarrow}{\rho_\uparrow + \rho_\downarrow} \tag{7.14}$$

Similarly, when the directions of the magnetisation in the two magnetic layers are anti-parallel to each other as shown in Figure 7.17c, then both the up-spin and down-spin electrons experience a small resistance in one layer and a large resistance in the other layer. The total resistivity of the anti-parallel magnetisation directions can be expressed as:

$$\rho_{\leftrightarrow} \propto \frac{1}{2}(\rho_\uparrow + \rho_\downarrow) \tag{7.15}$$

The overall change in the resistivity for the GMR is:

$$\Delta\rho = (\rho_\parallel - \rho_{\leftrightarrow}) \propto -\frac{1}{2}\frac{(\rho_\uparrow - \rho_\downarrow)^2}{(\rho_\uparrow + \rho_\downarrow)} \tag{7.16}$$

Figure 7.18 displays the relationship between the magnetoresistivity, the external magnetic field (H), and the magnetisation of a three-layer GMR specimen. As can be seen, a wide range of change in magnetoresistivity can be achieved by operating the specimen in parallel and anti-parallel magnetisation modes. For a sufficiently thin spacer (only a few atoms thick), there appears a strong "exchange coupling," which tends to favour anti-parallel alignment of the adjacent magnetic layers. Consequently, in the absence of an externally applied field, the magnetic layers alternate in magnetisation, resulting in a high resistance. When an external magnetic field is applied, it can overcome the interlayer coupling and force all of the layers to align with the field and reduce the resistance. Since a magnetic field in either direction will cause alignment of the magnetisation vectors, the resulting ρ vs. H plot is an even function, symmetric about zero.

A modified version of the multilayer GMR can be realised with two magnetic layers separated by a non-magnetic layer and a base anti-ferromagnetic "pinning" layer on to which the GMR structure is deposited (see Figure 7.19a). The anti-ferromagnetic layer does not have a net magnetisation of its own but tends to fix the polarisation of the adjacent ferromagnetic layer in a set direction. The polarisation of the other layer, however, is free to rotate in response to an applied field. The structure is termed as a "spin valve", due to the free rotation of the upper-layer magnetisation vector. When the spin value is properly biased, the rest position of the free layer is perpendicular to the pinned layer so that maximum sensitivity and signal swing is achieved. Consequently, a broad linear response can be achieved when the applied magnetic field is in the direction of the pinned layer. The change in resistance to the applied magnetic field can be approximated as follows:

$$\rho(\xi) \propto \rho(\xi = 0) + \frac{\Delta\rho}{\rho}\frac{(1 - \cos\,\xi)}{2} \tag{7.17}$$

where $\rho(\xi = 0)$ is the resistivity of the spin valve when the angle between the magnetisation of the pinned and the free layer is zero, $\Delta\rho/\rho$ is the maximum achievable magnetoresistivity (reaching between 15 and 20%) and ξ is the angle between the polarisation of the pinned layer and the free layer. Equation (7.17) is a phenomenological

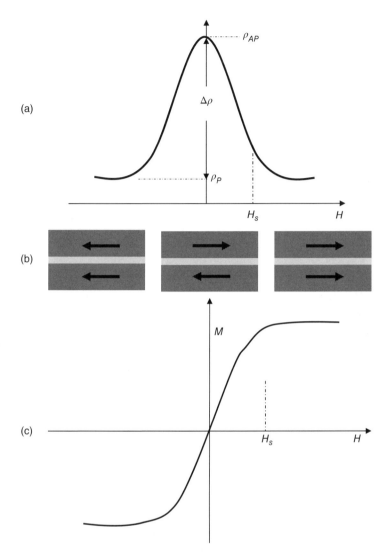

Figure 7.18 Relationship between magnetisation, applied magnetic field, and resistivity: (a) the change of resistance in the magnetic multilayer as a function of applied magnetic field; (b) parallel and anti-parallel magnetisation in the magnetic layers produces variable electron scattering thereby changing the magnetoresistance; (c) the relationship between the magnetisation and the applied magnetic field for the multilayer arrangement.

expression in that it yields a reproducible result that agrees with experimental observations but there is no theoretical foundation for it yet.

7.4 Tunnelling Magnetoresistance

Tunnelling magnetoresistance (TMR) structures are a recent development that produce a significant improvement in the magnetoresistance (Park et al. 2011;

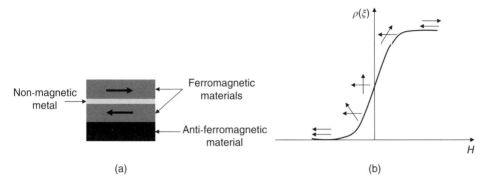

Figure 7.19 The spin valve GMR: (a) basic structure; (b) $\rho - H$ characteristics as a function of the angle between the magnetisation vectors of the two magnetic layers. Note that the magnetisation (polarisation) of the pinned layer is fixed; only the polarisation of the free layer varies in response to the external magnetic field.

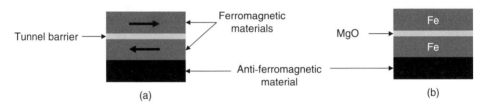

Figure 7.20 Tunnelling magnetoresistance: (a) two ferromagnetic materials are separated by a thin insulating material setting up a tunnelling magnetoresistance material; (b) a realisation of TMR using magnesium oxide as an insulating spacer.

Parkin et al. 2004). They are essentially similar to spin valves except instead of an intermediate conductive spacer, they use an insulating layer to separate the two magnetic layers (see Figure 7.20). Similar to the Josephson junction we discussed in Section 7.1, quantum electron tunnelling can occur from one of the ferromagnetic layers to the other through the insulating layer. The ease with which the tunnelling can take place depends on the alignment of the magnetic vectors in the two ferromagnetic materials. When the magnetisation vectors are aligned, there is a high probability of spin-polarised electrons tunnelling. On the other hand, when the magnetisation vectors are anti-parallel, the spin-polarised electrons are prevented from tunnelling, because of their opposing orientations. The magnetoresistance (ΔR) that can be achieved with this arrangement can exceed 200%.

Similar to the spin valve structure, TMR also uses a pinned magnetic layer and a free magnetic layer. In the absence of an applied field, the direction of magnetisation of the free layer is perpendicular to that of the pinned layer. An external magnetic field can reorient the free magnetisation vector in different directions. Often, it is sufficient to operate the TMR in two modes only: when the magnetisation vectors of the two vectors are parallel, the magnetoresistance is minimum; when they are anti-parallel, however, the magnetoresistance is maximum. Figures 7.21 and 7.22 show experimental results of the magnetoresistance that can be achieved with an Fe/MgO/Fe structure for different temperatures and thicknesses of the intermediate layer.

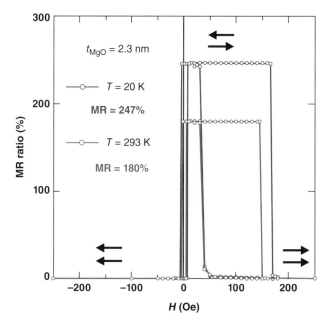

Figure 7.21 The tunnel magnetoresistance that can be achieved for an Fe/MgO/Fe structure for T = 293 K and 20 K measured at a bias voltage of 10 mV (the thickness of MgO (t_{MgO}) = 2.3 nm). The resistance- area product (RA) plotted here is the tunnel resistance for a 1μm × 1μm area. Arrows indicate the magnetisation configurations of the top and bottom Fe layers (electrodes). The magnetoresistance ratio is 180% at 293 K and 247% at 20 K. Reproduced with permission from Yuasa et al. (2004) (*Nature Material*, 2004).

MGR and TMR technologies are extensively used to store digital data in volatile and non-volatile memory (Chappert et al. 2007; Fert 2008). In these technologies, a single GMR or TMR structure is used to store a single bit. The external magnetic field (produced by the write head of a hard drive, for example) is used to align the magnetisation vectors of the ferromagnetic materials so that the overall magnetoresistance is a minimum (thus, the voltage across the GMR or TMR structure is zero, signifying a "0" bit); when, however, there is no applied magnetic field, the magnetoresistance is high, signifying a "1" bit. Figure 7.23 shows how a magnetoresistive random-access memory can be realised with a set of TMR structures.

Example 7.2 Explain how a spin valve sensor is used in contemporary disk drive heads to read data.

Figure 7.24 illustrates how a spin valve is used to read a bit of information from a disk drive. Each location beneath the spin valve stores a bit of information: either a "0" or a "1". The free layer of the spin value is initially magnetised perpendicular to the pinned layer, so that the current that is flowing through the circuit faces a minimum resistance. As the disk rotates, the bits on the hard disk pass beneath the read head and, depending on the information stored in these locations, the magnetisation of the free layer of the spin valve changes and as a result of this, the magnetoresistance of the spin valve changes. For example, if the bit on the hard disk stores a "1", then the polarisation of this bit is such

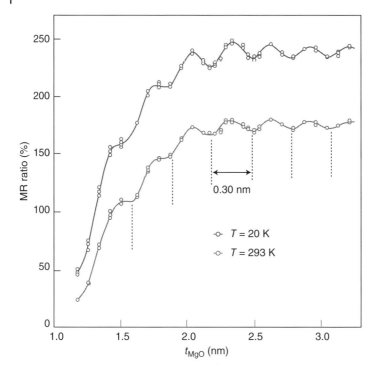

Figure 7.22 The magnetoresistance (MR) ratio of the Fe/MgO/Fe structure at T = 293 K and 20 K for variable t_{MgO} (measured at a bias voltage of 10 mV). Reproduced with permission from Yuasa et al. (2004) (*Nature Material*, 2004).

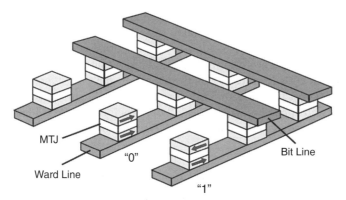

Figure 7.23 The conceptual architecture of a random access memory realised with TMR technology.

that when it passes beneath the read head, it aligns the magnetisation of the free layer of the spin value anti-parallel to the magnetisation of the pinned layer, in which case a maximum swing in the magnetoresistance is measured. When, on the other hand, a "0" bit passes, it aligns the magnetisation of the free layer parallel to the magnetisation of the pinned layer and, as a result, a minimum magnetoresistance is measured. The orientation of the free layer returns to its original position when the external magnetic field is withdrawn.

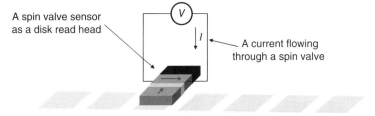

A spin valve sensor as a disk read head

A current flowing through a spin valve

Magnetic bits on a hard disk passing under the read head

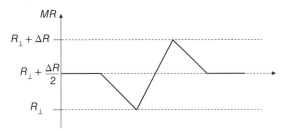

Figure 7.24 A spin valve is used to read a bit of information from a disk drive.

7.5 Hall-effect Sensing

When a charge carrier such as an electron moves at a velocity v inside a magnetic field, B, and perpendicular to it, it experiences a force which is perpendicular to both the magnetic field and its velocity, deflecting the electron from its path. If the electron continues travelling at the same velocity, the net effect of the induced force is to move it in a circular path. The force induced by the magnetic field is known as the Lorentz force and its magnitude and direction are determined by the following equation:

$$F_{L} = q(v \times B) \tag{7.18}$$

where q is the elementary charge and the "$\times$" sign inside the bracket indicates that the product is a cross product.

Suppose, we have a semiconductor slab as shown in Figure 7.25, where the charge carriers are electrons and holes. When a voltage is applied to the slab, the electrons and holes begin to flow in opposite direction along the slab. If an external magnetic field is applied to the slab perpendicular to the flow of the charge carriers, it tends to deflect them due to the Lorentz force. If the width of the slab is relatively long compared to the length of the slab, the charge carriers may experience a longer path but still manage to cross the slab. If, however, the width is relatively smaller, the effect of the magnetic field is such that their flow length is increased and a significant number of them will be accumulated on the two sides of the slab, as can be seen in the figure. This accumulation of charge carriers on the opposite sides of the slab tends to oppose the Lorentz force that gave rise to it. Once enough charge carriers are accumulated, the two forces become equal and no further charge carriers will be deflected. But due to the accumulated charge carriers on the opposite sides of the slab a voltage is induced, which is called the Hall voltage. It is directly proportional to the biasing current and the applied magnetic field

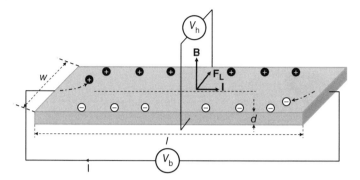

Figure 7.25 A magnetic field applied to moving charge carriers in a semiconductor slab induces a Lorentz force which is perpendicular to both the magnetic field and the direction of propagation. The Lorentz force deflects the charge carriers from the normal propagation path, giving rise to a Hall voltage, V_h.

and inversely proportional to the thickness of the slab and the charge density:

$$V_h = \frac{IB}{qnd} = R_H \frac{IB}{d} \tag{7.19}$$

where I is the current that flows in the slab due to the biasing voltage, V_b; n is the charge density; d is the thickness of the slab, and R_H is called the Hall coefficient. The extra length the charge carriers are forced to take can be viewed as an increment in resistance, and since this variation is due to the external magnetic field, the change in resistance is a magnetoresistance. It is proportional to the applied magnetic field and inversely proportional to the change density. Figure 7.26 shows the relationship between the Hall voltage and the applied magnetic field for different biasing voltages.

Hall-effect sensing can be realised with different semiconductor materials. The cheapest of them is an n-type silicon, but it is sensitive to temperature variations. GaAs

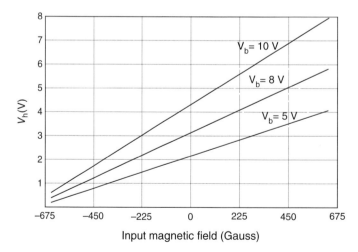

Figure 7.26 The Hall voltage (V_h) versus the input magnetic field for different biasing voltages for a typical Hall-effect sensor.

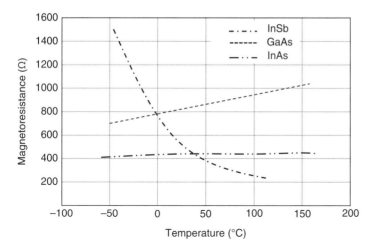

Figure 7.27 The magnetoresistance of different semiconductor materials as a function of temperature.

has a higher temperature capability due to its larger band gap. Due to relative high carrier mobility, InSb and InAs provide better sensitivity to field variations and robustness against temperature variations (particularly InAs). Figure 7.27 displays the effect of temperature on the magnetoresistance of different semiconductor materials for a fixed input magnetic field.

References

Andrew ER 2009 *Nuclear Magnetic Resonance* Cambridge University Press.

Azevedo A, Vilela-Leão L, Rodríguez-Suárez R, Santos AL and Rezende S 2011 Spin pumping and anisotropic magnetoresistance voltages in magnetic bilayers: Theory and experiment. *Physical Review B*, **83**(14), 144402.

Baibich MN, Broto JM, Fert A, Van Dau FN, Petroff F, Etienne P, Creuzet G, Friederich A and Chazelas J 1988 Giant magnetoresistance of (001) Fe/(001) Cr magnetic superlattices. *Physical Review Letters*, **61**(21), 2472.

Binasch G, Grünberg P, Saurenbach F and Zinn W 1989 Enhanced magnetoresistance in layered magnetic structures with antiferromagnetic interlayer exchange. *Physical Review B*, **39**(7), 4828.

Budden KG 2009 *Radio Waves in the Ionosphere*. Cambridge University Press.

Campbell JB and Wynne RH 2011 *Introduction to Remote Sensing*. Guilford Press.

Chappert C, Fert A and Van Dau FN 2007 The emergence of spin electronics in data storage. *Nature Materials*, **6**(11), 813–823.

Cheung S, Coleri S, Dundar B, Ganesh S, Tan CW and Varaiya P 2005 Traffic measurement and vehicle classification with single magnetic sensor. *Transportation Research Record: Journal of the Transportation Research Board*, **1917**, 173–181.

Dutta P, Grimmer M, Arora A, Bibyk S and Culler D 2005 Design of a wireless sensor network platform for detecting rare, random, and ephemeral events. *Proceedings of the 4th International Symposium on Information Processing in Sensor Networks*, p. 70.

Emery V and Kivelson S 1995 Importance of phase fluctuations in superconductors with small superfluid density. *Nature*, **374**(6521), 434–437.

Emsley JW, Feeney J and Sutcliffe LH 2013 *High Resolution Nuclear Magnetic Resonance Spectroscopy*, vol. 2. Elsevier.

Fagaly R 2006 Superconducting quantum interference device instruments and applications. *Review of Scientific Instruments*, **77**(10), 101101.

Fattori M, D'Errico C, Roati G, Zaccanti M, Jona-Lasinio M, Modugno M, Inguscio M and Modugno G 2008 Atom interferometry with a weakly interacting Bose–Einstein condensate. *Physical Review Letters*, **100**(8), 080405.

Fert A 2008 Nobel lecture: Origin, development, and future of spintronics. *Reviews of Modern Physics*, **80**(4), 1517.

Finlay C, Maus S, Beggan C, Bondar T, Chambodut A, Chernova T, Chulliat A, Golovkov V, Hamilton B, Hamoudi M *et al.* 2010 International geomagnetic reference field: the eleventh generation. *Geophysical Journal International*, **183**(3), 1216–1230.

Freitas P, Ferreira R, Cardoso S and Cardoso F 2007 Magnetoresistive sensors. *Journal of Physics: Condensed Matter*, **19**(16), 165221.

Fukazawa H, Yamada Y, Kondo K, Saito T, Kohori Y, Kuga K, Matsumoto Y, Nakatsuji S, Kito H, M. Shirage P *et al.* 2009 Possible multiple gap superconductivity with line nodes in heavily hole-doped superconductor KFe_2As_2 studied by 75As nuclear quadrupole resonance and specific heat. *Journal of the Physical Society of Japan*, **78**(8), 083712.

Geselowitz DB 1970 On the magnetic field generated outside an inhomogeneous volume conductor by internal current sources. *Magnetics, IEEE Transactions on*, **6**(2), 346–347.

Hayashi M, Thomas L, Bazaliy YB, Rettner C, Moriya R, Jiang X and Parkin S 2006 Influence of current on field-driven domain wall motion in permalloy nanowires from time resolved measurements of anisotropic magnetoresistance. *Physical Review Letters*, **96**(19), 197207.

Hore P 2015 *Nuclear Magnetic Resonance*. Oxford University Press.

Huettel SA, Song AW and McCarthy G 2004 *Functional Magnetic Resonance Imaging*, vol. 1. Sinauer Associates Sunderland.

Inoue A and Nishiyama N 2007 New bulk metallic glasses for applications as magnetic-sensing, chemical, and structural materials. *MRS Bulletin*, **32**(08), 651–658.

Kleiner R, Koelle D, Ludwig F and Clarke J 2004 Superconducting quantum interference devices: State of the art and applications. *Proceedings of the IEEE*, **92**(10), 1534–1548.

Knoll GF 2010 *Radiation Detection and Measurement*. John Wiley & Sons.

Koch R, Umbach C, Clark G, Chaudhari P and Laibowitz R 1987 Quantum interference devices made from superconducting oxide thin films. *Applied Physics Letters*, **51**(3), 200–202.

Koelle D, Kleiner R, Ludwig F, Dantsker E and Clarke J 1999 High-transition-temperature superconducting quantum interference devices. *Reviews of Modern Physics*, **71**(3), 631.

Lee S, Sauer K, Seltzer S, Alem O and Romalis M 2006 Subfemtotesla radio-frequency atomic magnetometer for detection of nuclear quadrupole resonance. *Applied Physics Letters*, **89**(21), 214106–216100.

Maki K 1968 The critical fluctuation of the order parameter in type-II superconductors. *Progress of Theoretical Physics*, **39**(4), 897–906.

Mazin I, Singh DJ, Johannes M and Du MH 2008 Unconventional superconductivity with a sign reversal in the order parameter of $LaFeAsO_{1-x}F_x$. *Physical Review Letters*, **101**(5), 057003.

McGuire T and Potter R 1975 Anisotropic magnetoresistance in ferromagnetic 3D alloys. *Magnetics, IEEE Transactions on*, **11**(4), 1018–1038.

Merrill RT and McElhinny MW 1983 *The Earth's magnetic field: Its history, origin and planetary perspective*, vol. 401. Cambridge Univ Press.

Moser J, Matos-Abiague A, Schuh D, Wegscheider W, Fabian J and Weiss D 2007 Tunneling anisotropic magnetoresistance and spin-orbit coupling in Fe/GaAs/Au tunnel junctions. *Physical Review Letters*, **99**(5), 056601.

Ogawa S, Lee TM, Kay AR and Tank DW 1990 Brain magnetic resonance imaging with contrast dependent on blood oxygenation. *Proceedings of the National Academy of Sciences*, **87**(24), 9868–9872.

Park B, Wunderlich J, Martí X, Holÿ V, Kurosaki Y, Yamada M, Yamamoto H, Nishide A, Hayakawa J, Takahashi H *et al.* 2011 A spin-valve-like magnetoresistance of an antiferromagnet-based tunnel junction. *Nature Materials*, **10**(5), 347–351.

Parkin SS, Hayashi M and Thomas L 2008 Magnetic domain-wall racetrack memory. *Science*, **320**(5873), 190–194.

Parkin SS, Kaiser C, Panchula A, Rice PM, Hughes B, Samant M and Yang SH 2004 Giant tunnelling magnetoresistance at room temperature with MgO (100) tunnel barriers. *Nature Materials*, **3**(12), 862–867.

Sander T, Preusser J, Mhaskar R, Kitching J, Trahms L and Knappe S 2012 Magnetoencephalography with a chip-scale atomic magnetometer. *Biomedical Optics Express*, **3**(5), 981–990.

Schmaus S, Bagrets A, Nahas Y, Yamada TK, Bork A, Bowen M, Beaurepaire E, Evers F and Wulfhekel W 2011 Giant magnetoresistance through a single molecule. *Nature Nanotechnology*, **6**(3), 185–189.

Spaldin NA 2010 *Magnetic Materials: Fundamentals and Applications*. Cambridge University Press.

Sze SM and Ng KK 2006 *Physics of Semiconductor Devices*. John Wiley & Sons.

Tonouchi M 2007 Cutting-edge terahertz technology. *Nature Photonics* **1**(2), 97–105.

Uchiyama T, Mohri K, Honkura Y and Panina L 2012 Recent advances of pico-tesla resolution magneto-impedance sensor based on amorphous wire CMOS IC MI sensor. *Magnetics, IEEE Transactions on*, **48**(11), 3833–3839.

Urushibara A, Moritomo Y, Arima T, Asamitsu A, Kido G and Tokura Y 1995 Insulator-metal transition and giant magnetoresistance in $La_{1-x}Sr_xMnO_3$. *Physical Review B*, **51**(20), 14103.

Van den Berg H, Coehoorn R, Gijs M, Grünberg P, Rasing T, Röll K and Hartmann U 2013 *Magnetic Multilayers and Giant Magnetoresistance: Fundamentals and Industrial Applications*, vol. 37. Springer Science & Business Media.

Yasui Y 2009 A brainwave signal measurement and data processing technique for daily life applications. *Journal of Physiological Anthropology*, **28**(3), 145–150.

Yuasa S, Nagahama T, Fukushima A, Suzuki Y and Ando K 2004 Giant room-temperature magnetoresistance in single-crystal Fe/MgO/Fe magnetic tunnel junctions. *Nature Materials*, **3**(12), 868–871.

Yun X, Bachmann ER, Moore IV H and Calusdian J 2007 Self-contained position tracking of human movement using small inertial/magnetic sensor modules *Robotics and Automation, 2007 IEEE International Conference on*, pp. 2526–2533.

8

Medical Sensing

Prince Hamlet's disdainful yet candid admission that man is endowed with infinite faculties is not merely a poetic hyperbole. When regarded as a measurand, the human body is a remarkable signal generator. The brain, the heart, the eyes, the lungs, the digestive system, the muscles, and different types of tissues in the body generate electrical and electromagnetic signals having various magnitudes and frequencies. These signals may give unique insights into the physical, emotional, cognitive, and psychological conditions of a person. Providing an exhaustive account of these signals is not the purpose of this chapter. Instead, it introduces the most important signals (action potentials) and how they are generated, measured, and processed. Readers interested in gaining more comprehensive insights into biomedical signals and sensors are referred to the following excellent books: (Hall 2015; Kandel et al. 2000; Levick 2013; Webster 2009).

The most important signals that can be measured from the human body are action potentials, which are generated by excitable cells. The media through which the action potentials propagate consist of extracellular fluid, blood, connective tissue, muscle, tendon, and bones containing different ions. Generally, the body can be regarded as an electrolyte solution and interfacing the sensing system with the human body using appropriate electrodes (Figure 8.1) enables the electrodes to chemically interact with the cations and anions in the electrolyte through the process of oxidation and reduction:

1) Some of the atoms of the electrode give up their electrons, become cations, and combine with the electrolyte:

$$C \rightleftharpoons C^{n+} + ne^-$$
(8.1)

In this case, the atoms from which the electrode is made should be the same as the atoms of the cations in the electrolyte.

2) Some of the anions in the electrolyte give up their excess electrons to the electrode and become electrically neutral atoms:

$$A^{m-} \rightleftharpoons A + me^-$$
(8.2)

The number of electrons (n and m) that can be transferred to the sensing system as a result of the oxidation and reduction processes depends on the concentration of ions within the electrolyte solution, which in turn depends on the action potentials generated

Principles and Applications of Ubiquitous Sensing, First Edition. Waltenegus Dargie.
© 2017 John Wiley & Sons, Ltd. Published 2017 by John Wiley & Sons, Ltd.
Companion Website: www.wiley.com/go/dargie2017

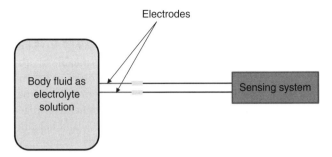

Figure 8.1 Action potentials generated by excitable cells in the body are responsible for the flow of cations and anions in body fluid, which can be regarded as an electrolyte solution. Interfacing the body with a sensing system through biopotential electrodes enables measurement of biopotentials (action potentials). The exchange of electrons at the interface takes place through the processes of oxidation and reduction.

by a particular set of excitable cells. The most widely studied and employed biopotential electrodes are wet adhesive Ag/AgCl electrodes (Chi et al. 2010; Cogan 2008; Hai et al. 2010; Webster 2009). In the subsequent sections of this chapter the two most significant types of action potentials, namely cardiac and brain action potentials, will be introduced.

Example 8.1 A pair of silver biopotential electrodes are used by a wireless ECG, which is attached to the chest of a patient. When a current flows through the anode of the electrode, it oxidises some of the silver atoms, as a result of which the atoms give up electrons, become silver ions, and join the body (the electrolyte). If the number of silver ions oxidised at the interface between the human body and the ECG electrodes is 6.25×10^{13}/s, how much ionic current flows through the electrode?

By definition, an electric current is the flow rate of electric charge – $dQ(t)/dt$ – in coulombs per second. During an oxidisation process, a single silver atom gives up a single electron:

$$Ag \longrightarrow Ag^+ + e^- \tag{8.3}$$

which will have a charge of 1.6×10^{-19} C. Therefore, the amount of charge flowing per second through the anode will be:

$$n \times 1.6 \times 10^{-19} \ C/s = 6.25 \times 10^{13} \times 1.6 \times 10^{-19} \ C/s = 16 \ \mu A$$

8.1 Excitable Cells and Biopotentials

Three types of cells in humans (as well as in many animals) are electrically excitable. These are nervous, muscular, and glandular tissue cells (De Biase et al. 2010; Hibino et al. 2010; Jack et al. 1975). When electrical impulses are given to these cells, they react by producing their own electrical pulses, in the shape and frequency of which vital physiological information is encoded. Some of these pulses induce emotional excitation, some of them inhibit emotional excitation, and some of them induce muscular contractions (Bean 2007).

The part of a cell that is primarily responsible for the generation and propagation of electrical pulses is the cell membrane. The cell membrane is made up of a lipoprotein complex – containing both proteins and lipids – and has a thickness of 7 – 15 nm (Webster 2009). It is selectively permeable to some ions, while it is impermeable to others. Moreover, its permeability (or conductance) depends on the concentration of ions within and outside the cell as well as on the magnitude and duration of the external excitation. The ions which are responsible for the existence of electrochemical reactions in excitable cells are sodium (Na^+), potassium (K^+), chlorine (Cl^-), and calcium (Ca^{2+}) ions (Bernstein and Morley 2006; Guevara 2003; Llinás 2012). There is a perpetual imbalance in the concentration of these ions within and outside of a cell, establishing a default potential difference. This imbalance is tirelessly maintained by an ion-pumping mechanism within a cell membrane even when it is externally disturbed (Webster 2009).

8.1.1 Resting Potential

When separately considered, both the inner and the outer regions of an excitable cell are electrically neutral. These is because the positive and negative ions in each region are naturally neutralised by the inherently negatively charged protein ions (anions) and by each other. However, since the cell membrane is impermeable to intracellular protein but selectively permeable to some of the ions, the electrical neutrality is broken when ions leave the cell or inter into the cell. To understand this better, consider Figure 8.2. Suppose the $^+$ balls represent the concentration of potassium ions in a frog's skeletal muscle. The intracellular concentration is 155 mmol l^{-1} and the extracellular concentration is 4 mmol l^{-1} per litre. In this arrangement, the collective positive charge built by the potassium ions within the cell is neutralised by the anions built by the negatively charged protein within the intracellular fluid (cytosol). The same is the case for the extracellular arrangement. However, if some of the potassium ions diffuse from within the cell into the extracellular medium (the bathing medium), then the internal medium will have more negatively charged ions because it has now given up some of its positive ions. Similarly, the extracellular medium has more positively charged ions, which brings it to

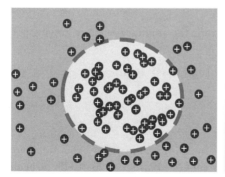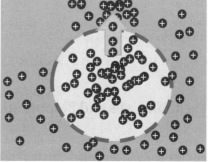

Figure 8.2 A difference in the concentration of ions between the intracellular and extracellular media produces an electrical potential difference. The chemical dynamic which is responsible for the creation of a membrane potential is the diffusion of ions – particularly, potassium ions (K^+) – from one medium to another. Since the cell membrane is impermeable to intracellular protein, the diffusion of ions creates a charge imbalance between the regions.

a higher potential than the internal medium. From this new state of things now emerges a potential difference between the internal and the external environment of the muscle cell. This potential difference is called the membrane potential.

In the example above, we mentioned that ions diffuse from the intracellular fluid to the bathing medium to establish the membrane potential. This does not happen naturally and a closer examination is required to understand the exact nature of the electrochemical dynamics. To begin with, in a single cell, there are different ion channels with different characteristics. Researchers have identified at least three types of voltage-dependent channel. These are sodium, potassium, and calcium channels (Alexander et al. 2011; Catterall 1985; Hille et al. 2001; Traynelis et al. 2010). There are at least 9 different families of sodium channels, 40 families of potassium channels, and 10 families of calcium channels (Goldin et al. 2000; Hille et al. 2001). Secondly, the concentration of sodium ions within a cell is significantly lower than the concentration of sodium ions outside of a cell; the concentration of potassium ions within a cell is significantly higher than the concentration of potassium ions outside of a cell. Likewise, in some nerve cells, the concentration of calcium ions in the external medium is higher than the concentration of calcium ions in the internal medium (Bean 2007; Bernstein and Morley 2006). Consequently, there is a diffusion of ions in and outside of the cell, even in a resting state. However, since the membrane is only slightly permeable to sodium ions but freely permeable to potassium ions, the mobility rates are different. The permeability coefficient, which is a measure of the flow of ions through a cell membrane in a unit time and under a uniform pressure, depends on the concentration gradient, the membrane's intrinsic permeability, and the ion's mass diffusivity. Generally, there is a free outward diffusion of K^+ and a feeble diffusion of Na^+ at resting condition.

Thirdly, due to the free outward diffusion of K^+ ions, the extracellular medium becomes more positive than the intracellular medium, creating an electric field directed inward. The tendency of this field is to repel any further diffusion of K^+ ions into the external medium. While the strength of the electric field is weaker than the force created by the diffusion gradient, the diffusion of K^+ will continue. However, the accumulation of K^+ in the external medium further strengthens the electric field and will eventually bring the diffusion force to equilibrium, at which point the diffusion stops. The result is a net potential difference between the inner and the external media, the inner medium being at a negative potential with respect to the outer medium. This transmembrane potential is called the resting potential; considered from the inner part of a cell, its value ranges from -50 to -80 mV, depending on the nature and location of the cell in the body.

As long as the cell is not externally excited to disturb the concentration imbalance, the resting potential remains unchanged. It is worth distinguishing two things:

- Even after the diffusion of some potassium ions (K^+), the internal concentration of potassium ions is still greater than the external concentration of potassium ions.
- However, the external medium is electrically at a higher potential than the internal medium. This is because the net charge difference inside the cell results in an excess of negative ions (because of the protein anions). At the same time, the addition of K^+ ions into the external medium creates an excess of positively charged ions. Figure 8.3 summarises the establishment of the membrane potential.

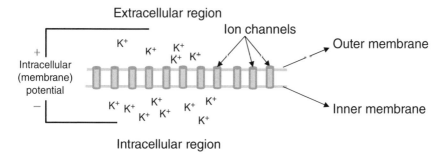

Figure 8.3 Establishment of the membrane potential.

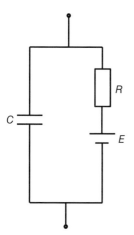

Figure 8.4 The cell membrane of an excitable cell modelled as an electric circuit consisting of a leaky capacitor, a resistor, and a DC electrical potential.

The cell membrane with its surrounding ions can be modelled as an electric circuit consisting of a leaky capacitor, a resistor, and an electric potential, as shown in Figure 8.4. The capacitor represents the inner and outer membranes, which are oppositely charged (a monolayer of cations being distributed on the outer surface of the membrane and a monolayer of anions being distributed on the inner surface of the membrane). The membrane is, however, leaky, on account of the permeable channels. The channels themselves can be modelled as resistors because of their selective and voltage-dependent nature; the function of the transmembrane potential is self-evident.

The transmembrane potential, after an equilibrium between the diffusion force and the inwardly electric field is achieved, can be approximated by the Nernst equation. For the potassium channel, this is given by (Webster 2009):

$$E_K = \frac{RT}{zF} \ln \left(\frac{[K^+]_o}{[K^+]_i} \right) \tag{8.4}$$

where $[K^+]_o$ and $[K^+]_i$ are the external and internal concentrations of potassium ions, respectively, R is the Rydberg gas constant, T is the temperature in Kelvin, z is the charge of a single ion (+1) for K^+, and F is Faraday's constant. For a human cell at body temperature (37°C), the above equation reduces to:

$$E_K = 0.0615 \log_{10} \left(\frac{[K^+]_o}{[K^+]_i} \right) V \tag{8.5}$$

More generally, the equilibrium potential is a result of the concentrations of different types of ions, most significantly, K^+, Na^+, and Cl^-. However, these ions have different permeability coefficients, as we have already discussed above. Given the outward and inward permeability coefficients for these ions, the equilibrium potential can be approximated by the following equation (Webster 2009) (see also (Hodgkin and Katz 1949)):

$$E = \frac{RT}{F} \ln \left(\frac{P_K[K]_o + P_{Na}[Na]_o + P_{Cl}K[Cl]_i}{P_K[K]_i + P_{Na}[Na]_i + P_{Cl}K[Cl]_o} \right) \tag{8.6}$$

where $[]_o$ denotes outward mobility and $[]_i$ denotes inward mobility. It must be noted that under the equilibrium condition, the net flow of current across the membrane is zero and the transmembrane electric field is assumed to be constant.

Example 8.2 The equivalent circuit shown in Figure 8.4 to model the resting membrane potential is simple but too abstract. A more realistic equivalent circuit would be the one shown in Figure 8.5. Explain the significance of the circuit elements.

As far as the resting potential is concerned, it is established as a result of the interplay of:

- the potential gradients established across the membrane due to the imbalance in the cation (Na^+, K^+) and anion (Cl^-) concentrations in and outside of the cell;
- the electric field established as a result of the inward and outward diffusion of ions; the tendency of the electric field is opposing the diffusion of ions;
- the existence of ion-selective permeability (channel);
- the existence of a tireless $Na^+ - K^+$ pumping mechanism within the cell, which is responsible for maintaining the ionic concentration imbalance.

These four aspects can be modelled by a capacitor, conducting channels (the inverse of resistance) with time-varying conductance (the inverse of resistivity), and three DC voltage sources signifying the potential gradients induced by the three dominant ionic concentration imbalances.

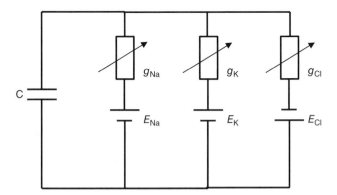

Figure 8.5 A more realistic electric equivalent circuit modelling the resting potential of an excitable cell. The capacitor represents the plasma membrane. The variable conductances represent the time-dependent ionic channels, and the DC voltage sources represent the potential gradients induced by the ionic concentration imbalance between the intracellular and extracellular media. Note that the polarity of the potential for the chlorine ions is reversed.

8.1.2 Channel Current

We have seen that the membrane potential has two important components or rather is a result of two important phenomena:

1) We have the resting potential, which is determined by the interplay of diffusion and the inwardly directed electric field.
2) During the action potential, the ionic channels open and close in a voltage- and time-dependent manner, generating inward- and outward-flowing currents. The variable membrane potential during this phase can be approximated by the Nernst potential, which depends on the ionic permeability coefficients of the channels.

If we assume that the ionic channels remain open during an action potential and their electric conductance is time invariant (that is, it does have a fixed value which is independent of time), then it is possible to approximate the transmembrane voltage using the equivalent circuit of Figure 8.4. For example, the current due to the flow of potassium ions can be approximated by:

$$I_K = g_K(V - E_K) \tag{8.7}$$

where g_K is the conductance of the potassium channels, V is the membrane potential, and E_K is the Nernst potential due to the flow of potassium ions across the membrane (Guevara 2003). When the two voltages in the bracket are equal, there will be no current flowing across the membrane; if the term is negative, current flow is inward; otherwise it is outward. We are interested in the condition when the membrane potential is not at equilibrium. During action potentials, there will be a flow of ionic current and therefore the capacitor charges; the net charge in the capacitor is given by:

$$Q = -CV \tag{8.8}$$

If we differentiate both sides of Eq. (8.8) with respect to time, we get a current term equal to:

$$I_K = \frac{dQ}{dt} = -C\frac{dV}{dt} \tag{8.9}$$

If we substitute Eq. (8.7) in Eq. (8.9) and rearrange terms, we get:

$$\frac{dV}{dt} + \frac{1}{\tau}(V - E_K) = 0 \tag{8.10}$$

where we have $\tau = \frac{g_K}{C} = \frac{1}{R_K C}$ is the time constant of the membrane. Clearly, Eq. (8.10) is a first-order differential equation, the solution of which is:

$$V(t) = E_K - (E_K - V(0))e^{-t/\tau} \tag{8.11}$$

Equation (8.11) is useful to understand the time-variant membrane potential, but it is inaccurate. This is because, the conductance of ionic channels during action potential is not constant; in reality, g_K is a function of time.

8.1.3 Action Potentials

A membrane's resting potential can be disturbed chemically or electrically. For example, a chemical solution of concentrated potassium ions can be injected into the cell body to neutralise the net negative charge; likewise, a solution of concentrated chlorine ions (Cl^-) can be injected into the external medium to neutralise the net positive charge due to the excess of potassium ions. Alternatively, the outer membrane can be connected to a negative electric potential while the internal membrane is connected to a positive electrical potential. When the resting potential is temporarily disturbed by an external stimulus, an interesting chain of electrochemical phenomena occurs.

When an external stimulus is applied to upset the ionic concentration, a process known as depolarisation begins to take place. If the stimulus is sufficiently large, the depolarisation process is regenerative; otherwise, nothing happens; as soon as the stimulus is removed, a tireless mechanism within the cell membrane pumps out sodium ions and pumps in potassium ion to maintain the status quo. But if the stimulus is sufficient to reduce the resting potential by approximately $-15\,mV$ (the threshold voltage), the sodium channels in the cell membrane quickly open to let a large number of sodium ions flow into the cell (recall that the concentration of sodium ions in the external medium is larger). This inward flow of sodium ions creates a sodium current (I_{Na}), which steadily increases the inner potential. Depending on the cell type, the influx of sodium ions raises the resting potential to several positive volts. After a while, however, the influx of sodium ions will stop, because the sodium channels are also time dependent and regardless of the stimulus, will automatically close. Thus the resting potential reaches a maximum. Since the precondition for the opening of the sodium ion channels (which is responsible for the triggering of the action potential) is the application of a stimulus large enough to exceed the threshold voltage, the creation of an action potential is an all-or-nothing event (Burns 2013; Kimura 2013; Levitan et al. 2015).

Meanwhile, at about the time the sodium channels are deactivated, voltage-dependent potassium channels open to allow the outward flow of potassium ions (a current due to potassium ions (I_K) begins to flow outwardly). The effect of the potassium current is to reduce the transmembrane potential back to its original level. This process is called repolarisation. However, even after the transmembrane potential reaches its resting potential, the potassium gates will remain open, allowing more potassium ions to leave the cell body. This phenomenon is called hyperpolarisation and its effect is to further reduce the transmembrane potential. Hyperpolarisation ends when the potassium channels close. After a while, the concentration difference returns back to its original level, thereby returning the membrane potential to its resting state. The time required by a cell to return the concentration imbalance to its original level is called the refractory period. Any further stimulus within this period produces no action potential. Figure 8.6 summarises the different phases of the action potential.

The shape and duration as well as the firing frequency of an action potential varies from cell to cell, as the function of cells is different from type to type as well as from location to location in the same body. Figure 8.7 displays different action potentials taken from different nerve cells. Figure 8.8 relates action potentials to transmembrane currents that flow and trigger different actions during the depolarisation and repolarisation phases for two different nerve cells.

8.1.4 Propagation of Action Potentials

In order to grasp the full significance of an action potential, one has to have a more complete picture of an excitable cell and its relationship with its environment. Figure 8.9

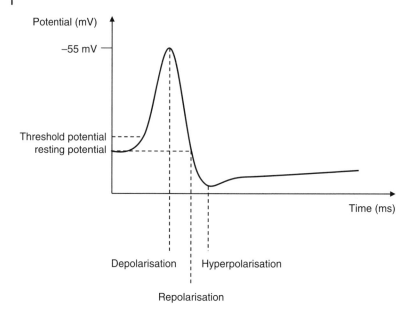

Figure 8.6 The creation of an action potential inside an electrically excitable cell. The time between the beginning of the inward flow of sodium ions and the arrival of the membrane potential to its maximum positive value constitutes the depolarisation phase. The time between the decline of the membrane potential and the membrane potential reaching its original resting potential level constitutes the repolarisation phase. The time from repolarisation to the closing of the time-dependent potassium channels constitutes the hyperpolarisation phase.

shows a typical neuron cell. The most important components of the cell for our discussion are:

- dendrites
- mitochondria
- axon hillock
- axon
- synaptic terminals.

The dendrites are components through which the cell receives excitation signals to generate an action potential. The excitation signals may come in the form of physical contractions, chemical releases (binding), or electrical pulses. In any case, their effect is to induce an action potential inside the neuron.

The mitochondria supply the energy required (adenosine triphosphate) to the sodium–potassium pumping mechanism within the cell membrane, so that it can tirelessly maintain the ionic imbalance between the internal and the external media against their electrochemical gradients. This mechanism continuously pumps sodium ions out of the cell and potassium ions into the cell at a ratio of $3Na^+ : 2K^+$ (Webster 2009). Consequently, the net effect is increasing the negativity of the potential of the inner medium.

The axon hillock is the region at which the axon of a neuron outwardly extends from its cell. It is a region with the highest density of sodium channels, and therefore where action potentials are first generated.

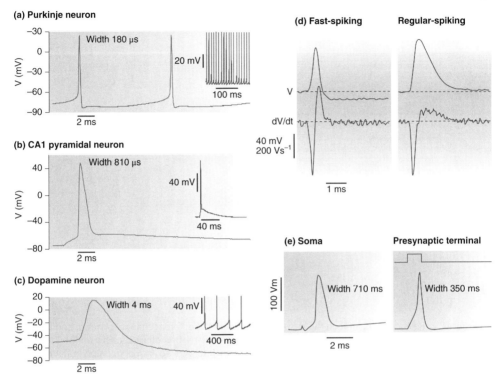

Figure 8.7 Action potentials of different nerve cells: (a) spontaneous action potentials taken from an acutely dissociated mouse cerebral Purkinje neuron; (b) action potentials taken from hippocampal CA1 pyramidal neuron in brain slice; (c) spontaneous action potentials in a midbrain dopamine neuron; (d) comparison of action potentials of fast-spiking and regular-spiking cortical neurons; (e) different action potential widths in the soma (cell body) of dentate gyrus granule neurones and in the mossy fibre bouton (a presynaptic terminal made by a granule neurone). Reproduced with permission from Bean 2007 (*Nature Review*, 2008)

The axon is a cellular extension that is responsible for the propagation of action potentials. In humans, it may have a length of up to 1 m, but in other species (such as giraffes), it may extend to several meters. There can be multiple dendrites in a single neuron, but there can never be more than a single axon, even though the axon may branch multiple times before it terminates. Synapses interface a neuron with the soma, dendrite, or axon of another neuron, or a blood vessel or an extracellular fluid. There are chemical and electrical synapses. During interaction between two neurons, the neuron initiating the interaction is called a presynaptic neuron, while the recipient is called the postsynaptic neuron. Since the propagation of action potentials within an axon is always unidirectional, presynaptic and postsynaptic neurons do not interchange roles within the same setting. A communication between a presynaptic neuron and a postsynaptic neuron may take place through chemical or electrical synapses. When chemical synapses are employed, they translate the electrical activities within the presynaptic neuron into chemical activities in the postsynaptic neuron. This is done when action potentials activate voltage-dependent calcium channels, which induce the release of neurotransmitters

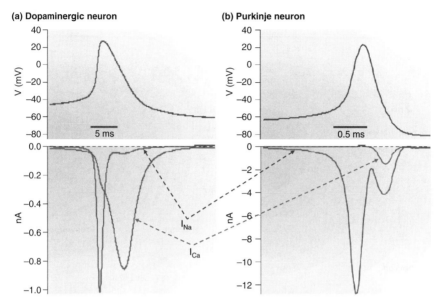

Figure 8.8 Relationship between the action potential and the different channel currents (sodium and calcium currents) generated during depolarisation and repolarisation phases: (a) in a midbrain dopaminergic neuron; (b) in a cerebellar Purkinje neuron. Reproduced with permission from Bean 2007 (*Nature Review*, 2008)

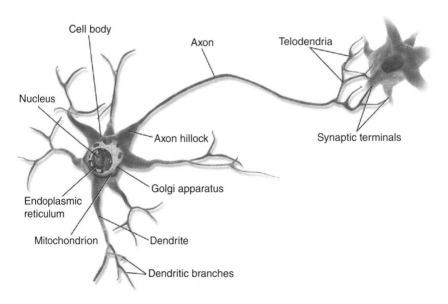

Figure 8.9 A typical neuron cell consisting of a cell body (soma), dendrites, an axon, and multiple synaptic terminals. Courtesy of Blausen Medical Communications, Inc.

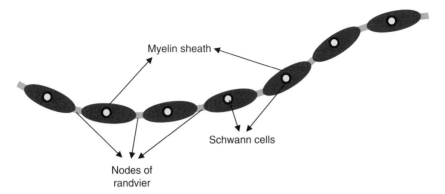

Figure 8.10 The structure of a myelinated axon or nerve fibre.

(noradrenaline, serotonin, dopamine, acetylcholine, and histamine) that bind to the receptors of the postsynaptic neurons. The tendency of some of the neurotransmitters is to excite (glutamatergic) or inhibit (GABAergic) the postsynaptic neurons. When electrical synapses are employed, they allow the flow of an electric current into the postsynaptic neuron, which then causes a change in the potential of the postsynaptic neuron.

Some axons are myelinated while others are not. Myelinated axons exclusively belong to nerve cells and are called nerve fibres (see Figure 8.10). The myelin sheaths in these axons serve as an electrical insulation and prevent the generation of action potentials in those regions. Therefore, transmembrane ionic flow occurs only at the Randvier nodes that are not myelinated. A myelinated axon typically transmits action potentials faster than an unmyelinated axon having the same dimension.

Now that we have an overview of the anatomy of a neuron cell, we are ready to discuss the propagation of action potentials. As we have already mentioned, the axon hillock is the part of a neuron with the highest density of sodium channels. So, when a neuron is stimulated, these are the channels which quickly open and allow the flow of sodium current into the cell. This initiates the generation of an action potential. At the same time, positive charges begin to flow inside the axon, depolarising adjacent areas of the membrane and inducing the generation of an action potential. The action potential thus created propagates along the axon, away from the cell body and towards the synapses. In myelinated axons, the action potential jumps from node to node since it is forbidden to occur along the regions that are insulated by myelin sheaths. This type of propagation is called saltatory conduction. Due to the jumping nature, a myelinated axon is typically faster in the conduction of action potentials than an unmyelinated axon of the same dimension. Figure 8.11 demonstrates how the activation of voltage-dependent channels induces the generation of action potentials and how the flow of positive charges (ions) inside the axon activates adjacent channels to generate and thereby propagate action potentials. Following the depolarisation-repolarisation-hyperpolarisation phases, the sodium–potassium pumping mechanism returns the transmembrane potential to its resting state. This is depicted in Figure 8.12.

8.1.5 Measuring Action Potentials

The first measurement set up for measuring the action potential of a giant axon of a squid was made by the joint endeavour of Hodgkin and Huxley (Hodgkin and Huxley

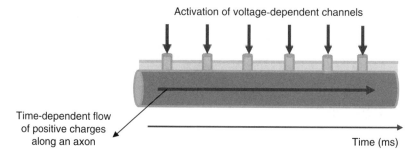

Activation of voltage-dependent channels

Time-dependent flow
of positive charges
along an axon

Time (ms)

Figure 8.11 The propagation of action potentials follows the activation of adjacent ion channels by the internal flow of positively charged ions inside the axon. The flow of ions as well as the propagation of action potentials inside the axon is unidirectional.

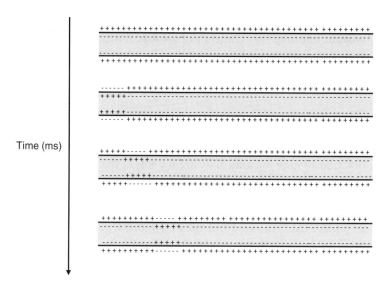

Time (ms)

Figure 8.12 The transmembrane potential returns to its resting state once adjacent channels are activated and the action potential propagates forward.

1952a, 1952b; Hodgkin et al. 1952). In their setting, they inserted one electrode inside the axon while the other was placed inside the extracellular fluid. Then, the axon was given a stimulus externally. A sensitive differential amplifier was used to sense and amplify the output voltage. As can be seen in Figure 8.13, the output of the amplifier equals the difference of the comparator inputs multiplied by the amplification factor of the amplifier: $V_o = A_v(V_+ - V_-)$. In the absence of the external stimulus, the output voltage is a function of the resting potential. When a stimulus signal is applied to the axon, however, it changes the resting potential, which initiates the generation of an action potential. The effect can be observed in the change of the output voltage's waveform (both magnitude and shape).

The measurement set up in Figure 8.13 is unstable due to the sensitivity of the amplifier to internal noise. A more stable set up can be realised by a closed-loop amplifier with

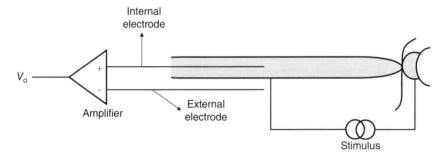

Figure 8.13 Set up for measuring an action potential.

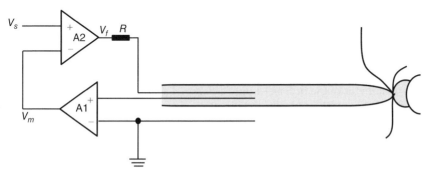

Figure 8.14 A stable measurement set up consisting of a negative feedback amplifier for measuring an action potential.

a negative feedback. This is displayed in Figure 8.14. The additional advantage of the feedback amplifier is that it is possible to:

- condition the stimulus voltage (V_s) to a desirable level
- measure channel current by measuring the voltage drop across the resistor R.

Alternatively, a sensitive ammeter can be inserted between the resistor and the electrode to directly measure the channel current.

In order to explain the operational principle of the feedback amplifier, assume that the potential of the extracellular electrode is zero. This is not a necessary precondition to the operation of the amplifier, but simplifies the explanation. Moreover, assume that both amplifiers have unity gain. Hence,

$$V_m = A_{v1}(V_+ - V_-) = V_+ \tag{8.12}$$

Furthermore, we have,

$$V_f = A_{v2}(V_m - V_s) = V_s - V_m \tag{8.13}$$

If we condition the stimulus voltage to be twice the intracellular potential, $V_s = 2V_+$, then, we have:

$$V_f = V_s - V_m = 2V_+ - V_+ = V_+ \tag{8.14}$$

Furthermore, if we neglect the voltage drop across the resistor, then the two electrodes within the same intracellular fluid have the same potential, and that is what we wish to have. Suppose, however, the stimulus voltage is changed so that $V_s = 3V_+$. How does the feedback amplifier react to this change?

$$V_f = V_s - V_{in} = 3V_+ - V_+ = 2V_+ \tag{8.15}$$

Clearly, the two electrodes within the same intracellular fluid are at two different potentials: one is at $2V_+$ mV and the other at V_+ mV. How does the amplifier react to this imbalance? As a result of the intracellular potential difference, current begins to flow from the higher potential to the lower potential until the potential difference is removed. In other words, current begins to flow through the amplifier to balance the potential difference. Similarly, when the ionic channels open and current begins to flow in and out of the axon, the membrane potential changes and, as a result, the feedback voltage changes, inducing a flow of current within the resistor, which can be sensed.

One of the difficulties with characterising action potentials in excitable cells is the difficulty of measuring current from individual channels. Given the high density of channels and the small spaces between them, measuring the current of individual channels is a formidable challenge. Scientists use the patch-clamp technique (Davie et al. 2006; Hamill et al. 1981; Sakmann 2013) in which a glass microelectrode (or pipette), the tip of which has a diameter of the order of 1 μm, is clamped against the membrane of a cell to make a tight contact with it. Ideally, the pipette makes contact with a single channel and the ions that flow through the channel also flow through the pipette. Figure 8.15 illustrates the idea.

In addition to the measurement challenge, the opening and closing of a single ionic channel exhibits randomness. This is true in two respects:

1) No two channels have an identical current-flow pattern during the generation or propagation of one and the same action potential.
2) Repeated measurements from the same channel during the generation and propagation of successive action potentials result in dissimilar patterns, but the ensemble averages of the channel current measurements exhibit similar statistical properties.

Figure 8.16 illustrates the result of repeated measurements of the current flowing through a single sodium channel. Figure 8.17 shows the ensemble average of the current flowing through a single sodium channel. This plot represents the expected current flow through a single sodium channel as a function of time.

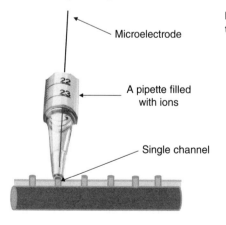

Microelectrode

22
23

A pipette filled with ions

Single channel

Figure 8.15 An illustration of the patch-clamp technique to measure a single channel current.

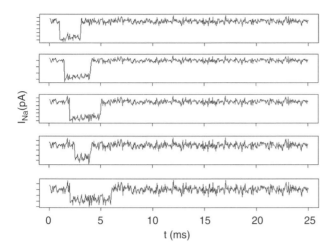

Figure 8.16 Computer simulation of repeated measurements of the current flowing through a single sodium channel during the depolarisation of action potentials.

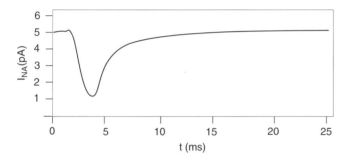

Figure 8.17 The construction of the ensemble average of the current flowing through a single sodium channel.

8.2 Cardiac Action Potentials

The heart requires and produces its own action potentials to function properly (to enable beating more than 100,000 times a day). The excitable cells that are responsible for the generation and propagation of these action potentials are millions of cardiac muscle cells or myocardiocytes, which spread mostly across the free walls of the left and right ventricles and the intraventricular septum. Each of these cells contains specialised organelles called myofibrils, which make up long chains of sarcomeres, the fundamental contractile units of a muscle cell. The muscle cells are also endowed with a high density of mitochondria for providing an ample supply of adenosine triphosphate, both to the sodium–potassium pumping mechanism (as well as the Na^+, Ca^{2+} exchanger)) inside the membrane and to the myofibrils, so that they can contract and relax without fatigue (refer also to (Klabunde 2011; Levick 2013)). The anatomy of the cardiac cells is suitable for fast propagation of action potentials: the cells are long and thin and interface each other tightly at the intercalated disks, forming tight, electrically conductive muscle fibres. There is also a considerable branching to and interconnecting with parallel

Figure 8.18 A myocardial fibre consisting of tightly connected myocardiocytes. The cells branch and interconnect with both adjacent and parallel cells for a rapid mechanical contraction and the propagation of action potentials.

fibres (see Figure 8.18). The cells are surrounded by a plasma membrane that makes end-to-end contact with adjacent cells.

The ions which are responsible in humans for producing cardiac action potentials are K^+, Na^+, and Ca^{2+} ions. These ions flow into and outside of multiple cells simultaneously in a coordinated fashion to produce action potentials in multiple locations. Like most excitable cells, sodium ions flow inward through sodium channels producing sodium current (I_{Na}) and potassium ions flow outward through potassium channels. In addition, there is an inward calcium current (I_{CaL}) flowing through calcium channels during action potential. There is also a channel, designated I_{K1}, which is impermeable to Na^+ and Ca^{2+} but permeable to K^+ even at resting potential. The flow of potassium ions exhibits four distinct features during action potential:

- transient outward current (I_{to})
- ultra-rapid current (I_{Kur})
- rapid current (I_{Kr})
- slow current (I_{Ks}).

Figure 8.19 displays a typical ventricular/atrium action potential. When the transmembrane potential exceeds the crucial threshold voltage, sodium channels (in the case of the atrium and ventricle) and calcium channels (in the case of SA and AV nodes) open to admit the inward flow of positive ions, thereby rapidly depolarising the membrane potential. This phase is known as action potential (AP) upstroke or AP Phase 0. Sequentially, first the Na^+ ions quickly depolarise the membrane potential; then the Ca^{2+} ions enable the cell to:

- maintain the depolarisation phase in the presence of outwardly flowing potassium current, and,
- mechanically contract.

After the rapid depolarisation, the time-dependent potassium channels open to enable an outward potassium current (I_{to}), but the inward calcium current (I_{CaL}) counterbalances this flow and establish a "plateau" in the action potential. This is known as Phase 2.

Following this phase, different potassium currents flow out of the cell to repolarise the membrane potential. The I_{to} and I_{Kur} are responsible for an initial rapid repolarisation (Phase 1), while I_{Kr} and I_{Ks} are primarily responsible for bringing the membrane potential to its resting state (Phase 3). The time elapsed between the AP upstroke and the return of the transmembrane potential to its resting state is referred to as the AP duration. The refractory period is approximately equal to the AP duration.

The rapid I_{Na} current during depolarisation provides a flow of positive ions between adjacent cardiomyocytes. The intercalated discs, located at the ends of the cells and which are rich in low-resistance gap junctions, provide large, non-selective channels

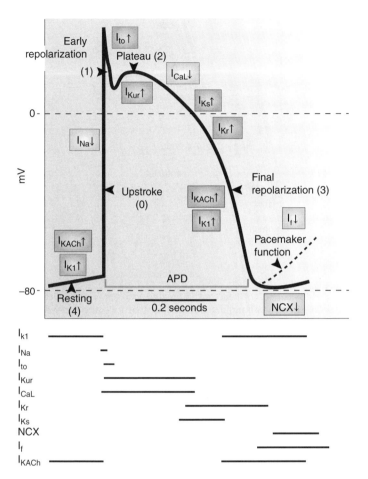

Figure 8.19 The action potential of a typical atrial or ventricular myocardiocyte. The action potential has four phases depicted as (1), (2), (3), and (4). Phase (4) depicts a resting or diastolic state. The upstroke (also labelled sometimes as Phase 0) is marked by a significant depolarisation of membrane potential. The voltage dependent inward flow of Na^+ ions is responsible for the upstroke. Phase 2 is marked by an early repolarisation due to the closing of Na^+ channels and the opening of potassium channels to enable an outward flow of K^+ ions. However, this activity is counterbalanced by the opening of calcium channels which enable an inward flow of Ca^{2+} ions creating a tentative plateau in the action potential (Phase 2). But the steady outward flow of potassium ions prevails to complete the repolarisation process (Phase 3). The time duration between the initial depolarisation and the eventual returning of the membrane potential to its resting state is known as the action potential duration (APD). The time a cell requires to be ready for the next action potential is called a refractory period, which is approximately equal to the APD. Reproduced with permission from (Nattel and Carlsson 2006) (*Nature Review*, 2006).

allowing the sodium ions and other small molecules to diffuse freely between cells, thus maintaining a coordinated forward propagation of action potentials.

8.2.1 Propagation of Cardiac Action Potentials

The bioelectrical system responsible for the generation and propagation of cardiac action potentials consists of the SA node, the atrial muscles, the AV node, the His

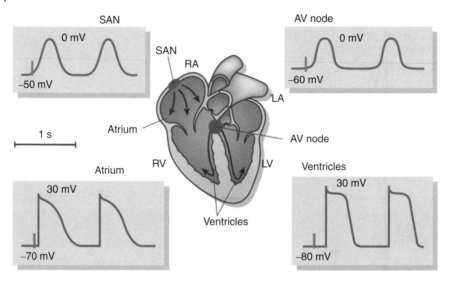

Figure 8.20 The electrical conduction system of the heart. A coordinated action potential which is responsible for the proper contraction of the heart originates at the sinoatrial node (the pacemaker) at the top-right part of the heart near the top of the right atrium and propagates through the atrial muscles to the atrioventricular (AV) node, the bundle of His, the left and right bundle branches, the Purkinje fibres, and finally diffuses into the ventricular muscles. Reproduced with permission from (Nattel and Carlsson 2006) (*Nature Review*, 2006).

bundle, the right and left bundle branches, the Purkinje fibres, and the ventricular muscles. To put this system into perspective, consider Figure 8.20. The heart consists of four chambers: two atria and two ventricles. The ventricles are responsible for pumping blood while the atria store blood temporarily and flash it into the ventricles when they are filled. Between the two pumping phases the ventricles rest while the atria contract to be filled with blood. The resting phase is called diastole and the pumping phase is called systole. The bioelectrical system encompasses the regions designated by the black arrows in the figure (however, the His bundle, which is located between the AV node and the right and left bundle branches, is not shown here).

Cardiac action potentials originate in the SA node and terminate at the ventricular muscles. In humans, the end-to-end propagation duration is about 0.2 s. While the normal contractual cells (cardiomyocytes) contribute to the propagation of action potential, they are, however, not fast enough to enable the end-to-end (organ-level) propagation within this time. Instead, the predominant conduction takes place through specialized, non-contractile cardiomyocytes (also known as conducting cardiomyocytes), which can propagate action potentials at a much faster speed.

The propagation of action potentials is both a micro- and a macro-scale process. At a micro-scale, the relationship between adjacent cells, and at a macro-scale the relationship between the specialized conduction tissues of different regions, determine the speed of propagation as well as the evolution of the waveforms of action potentials. At a micro-scale, the low-resistance gap junctions in the intercalated discs provide densely populated and non-selective channels for the free longitudinal flow of positive ions into adjacent cells, which then trigger further depolarisation. The gap junction channels are made up of specialised proteins known as connexin (Cx). There are different subtypes of connexin, but the most significant ones are Cx40, Cx43, and Cx45. These protein subtypes render unique electro-physiological properties to the gap junction channels such as conductance, permeability, and voltage dependence. Cx40 is present predominantly in the cells of the atria, the His bundle, and the Purkinje fibres; Cx43 is the key component of the working myocardiocytes of both the atria and ventricles; and Cx45 is primarily confined within the SA and AV nodes (see Bernstein and Morley 2006 and references therein).

The micro-scale action potential propagation is also influenced by morphological factors such as the size, shape, and excitability of individual cells as well as the source–sink relationship (such as mismatch in excitatory current) between adjacent cells. Moreover, the curvature of the wavefront influences the conduction velocity. As the curvature of the wavefront becomes more convex (see Figure 8.19), the velocity decreases, finally vanishing when the curvature takes some critical shape. For example, an abrupt change in wavefront curvature produces non-uniform and anisotropic conduction. At a macro-scale, factors such as "the distribution of non-excitable tissues, (fibrocytes, collagen, veins, arteries, nerves), the presence of specialized conduction tissues, and the macroscopic formations of [myocardiocytes] into larger structures (ventricles, atria)" influence action potential propagation (Bernstein and Morley 2006).

The SA node, located in the wall of the right atrium, consists of a group of specialised cardiomyocytes. Even though the cells resemble normal cardiomyocytes and possess contractile units, they do not contract with the same robustness of normal cardiomyocytes. They are also thinner in structure and richly innervated by sympathetic and parasympathetic nerves. This node is responsible for generating the impulses (slowly increasing positive potential) that regularly fire action potentials. Therefore, it is also known as the pacemaker node.

The SA node is directly connected to contractile atrial cardiomyocytes and additional conducting fibres known as the "interatrial band", the latter being responsible for rapidly spreading action potentials from the right to the left atrium, during which time the action potentials induce the atrial muscles to contract, thereby forcing the blood they accumulated to flow into their corresponding ventricles. Then action potentials reach the AV node, which is located between the atrioventricular valves and the aortic valve and connects the atria and the ventricles electrically. A set of highly specialized conducting cardiomyocytes extend from the AV node via the intraventricular septum to the ventricular walls.

The cells in the AV node conduct action potentials relatively slowly, producing a considerable delay to their propagation from the atria to the ventricles and providing the atria with sufficient time to contract and fill the ventricles prior to ventricular contraction. Unlike the cardiomyocytes of the AV node, the fibres extending from it are excellent conductors and enable a fast propagation into the outer walls of both ventricles. The fibres in the intraventricular septum are organized into bundles known as the His bundles, which then branch off into right and left bundles and extend further down the intraventricular septum, towards the apex of the heart and into the outer walls of both ventricles, whereto also the action potentials diffuse. Figure 8.21 (upper part) shows the waveform evolution and timing of an action potential as it propagates from the SA node to the right and left ventricular muscles through the atrium muscles, AV node, His bundle, bundle branches, and Purkinje fibres.

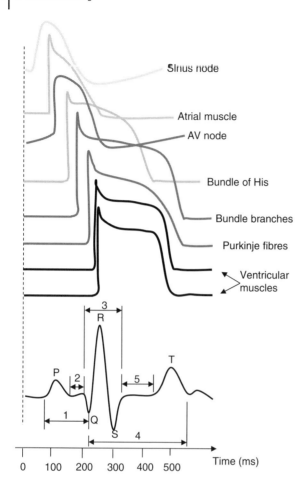

Figure 8.21 The generation of the electrocardiogram wave complex. Different parts of the heart can be activated at the same time during the propagation of action potentials. At any given point in the thoracic region, the scalar potential that can be picked up by the electrode of an electrocardiogram is the superposition of waves originating from different regions. The aggregate wave complex has three distinct regions: the P wave, the QRS wave, and the T wave. The P wave is predominantly a result of atrial depolarisation, the QRS complex primarily a result of ventricular depolarisation, and the T wave a result of ventricular repolarisation. Sometimes there is a small wave coming after the T wave called the U wave; it is a result of the repolarisation of ventricular muscles. The numbers 1, 2, 3, 4, and 5 depict notable durations that reveal particular cardiac symptoms. Often cardiac specialists investigate these durations to determine various health issues.

8.2.2 The Electrocardiogram

As action potentials propagate within the specialised electrical conduction system, they activate and deactivate multiple regions simultaneously in the atria and ventricles (since the atria muscles, Purkinje fibres, and ventricular muscles branch off in different directions). Due to the difference in propagation characteristic of the various conduction regions (tissues), the action potentials of the different regions will have different waveforms and durations. If we measure the potential of a single point (a "scalar lead") at the surface of the outer body in the thoracic region as a function of time, what we observe is the superposition of the various waveforms as the atria and ventricles depolarise and repolarise.

Figure 8.21 (below) shows the magnitude of a single-body surface potential difference as a function of time. The waveform consists of three noteworthy regions:

- the P wave
- the QRS wave complex
- the T wave.

The P wave is generated as a result of atrial depolarisation, the QRS complex primarily as a result of ventricular depolarisation, and the T wave as a result of ventricular repolarisation. The QRS complex also overlaps with the atrial repolarisation, but due to its dominance obscures it. Hence, the waveform produced by the atrial repolarisation cannot be shown distinctly. The small wave that comes after the T wave is called the U wave; it is not always present, but when it is, it is believed to be a result of the repolarisation of the ventricular muscles. The conduction delay in the AV nodes results in a zero potential region between the P and QRS regions. Likewise, the zero potential between the QRS and T regions is due to the plateau regions of individual ventricular cells.

The heart's beating rate is attuned to the body's metabolic needs because the SA node belongs to a feedback system, consisting of the sympathetic and parasympathetic nervous systems. The responsibility of the feedback system is maintaining adequate blood flow and blood pressure in the body. When the body is resting or sleeping, the heart beats slower than normal but when it is exercising or perceived to be in danger, the heart beats faster than normal. Similarly, when there is a decrease in blood pressure due to a loss of blood or dehydration, the heart beats faster so as to increase blood pressure. In a normal adult, the changes in the heartbeat are between upper and lower bounds and the heartbeat returns to a normal state when any stimulus is removed. Abnormal conditions in the cardiac electrical conduction system may result in a heartbeat which is too fast (above 100 beats per minute) or too slow (below 60 beats per minute). It is called tachycardia when it is too fast and bradycardia when it is too slow. Additional abnormal conditions produce irregular and uncoordinated atrial and ventricular contractions, resulting in quiver (fibrillation) as opposed to proper contractions.

There are three conditions that give raise to abnormal cardiac rhythms (arrhythmia):

- damage in or around the propagation path of action potentials (re-entry)
- loss of membrane potential
- afterdepolarisations.

8.2.2.1 Re-entry

Damage in the electrical conduction system (for example, a damage in the His bundle or the bundle branches or in the atrial or ventricular muscles due to myocardial ischemia or infarction) potentially block or alter the proper and coordinated propagation of action potentials. The alteration can be in:

- the propagation velocity of action potential
- the length of the refractory period of individual cells in the surrounding regions.

When action potentials travel at a significantly low speed, neighbouring cells can get sufficient time to complete their refractory period and be ready to be fired (depolarised) in the wrong direction. Similarly, when the refractory period of excitable cells is shorter, then they can be fired at a speed that is considerably higher than normal. The cumulative effect of these phenomena is that the action potentials, instead of following their normal propagation path that originates at the SA node and terminates at the ventricular muscles, can be trapped within a region in the atria or the ventricles, circulating there in a closed circuit. This condition is called re-entry and leads to atrial and ventricular fibrillation.

The effect of re-entry on the atrium and ventricles is different. In the atrium, re-entry gives rise to atrial fibrillation and manifests itself in the form of a sustained, very rapid (400–600 per minute), and irregular firing of action potentials. However, due to the inherent slowness of myocardiocytes in the AV node in conducting action potentials, the AV node acts like a filter and thus protects the ventricles from absorbing action potentials at the same rate. Nevertheless, action potentials still reach the ventricles at a considerably higher rate (typically about 150 per minute). Meanwhile, the action potentials in the atria induce it to contract and flutter rapidly. The contraction, however, is not effective and the ventricles will be filled with insufficient blood. Likewise, when action potentials are trapped inside the ventricles, they produce rapid and uncoordinated contractions of the cardiac muscles, giving rise to ventricular fibrillation. Consequently, the ventricles quiver rather than contract properly. Sustained ventricular fibrillation is fatal. Figure 8.22 compares normal action potential propagation with atrial and ventricular fibrillations. It also shows the change in the frequency of firing of action potentials during fibrillation.

8.2.2.2 Loss of Membrane Potential

The normal cardiac rhythm is regulated by the SA node, which is also called the cardiac pacemaker. There are, however, other cells in various regions of the heart (see Figure 8.23) that spontaneously initiate action potentials. Under normal circumstances, their effect is undermined by the SA node because it is inherently the fastest pacemaker. But under abnormal circumstances, some myocardial cells (for example, hypoxic myocardiocytes surrounding a scar) become increasingly irritable (hyperirritable), in that they can easily be depolarised and assert themselves as dominant pacemakers. One of the causes of irritability is a loss of transmembrane resting potential. In other words, the diastolic (the resting state of the heart) intracellular potential is less negative than normal, as a result of which a low triggering impulse is sufficient to depolarise the membrane potential above the threshold potential and to initiate an action potential.

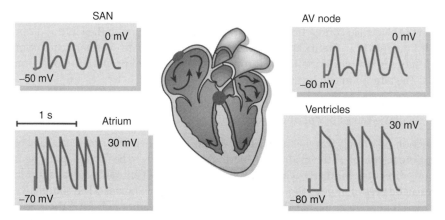

Figure 8.22 When the electrical conduction system is damaged or affected by surrounding tissues that are damaged, it may lead to atrial and ventricular re-entry, which in turn leads to atrial and ventricular fibrillation. The figure of the heart at the top summarises the proper conduction of action potentials and shows the expected waveforms whereas the figure at the bottom shows a re-entry in different regions and how this modifies the shape and magnitude of atrial and ventricular action potentials. Reproduced with permission from Nattel (2002) (*Nature*, Macmillan Magazines Ltd. 2002).

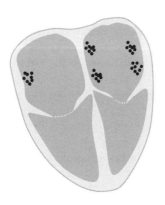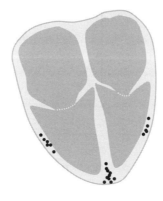

Figure 8.23 Different hyperirritable cells in different regions of the heart lead to atrial and ventricular fibrillation. A typical cause of irritable cells is hypoxic myocardiocytes around a scar.

The loss in membrane potential can be due to:

1) an increase in the concentration of potassium ions (K^+) in the extracellular medium
2) a decrease in the concentration of sodium ions (Na^+) in the intracellular medium
3) an increase in the permeability of a cell membrane to sodium ions (Na^+)
4) a decrease in the permeability of a cell membrane to potassium ions (K^+)

The cumulative effect of the above conditions is a net gain of cations inside the cell, which changes the intracellular potential from approximately -85 mV to approximately -50 to -60 mV. In case (1), when the concentration of K^+ in the external medium increases, the outward diffusion of K^+ decreases, making the intracellular potential more positive than normal. In case (2), when the concentration of Na^+ ions in the intracellular medium increases, there will be a high inward Na^+ flow, which tends to increase the concentration of positive ions within the cell thereby reducing the negativity of the intracellular potential. The effect of case (3) is similar to case (2) whereas the effect of case (4) is similar to case (1).

8.2.2.3 Afterdepolarisations

The term "afterdepolarisation" is used to characterise the occurrence of an abnormal depolarisation in a membrane potential before the firing of an action potential or a refractory period is properly completed. Afterdepolarisation can occur in Phase 2, 3, or 4 (see Figure 8.19) and give rise to cardiac arrhythmia. When it occurs during Phase 2 and 3, it is called early afterdepolarisation and when it occurs during Phase 4, it is called delayed afterdepolarisation. Early afterdepolarisation is a depolarisation of a membrane potential after the membrane has begun to repolarise but before this phase is completed. This is typically the case when calcium channels suddenly open during Phase 2 and when sodium channels open during Phase 3. Delayed afterdepolarisation, on the other hand, refers to a depolarisation in membrane potential before a cell's proper refractory period is completed. The inward flow of excessively large Ca^{2+} current due to the Na^+/Ca^{2+} exchanger is regarded as the cause of delayed afterdepolarisation.

Figure 8.24 describes how afterdepolarisation occurs and its contribution to cardiac conditions. In (a), the solid lines represent the proper firing of an action potential by a

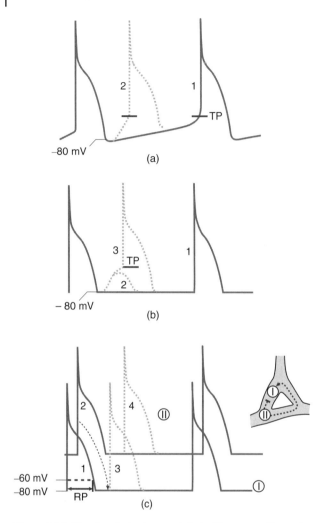

Figure 8.24 Afterdepolarisations and their effect on cardiac action potentials. In (a) cardiac action potentials are initiated by the spontaneous depolarisation of the transmembrane potential of myocardiocytes. The existence of a refractory period normally regulates the frequency of firing. When a rate of depolarisation increases, it leads to abnormal firing. In (b) spontaneous depolarisation exceeds the threshold voltage leading to early afterdepolarisation and thereby introducing an extra atrial or ventricular contraction. In (c), the existence of an afterdepolarisation and a change in the propagation velocity of action potentials in one region leads to a re-entry which produces the circulation of action potentials within an enclosed region. The cumulative effect is atrial or ventricular fibrillation. Reproduced with permission from Nattel (2002) (*Nature*, Macmillan Magazines Ltd, 2002).

normal pacemaker, while the dashed line shows an abnormal occurrence of depolarisation in membrane potential during a refractory period. In (b), depolarisation exceeds the threshold voltage due to an inward flow of Ca^{2+} during a refractory period, as a result of which the membrane potential exceeds the threshold voltage to fire an action potential. In (c), the occurrence of re-entry due to a delayed afterdepolarisations is demonstrated. A delayed repolarisation in Zone II (depicted by (2)) initiates an early afterdepolarisation

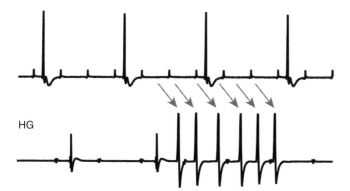

Figure 8.25 ECG measurement of glucose-induced arrhythmias: (top) during baseline conditions; (bottom) high glucose concentrations (HG). Reproduced with permission from Erickson et al. (2013) (*Nature*, Macmillan Magazines Ltd., 2002).

in Zone I (depicted by (3)), which in turn propagates to Zone II to initiate an early after-depolarisation there (depicted by (4)) and this process can continue indefinitely, creating a re-entry.

Example 8.3   We are monitoring a diabetic patient using an ECG. We suspect that an excess amount of glucose circulating in the blood plasma modifies the molecular structure of an important regulatory enzyme in the heart and the brain (CaMKII), which in turn results in the hyper-activation of Ca^{2+} channels in the heart. If the ECG measurement we obtain is the one shown in Figure 8.25, explain the phenomenon that must have taken place in the heart to produce the ECG measurement.

The figure describes a cardiac condition known as arrhythmias. It occurs when a portion of the myocardial cells become irritable and fire (discharge) independently. The site in which the irritable cells are located is called an ectopic focus and may belong to the AV node, the specialised conduction system, the atrial wall, or the ventricular wall. Depending on the ectopic focus, arrhythmias may manifest themselves in different ways. The case we are considering is a paroxysmal tachycardia, in which the ectopic focus fires rapid and regular pulses lasting for minutes, hours, or even days (Webster 2009). As can be seen, the waves consist of only the QRS-wave complex suggesting that the irritable nodes fire at a much faster rate than the SA node.

8.3   Brain Action Potentials

The human brain is an essential component of the central nervous system (the other being the spinal cord), which, together with the peripheral nervous system, controls every aspect of mental and physical activity. Nerve fibres extend from the brain to the face, ears, eyes, and nose as well as to the spinal cord, from which they further spread to the rest of the body. Sensory nerves interface the body with the physical environment to collect surrounding information and to transmit it to the spinal cord. The spinal cord aggregates and filters the information before forwarding it to the brain. The brain interprets the information and, when necessary, directs motor responses. Likewise, motor neurons convey instructions from the brain to the rest of the body.

The electrical activities of neurons in a developing human brain begin to take place between the 17th and the 23rd weeks of prenatal development. It is estimated that the full number of neural cells are already developed at birth, which amounts to approximately 10^{11} neurons with an average density of 10^4 neurons per cubic millimeter (Kandel et al. 2000; Teplan 2002). These neurons will have approximately 5.10^{14} synapses by the time the child is well developed. Brain neurons are densely interconnected to form robust and resilient networks that enable rapid and economical communication between the different parts of the brain (Braitenberg and Schüz 2013; Huttenlocher et al. 1979; Llinás 2012).

The brain consists of three main parts:

- the forebrain, consisting of the cerebrum, thalamus, and hypothalamus
- the midbrain, consisting of the tectum and tegmentum
- the hindbrain, consisting of the cerebellum, pons, and medulla.

In the literature, however, the midbrain, pons, and medulla are collectively referred to as the brainstem. The cerebellum coordinates voluntary movements of muscles and maintains balance. The brain stem controls respiration, heart regulation, biorhythms, neurohormone and hormone secretion. The cerebrum makes up the largest part of the human brain and is associated with complex functions such as thought and action. Functionally, it is divided into four sections: the frontal lobe, parietal lobe, occipital lobe, and temporal lobe, each of which is associated with particular functions. These are summarised in Table 8.1.

Figure 8.26 displays the four functional sections of the cerebrum. In terms of arrangement, the cerebrum is divided into two hemispheres: the left and the right hemispheres. These exhibit a certain symmetry, but recent studies suggest that each side functions slightly differently. For example, in some individuals, the right hemisphere is associated with creativity whereas the left hemisphere is associated with the ability to reason logically (Duvernoy 2012; Mai et al. 2016). The two hemispheres are interconnected via the corpus callosum, which consists of a bundle of axons. The grey surface of the cerebrum (distributed at the cerebral hemispheres or the cerebral cortex) is densely populated by highly interconnected neurons, 85 % of which are pyramidal neurons (Väisänen 2008). The neurons are tightly placed parallel to each other and perpendicular to the local cortical surface (Bullmore and Sporns 2012). The nerve fibres beneath the grey surface of the cerebellum transport signals between the neurons and the other parts of the brain and body.

Table 8.1 A summary of the different sections of the cerebrum and their functions.

Lobe	Function
Frontal	Reasoning, planning, parts of speech, movement, emotions, and, problem solving
Parietal	Orientation, recognition, perception of stimuli
Occipital	Visual processing
Temporal	Perception and recognition of auditory stimuli, memory, and speech

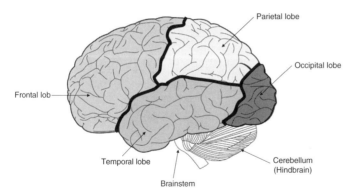

Parietal lobe

Occipital lobe

Frontal lob

Temporal lobe

Cerebellum
(Hindbrain)

Brainstem

Figure 8.26 Functionally, the cerebral cortex consists of four sections: the frontal lobe, the parietal lobe, the occipital lobe, and the temporal lobe. Each hemisphere of the brain contains these sections; the hemispheres themselves are interconnected via the corpus callosum, a flat bundle of neural fibres (200–250 million contralateral axonal projections).

When neuronal cells in the brain are depolarised, voltage- and time-dependent Na^+, K^+, Ca^{2+}, and Cl^- flow into and out of the cells, giving rise to intracellular and extra-cellular currents. Given the high density of neurons and the complexity of neuronal interactions and data-processing activities, studying the causes and effects of action potentials in a single neuron is difficult; technically as well as conceptually. However, neuronal activities in the brain are highly coordinated and synchronised, as a result of which the electrical potential to which the activities of a collection of neurons give rise within a volume of a brain tissue can be measured and reasoned about. When this potential is measured by penetrating depth electrodes into the brain, it is known as a local field potential (LFP); when it is measured at the cortical surface by inserting strips and grids of subdural electrodes, it is known as an electrocorticogram (ECoG); when it is measured at the surface of the scalp non-intrusively, it is known as electroencephalogram (EEG). Regardless of the mechanism employed, the electrical potential (V_e) can be a scalar or a vector quantity. When measured with respect to a reference potential, it is a scalar quantity. Often, however, the difference in potential between two locations on the scalp, the cortical surface, or an extracellular medium is a matter of interest, in which case, the potential difference is a function of the distance between the two locations and the underlying physiology. In this case, the potential difference is a vector quantity and gives rise to an electric field that is defined as the negative spatial gradient of V_e (Buzsáki et al. 2012).

As can be seen in Figure 8.27, the electric field that can be measured (in volts-meter) by employing electrodes is the effect of the superposition of different extracellular currents generated at different regions of the brain. The three measurement approaches have advantages and disadvantages. Understandably, the quality of signal that can be sensed reduces as the distance between the sources of membrane current and the measurement point increases. On the other hand, the level of intrusion increases as the distance reduces. Another perspective on the quality of the sensed signal comes by considering spatial and temporal resolution. Experimental results show that EEG measurements have excellent temporal resolution (in that the change in the potential waveform as a function of time can be detected in a millisecond range) but exhibit poor

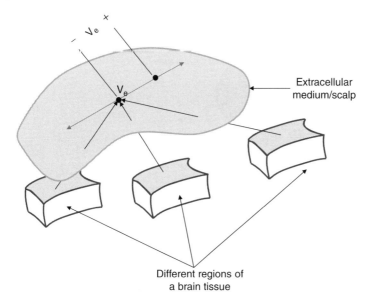

Figure 8.27 Extracellular currents generated in different parts of the brain as a result of neuronal activities (action potentials) will have an aggregate or a superposition effect on the electric field, which can be measured within or outside of the brain. The figure shows a scalar potential and a vector potential field (V_e). The scalar electric potential measures the potential of a single location with respect to a reference potential whereas the vector field measures the potential difference between two points. In the latter case, the potential difference is a function of the distance between the two points.

spatial resolution (potential waveforms become blurry as the distance between two measurement points diminishes). The invasive techniques have better spatial resolution but their temporal resolution is poor. Figure 8.28 compares the waveforms obtained by LFP and ECoG.

During action potentials, the components of excitable cells (the dendrites, soma, axons, and synapses) generate transmembrane currents, each of which contributes to the extracellular field. However, the most important components are the postsynaptic currents, which are slower to fluctuate than the rest as a result of which the superposition effect in postsynaptic currents can be distinguishable (Buzsáki et al. 2012; Michel and Murray 2012; Smith 2005). The amplitude and frequency evolutions of the measurable extracellular potential can easily be detected as the distance of the measurement point increases from an active neuron in all directions (Lindén et al. 2010). This is illustrated in Figure 8.29, where the amplitude of the extracellular potential rapidly decreases from 100 mV to 0.1 mV as the the distance from the soma of a neuron increases.

8.3.1 Electroencephalography

A portion of the extracellular current that is generated by postsynaptic potentials in the dendrites of large pyramidal neurons flows through different hard and soft tissues of the brain, the cerebrospinal fluid, the skull, and the skin and produces a voltage drop at the surface of the scalp. When this voltage is read by a single electrode, it is called electroencephalography. The difference in potential between two electrodes attached to the

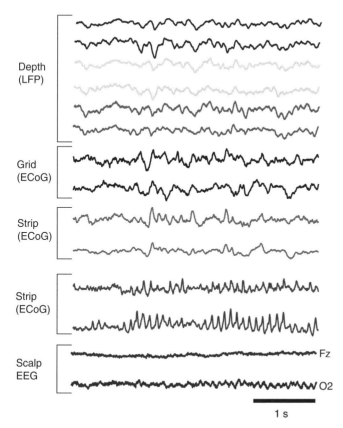

Figure 8.28 Simultaneous recordings of V_e from different regions of the human brain. From top to bottom: three depth electrodes were used in the left amygdala and hippocampus to measure LFP (three measurement for each site); a 3×8 subdural grid electrode array was placed over the lateral left temporal cortex to measure the ECoG; two four-contact strips were placed under the inferior temporal surface for measuring the ECoG; an eight-contact strip was placed over the left orbitofrontal surface for measuring the ECoG; two electrodes were placed on both hemispheres on the scalp to measure the EEG. The measurements were taken from a patient with drug-resistant epilepsy. As can be seen the measurements from the intracerebral (LFP) and ECoG have larger amplitudes and higher frequency components and greater resolution compared to scalp EEG. Reproduced with permission from Buzsáki et al. (2012), (*Nature Reviews Neuroscience*, 2012).

surface of the scalp strongly depends on the underlying structure as well as the relative location of the electrodes with respect to the firing neurons. Due to signal distortions at various junctions, there is little discernible resemblance between the potential difference of an electroencephalogram and the firing pattern of the contributing individual neurons (Smith 2005). However, electroencephalography plays a vital role in the study of brain activities in medicine, neurophysiology, and cognitive science. Teplan (2002) summarises some of the essential roles of electroencephalography (refer also to Sanei and Chambers 2013 and Niedermeyer and da Silva 2005). These are:

- monitoring alertness, coma and brain death
- locating areas of damage following head injury, stroke, tumour, etc.
- testing afferent pathways (by evoked potentials)

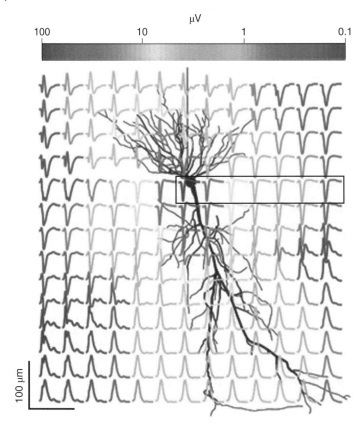

Figure 8.29 The distribution of a local field potential during an action potential in a pyramidal cell. The field exhibits a change in amplitude and frequency as the distance between the soma and the measurement point increases. Reproduced with permission from Buzsáki et al. (2012), (*Nature Reviews Neuroscience*, 2012).

- monitoring cognitive engagement (alpha rhythm)
- producing biofeedback situations
- controlling the depth of anaesthesia ("servo anaesthesia")
- investigating epilepsy and the origin of seizure
- testing epilepsy drug effects
- assisting in experimental cortical excision of epileptic focus
- monitoring human and animal brain development
- testing drugs for convulsive effects
- investigating sleep disorder and physiology.

Depending on the activity and resting level of the brain, the frequency of EEG waveforms vary from a fraction of a hertz to several kilohertz, even though the frequency range that is relevant for clinical investigation is usually from 0.1–100 Hz. In some cases, a stricter range from 0.3 –70 Hz is considered to be of particular interest (Väisänen 2008). Similarly, the amplitude of the signal that can be measured from the surface of the scalp lies between 10 and 100 mV. In contrast, the amplitude of the signal that can be measured directly from the surface of the cortex lies between 500 and 1500 µV

(Buzsáki et al. 2012). Approximately $6\,cm^2$ of the cortical gyri need to be activated synchronously to produce a measurable potential at the surface of the scalp (Väisänen 2008). When there are restricted lesions (damages) such as tumour, haemorrhage, and thrombosis, the EEG waveforms exhibit dominant low frequencies and an overall reduction in amplitude. Epileptic conditions are known for stimulating the cerebral cortex and thereby producing large amplitude signals (up to $1000\,mV$), which are referred to as "spikes". Similarly, EEG patterns can be modified by a wide range of biochemical, metabolic, circulatory, hormonal, neuroelectric, and behavioural factors (Teplan 2002).

EEG can be measured without interfering with the activities of the brain (passive observation) or by intentionally inducing it to do certain activities. In the second case, the EEG readings are called evoked potentials. The stimuli for induced activities can be sensory, motor, or cognitive. The purpose of evoked potentials is to study the brain's response (such as alertness, span of attention, or selective attention) to the stimuli. The most significant EEG frequency bands that give insight into different brain activities are summarised in Table 8.2.

Generally, the EEG waveforms tend to contain high-frequency components when the brain's activity or engagement level is high, and low-frequency components when the brain relaxes or when there are lesions in some of its regions. Furthermore, whether or not some of the frequency bands can be observed depends on the location on the scalp from which measurements are obtained, because there is a

Table 8.2 Classification of brain waves and a description of their significance.

Brain wave	Frequency (Hz)	Prominent location	Significance
δ-waves	< 4	Frontal (adults), occipital (children)	Deep sleep, Lesion in subcortical regions
θ-waves	$4-7$	Parietal, temporal	Drowsiness, arousal, stress, and disappointment, lesion in subcortical regions
α-waves	$8-15$	Occipital	Eyes closed, relaxation
β-waves (I)	$14-28$	Frontal	Serenity, relaxed motor activity
β-waves (II)	$16-31$	Frontal	Intense brain activity

Table 8.3 Electrode placement in the 10-20 international system and its association with particular brain functions.

Electrode placement	Brain region	Brain functions
F7	Frontal	Rational activities
Fz	Frontal	Intentional and motivational functions
F8	Frontal	Emotional impulses
C3, C4, Cz	Central	Sensory and motor functions
P3, P4, Pz	Parietal	Perception and differentiation
T3 and T4	Temporal	Emotion processor
O1 and O2	Occipital	Primary visual areas

coarse-grained correspondence between the EEG measurements of specific locations and the underlying activity regions in the brain.

The frequencies in the δ-band are the highest in amplitude. In adults, they are prominently observed in the frontal region of the scalp when they are in deep sleep and in children they are prominently observed in the occipital region. They also occur when there are lesions in subcortical regions of the brain (hippocampus, amygdala, claustrum). The frequencies in the θ-band are prominent in the parietal and temporal regions of the scalp of young children but may also be observed in states of drowsiness, arousal, stress, and disappointment in older children and adults. Typically, observation of excess θ-waves in adults implies abnormal brain activity, for example, lesions in subcortical regions.

Figure 8.31 displays one of the most prominent EEG events during deep sleep, namely, small waves (δ-waves) having frequencies below 1 Hz. These waves occur as a result of the oscillation of cortical neurons between active (up) and inactive (down) states during non-rapid eye movement phase. Scientists have not to date determined whether these waves are the result of synchronised activities across the majority or the minority of regions of the brain. According to Nir et al. 2011, they are a result of synchronised local activities, usually involving propagation from the medial prefrontal cortex to the medial temporal lobe and hippocampus. The study suggests that intracerebral communication during sleep is constrained because slow and spindle oscillations often occur out of phase in different brain regions.

The frequencies in the α-band are the most ubiquitous waves that can be observed in normal human beings when they are awake and in a serene and resting state. They are best observed in the occipital region of the scalp of both hemispheres, although they can sometimes be observed in the parietal and frontal regions as well. The magnitude of these waves ranges from 20 to 200 mV (peak to peak). α-waves can be induced by simply closing the eyes and relaxing; they disappear when the eyes are opened or when the brain is focused, for example, during thinking or calculating. The precise origin of α-waves is still not sufficiently understood. It is assumed that dendrite action potentials are predominantly responsible (Hughes and Crunelli 2005; Klimesch et al. 2007). In contrast, evoked potentials generated in the brain stem are often attributed to axonal and synaptic currents. A special band of frequencies that overlaps with α-waves but is believed to be induced by the firing of motor neurons in a resting state is known as the μ-band. The suppression of the μ-band occurs when a person observes a motor action at a resting state, [1] suggesting the firing of mirror neurons that are out of sync with the motor neurons and, therefore, cancel the μ-waves (Oberman et al. 2005; Pineda 2005).

β-waves have two components: the frequencies that are about twice the α-frequencies (classified as β-I) and the higher frequencies, which are classified as β-II. Both components are closely linked to motor and intense mental activity. β-I components occur in a relaxed states in a similar way that α-waves appear during a relaxed state and disappear when the brain starts to concentrate; β-II components are induced by intense activity in the central nervous system and during tension. Hence, β-II are induced whilst β-I

---

1 Many primates and birds posses mirror neurons that fire when they act or when they observe other animals engaged in similar actions, in which case the neurons mirror the actions or behaviours of the animals they observe. In human beings, the activities inducing mirror neurons to fire are associated with the premotor cortex, the supplementary motor area, the primary somatosensory cortex, and the inferior parietal cortex (Cozolino 2014; Gazzola et al. 2007; Molenberghs et al. 2009).

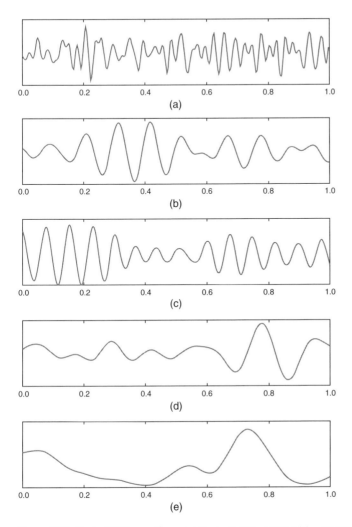

Figure 8.30 Typical EEG waveforms: (a) β-waves; (b) α-waves; (c) μ-waves; (d) θ-waves; (e) δ-waves. Courtesy of Hugo Gambo, 2005.

are inhibited by brain activity (Subasi and Ercelebi 2005). Figure 8.30 summarises the typical EEG waveforms.

8.3.2 Volume Conduction

In clinical and research settings, the investigation of brain action potentials can be considered either as a forward problem or as an inverse problem (Nunez and Srinivasan 2006; Väisänen 2008). In the forward problem, the task is to study how the activity of neurons in a particular region in the brain can be manifested in the form of a waveform in particular regions of the scalp. In the inverse problem, as the name implies, the task is the opposite: a manifested waveform at a particular point in the scalp is mapped to a particular region in the brain in order to reason about the causes of the waveforms. In

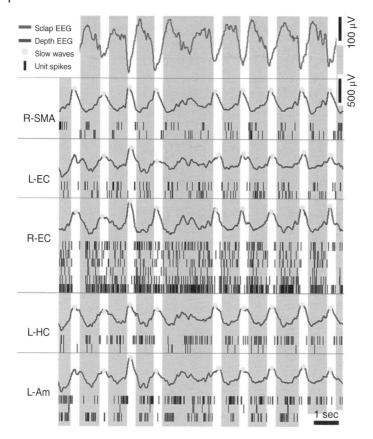

Figure 8.31 Scalp and depth EEG measurements taken from different regions of the brain during 11.5 s of deep NREM sleep in one individual. Except the topmost plot, which is taken from a scalp EEG (Cz), all the others are taken from a depth EEG. Hence they depict measurements from (top to bottom): the right supplementary motor area (R-SMA), left entorhinal cortex (L-EC), right entorhinal cortex (R-EC), left hippocampus (L-HC), and left amygdala (L-Am). The dark rasters depict unit spikes. The dots on the waveforms indicate individual slow waves detected automatically in each channel separately. The grey and white vertical stripes (bars) mark the "ON" and "OFF" periods occurring in unison across multiple brain regions. Figure reprinted with permission from Nir et al. (2011) (Elsevier, *Neuron*, 2011).

either case, a closer understanding of the nature of wave propagation between the brain and the scalp is vital.

When the concentration of ions in the extracellular medium varies (for example, through the outward flow of intracellular K^+ ions), it creates a repulsion force on the surrounding ions (because like charges repel each other). The magnitude of the repulsion force depends on the magnitude of the flow as well as the ease with which the ions can travel through the surrounding tissues. The ions upon which the repulsion force is exerted in turn exert a repulsion force on their neighbouring ions and so propagate the repulsion force outward, towards the scalp.

Figure 8.32 shows the different tissues that make up the region between the brain (the cerebrum) and the scalp. Because of the difference in the nature of the tissues, the propagating waveform experiences different resistivity as it propagates towards the

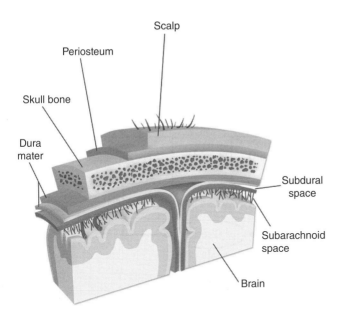

Figure 8.32 The current responsible for the electric potential that can be measured at the surface of the scalp is generated in the extracellular medium in the brain and propagates though the cerebrospinal fluid, the different soft and hard tissues, the skull, the skin, and the scalp. The collective impact of these structures is modelled as a volume conductor. (Courtesy of Blausen Gallery, 2014)

scalp. Indeed, the waveform experiences a non-linear distortion that it is not possible to establish a one-to-one mapping between the initial source of the waveform and the waveform that is measured by electrodes at the surface of the scalp. However, modelling the intermediate region as a whole or as a unified conductive medium is useful for addressing both the forward and the inverse problems. When this is done, the conductive region is called a volume conductor. Figure 8.33 shows the role of the volume conductor in solving the forward and backward problems.

8.3.3 Electrode Placement

As we have seen previously, different brain areas can be associated with different functions. Hence, it is reasonable to place electrodes in specific locations in the scalp to record and reason about the electrical potentials that can be generated by specific activities. Figure 8.34 shows the international 10-20 system for electrode placement in the surface of the scalp during EEG recording. From the figure, it is possible to observe that some of the electrodes are placed near to specific brain regions. For example:

- F7 is located near the brain region which is responsible for rational activities
- Fz is located near the brain region which is responsible for intentional and motivational functions
- F8 is located close to the brain region which is the source of emotional impulses.

Similarly, C3, C4, and Cz are close to the brain region that deals with sensory and motor functions; P3, P4, and Pz are close to the brain regions which contribute to perception and differentiation; T3 and T4 are close to the brain regions which are

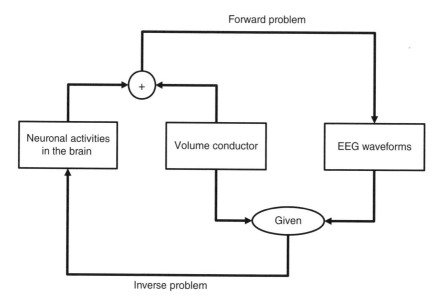

Figure 8.33 Modelling the region between the brain and the scalp as a volume conductor enables establishment of a relationship between neuronal activities in the brain and the EEG waveforms measured by EEG electrodes at the surface of the brain both in addressing the forward and the inverse problems.

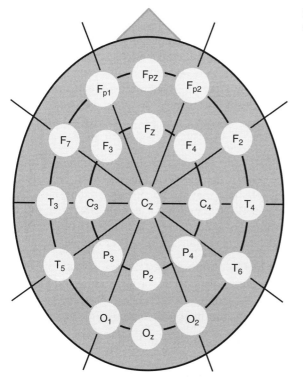

Figure 8.34 The standard 10-20 electrode placement.

responsible for processing emotions; T5 and T6 are placed near the region of the brain which is associated with certain memory functions. Similarly, O1 and O2 are placed above the brain region which is primarily responsible for visual functions (Teplan 2002). Having said this, it must be remarked that establishing direct relationships between specific brain activities and the EEG electrodes is a difficult task, not only due to the volume conduction problem but also the exact location of the activity sources is still an open research issue.

The spatial resolution that can be achieved by the 10-20 system is not sufficient for many scientific and research investigations, even though it may be sufficient for clinical diagnosis. There are several non-standard constellations that can be considered, depending on the desired spatial resolution. Taking the average size of the surface area of the scalp of an adult, the interelectrode distance in the 10-20 system is approximately 6 cm. This distance can be reduced to 3 cm when 64 electrodes are used, to 62.25 cm when 128 electrodes are used, and to 1.6 cm when 256 electrodes are used (Väisänen 2008). The existence of electrode caps has enabled the equidistant placement of electrodes in larger constellations (256 electrodes) and simplified their use.

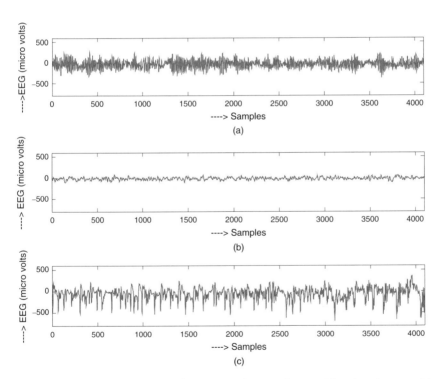

Figure 8.35 EEG measurements: (a) from a normal person during a relaxed awake state with eyes-open; and from an epileptic patient (b) during pre-ictal and (c) ictal states. Each measurement has a duration of 23.6 s. The measurements are sampled at 173.61 Hz. Reprinted with permission from Acharya et al. (2012), (Elsevier, *Biomedical Signal Processing and Control*).

Example 8.4 One of the applications of EEG is the determination of an epileptic seizure. Experienced neurophysiologists often inspect EEG readings to determine the nature of a seizure. Figure 8.35 shows EEG measurements for three different states. The top measurement is taken during a normal state, but the middle and the bottom measurements are taken during pre-ictal (referring to the physiological state immediately preceding an actual seizure) and ictal (referring to the physiological state of a seizure) stages, respectively. If you were asked to visually inspect the measurements and infer the nature of the seizure, how would you proceed with the task?

Epilepsy is a neurophysiological disorder arising from the generation of excessive action potentials by hyper-excitable neurons of the cerebral cortex. Epileptic seizures can be focal or generalised. During a focal seizure, only a particular region of the brain is affected, whereas in the generalised case the whole brain is affected. In both cases, epilepsy manifests itself by recurrent seizures. There are, however, different underlying causes for a seizure besides epilepsy. Indeed, episodic events arising from low blood sugar, low oxygen, abnormal sodium, calcium, and potassium in the blood can cause seizures. These are not related to epilepsy; other similar causes include dementia, head injury, brain infections, congenital birth defects, birth-related brain injuries, tumours and other space occupying lesions (Acharya et al. 2013). Even though the study of epileptic seizures is an active research area, unlike other seizures, epileptic seizures are results of a coordinated or synchronised neuronal firing (there are fewer independent functions during seizure, and processes are active in the brain).

References

Acharya UR, Molinari F, Sree SV, Chattopadhyay S, Ng KH and Suri JS 2012 Automated diagnosis of epileptic EEG using entropies. *Biomedical Signal Processing and Control*, 7(4), 401–408.

Acharya UR, Sree SV, Swapna G, Martis RJ and Suri JS 2013 Automated EEG analysis of epilepsy: a review. *Knowledge-Based Systems*, 45, 147–165.

Alexander SP, Mathie A and Peters JA 2011 Ion channels. *British Journal of Pharmacology*, 164(s1), S137–S174.

Bean BP 2007 The action potential in mammalian central neurons. *Nature Reviews Neuroscience*, 8(6), 451–465.

Bernstein S and Morley G 2006 Gap junctions and propagation of the cardiac action potential. In: *Cardiovascular Gap Junctions*, vol. 42. Karger.

Braitenberg V and Schüz A 2013 *Cortex: Statistics And Geometry of Neuronal Connectivity*. Springer Science & Business Media.

Bullmore E and Sporns O 2012 The economy of brain network organization. *Nature Reviews Neuroscience*, 13(5), 336–349.

Burns N 2013 *Cardiovascular physiology*. Presentation, retrieved from School of Medicine, Trinity College, Dublin.

Buzsáki G, Anastassiou CA and Koch C 2012 The origin of extracellular fields and currents–EEG, ECoG, LFP and spikes. *Nature Reviews Neuroscience*, 13(6), 407–420.

Catterall WA 1985 Structure and function of voltage-sensitive ion channels. *Annual Reviews of Biochemistry*, 64, 493–531.

Chi YM, Jung TP and Cauwenberghs G 2010 Dry-contact and noncontact biopotential electrodes: Methodological review. *IEEE Reviews in Biomedical Engineering*, **3**, 106–119.

Cogan SF 2008 Neural stimulation and recording electrodes. *Annual Reviews of Biomedical Engineering*, **10**, 275–309.

Cozolino L 2014 The Neuroscience of Human Relationships: Attachment and the Developing Social Brain *(Norton Series on Interpersonal Neurobiology).* WW Norton & Company.

Davie JT, Kole MH, Letzkus JJ, Rancz EA, Spruston N, Stuart GJ and Häusser M 2006 Dendritic patch-clamp recording. *Nature Protocols*, **1**(3), 1235–1247.

De Biase LM, Nishiyama A and Bergles DE 2010 Excitability and synaptic communication within the oligodendrocyte lineage. *The Journal of Neuroscience*, **30**(10), 3600–3611.

Duvernoy HM 2012 *The Human Brain: Surface, Three-dimensional Sectional Anatomy with MRI, and Blood Supply*. Springer Science & Business Media.

Erickson JR, Pereira L, Wang L, Han G, Ferguson A, Dao K, Copeland RJ, Despa F, Hart GW, Ripplinger CM *et al.* 2013 Diabetic hyperglycaemia activates camkii and arrhythmias by o-linked glycosylation. *Nature*, **502**(7471), 372–376.

Gazzola V, Rizzolatti G, Wicker B and Keysers C 2007 The anthropomorphic brain: the mirror neuron system responds to human and robotic actions. *Neuroimage*, **35**(4), 1674–1684.

Goldin AL, Barchi RL, Caldwell JH, Hofmann F, Howe JR, Hunter JC, Kallen RG, Mandel G, Meisler MH, Netter YB *et al.* 2000 Nomenclature of voltage-gated sodium channels. *Neuron*, **28**(2), 365–368.

Guevara MR 2003 Dynamics of excitable cells. In: *Nonlinear Dynamics in Physiology and Medicine*. Springer, pp. 87–121.

Hai A, Shappir J and Spira ME 2010 In-cell recordings by extracellular microelectrodes. *Nature Methods*, **7**(3), 200–202.

Hall JE 2015 *Guyton and Hall Textbook of Medical Physiology*. Elsevier Health Sciences.

Hamill OP, Marty A, Neher E, Sakmann B and Sigworth F 1981 Improved patch-clamp techniques for high-resolution current recording from cells and cell-free membrane patches. *Pflügers Archiv*, **391**(2), 85–100.

Hibino H, Inanobe A, Furutani K, Murakami S, Findlay I and Kurachi Y 2010 Inwardly rectifying potassium channels: their structure, function, and physiological roles. *Physiological Reviews*, **90**(1), 291–366.

Hille B *et al.* 2001 *Ion Channels of Excitable Membranes*, vol. 507. Sinauer.

Hodgkin AL and Huxley AF 1952a Currents carried by sodium and potassium ions through the membrane of the giant axon of loligo. *The Journal of Physiology*, **116**(4), 449.

Hodgkin AL and Huxley AF 1952b A quantitative description of membrane current and its application to conduction and excitation in nerve. *The Journal of Physiology*, **117**(4), 500.

Hodgkin AL and Katz B 1949 The effect of sodium ions on the electrical activity of the giant axon of the squid. *The Journal of Physiology*, **108**(1), 37.

Hodgkin AL, Huxley A and Katz B 1952 Measurement of current-voltage relations in the membrane of the giant axon of loligo. *The Journal of Physiology*, **116**(4), 424–448.

Hughes SW and Crunelli V 2005 Thalamic mechanisms of EEG alpha rhythms and their pathological implications. *The Neuroscientist*, **11**(4), 357–372.

Huttenlocher PR *et al.* 1979 Synaptic density in human frontal cortex-developmental changes and effects of aging. *Brain Research*, **163**(2), 195–205.

Jack JJB, Noble D and Tsien RW 1975 *Electric Current Flow in Excitable Cells*. Clarendon Press.

Kandel ER, Schwartz JH, Jessell TM *et al.* 2000 *Principles of Neural Science*, vol. 4. McGraw-Hill.

Kimura J 2013 *Electrodiagnosis in diseases of nerve and muscle: principles and practice*. Oxford University Press.

Klabunde R 2011 *Cardiovascular Physiology Concepts*. Lippincott Williams & Wilkins.

Klimesch W, Sauseng P and Hanslmayr S 2007 EEG alpha oscillations: the inhibition–timing hypothesis. *Brain Research Reviews*, **53**(1), 63–88.

Levick JR 2013 *An Introduction to Cardiovascular Physiology*. Butterworth-Heinemann.

Levitan IB, Kaczmarek LK *et al.* 2015 *The Neuron: Cell and Molecular Biology*. Oxford University Press.

Lindén H, Pettersen KH and Einevoll GT 2010 Intrinsic dendritic filtering gives low-pass power spectra of local field potentials. *Journal of Computational Neuroscience*, **29**(3), 423–444.

Llinás RR 2012 Commentary on 'Electrophysiological properties of in vitro Purkinje cell dendrites in mammalian cerebellar slices', *J. Physiol* 1980; 305: 197–213. *The Cerebellum*, **11**(3), 629.

Mai JK, Majtanik M and Paxinos G 2016 *Atlas of the Human Brain*. Academic Press.

Michel CM and Murray MM 2012 Towards the utilization of EEG as a brain imaging tool. *Neuroimage*, **61**(2), 371–385.

Molenberghs P, Cunnington R and Mattingley JB 2009 Is the mirror neuron system involved in imitation? A short review and meta-analysis. *Neuroscience & Biobehavioral Reviews*, **33**(7), 975–980.

Nattel S 2002 New ideas about atrial fibrillation 50 years on. *Nature*, **415**(6868), 219–226.

Nattel S and Carlsson L 2006 Innovative approaches to anti-arrhythmic drug therapy. *Nature Reviews Drug Discovery*, **5**(12), 1034–1049.

Niedermeyer E and da Silva FL 2005 *Electroencephalography: Basic Principles, Clinical Applications, and Related Fields*. Lippincott Williams & Wilkins.

Nir Y, Staba RJ, Andrillon T, Vyazovskiy VV, Cirelli C, Fried I and Tononi G 2011 Regional slow waves and spindles in human sleep. *Neuron*, **70**(1), 153–169.

Nunez PL and Srinivasan R 2006 *Electric Fields of the Brain: The Neurophysics of EEG*. Oxford University Press.

Oberman LM, Hubbard EM, McCleery JP, Altschuler EL, Ramachandran VS and Pineda JA 2005 EEG evidence for mirror neuron dysfunction in autism spectrum disorders. *Cognitive Brain Research*, **24**(2), 190–198.

Pineda JA 2005 The functional significance of mu rhythms: translating 'seeing' and 'hearing' into 'doing'. *Brain Research Reviews*, **50**(1), 57–68.

Sakmann B 2013 *Single-channel Recording*. Springer Science & Business Media.

Sanei S and Chambers JA 2013 *EEG Signal Processing*. John Wiley & Sons.

Smith JS 2005 The local mean decomposition and its application to EEG perception data. *Journal of the Royal Society Interface*, **2**(5), 443–454.

Subasi A and Ercelebi E 2005 Classification of EEG signals using neural network and logistic regression. *Computer Methods and Programs in Biomedicine*, **78**(2), 87–99.

Teplan M 2002 Fundamentals of EEG measurement. *Measurement Science Review*, **2**(2), 1–11.

Traynelis SF, Wollmuth LP, McBain CJ, Menniti FS, Vance KM, Ogden KK, Hansen KB, Yuan H, Myers SJ and Dingledine R 2010 Glutamate receptor ion channels: structure, regulation, and function. *Pharmacological Reviews*, **62**(3), 405–496.

Väisänen O 2008 *Multichannel EEG Methods to Improve the Spatial Rresolution of Cortical Potential Distribution and the Signal Quality of Deep Brain Sources*. Thesis, Tampere University of Technology, Finland.

Webster J 2009 *Medical Instrumentation: Application and Design*. John Wiley & Sons.

9

Microelectromechanical Systems

A book about sensors will not be complete without consideration of microelectromechanical systems (MEMS). However, due to the vastness of the subject and its multi-disciplinary nature, a single chapter may not do it justice. Devoting to the subject more than a single chapter, on the other hand, makes the book unnecessarily disproportional. Moreover, there are a number of books (for example, by Bao 2005; Korvink and Paul 2010; Liu 2010; Lyshevski 2002; Maluf and Williams 2004) and articles (for example by Fedder et al. 2008; Judy 2001; Kaajakari et al. 2009; Reina et al. 2008) that readers wishing to get an in-depth insight may refer to. With this cautionary remark, this chapter is intended to give an introduction to MEMS and the most important design, fabrication, and integration processes. It wil also introduce some of the most representative and state-of-the-art MEMS sensors.

MEMS represent, as their name suggests, the miniaturised version of macro systems that have mechanical and electrical components. However, this description is not adequate, since the systems may also have optical, thermal, chemical, magnetic, biological, and other components (Challa et al. 2008; Krüger et al. 2002; Ma and Kuo 2003; Staples et al. 2006; Waggoner and Craighead 2007). The most pervasive MEMS are sensors, actuators, and optical devices, (such as mirrors, waveguides, and splitters. These systems have a wide range of applications owing to their small size and additional desirable features that come along with small size. Some of the desirable features are the ease with which they can be mass produced and embedded into physical objects and processes as well as their low power consumption. The advantage of mass production is that MEMS are cheap, but the most desirable features are the ones that are directly associated with the performance improvement that can be achieved with physical scaling (Arlett et al. 2011; Cook-Chennault et al. 2008). In the subsequent sections these features and the design and fabrication processes will be addressed in some detail.

9.1　Miniaturisation and Scaling

Miniaturisation has important merits as well as some challenges. To understand both, it is useful to consider first the magnitude of scaling we are concerned with. Typically, a miniaturised system is more than 1000 times smaller than its macro counterpart. When a sensing or actuation material is reduced to this dimension, its physical and mechanical properties, among other things, undergo considerable change. Depending on the specific task of the sensor or actuator, this change may be desirable or undesirable.

Principles and Applications of Ubiquitous Sensing, First Edition. Waltenegus Dargie.
© 2017 John Wiley & Sons, Ltd. Published 2017 by John Wiley & Sons, Ltd.
Companion Website: www.wiley.com/go/dargie2017

Regardless of the task, the change also significantly affects the design, fabrication, and integration processes.

9.1.1 Physical Properties

In macro sensors and actuators, the size of the transduction elements (piezoelectric, photoelectric, piezoresistive, and so on) makes it plausible to assume that they are homogeneous even though there may be grain boundaries (defects) in the structures and microscopic fluctuations in the material composition that tend to reduce the electrical and thermal conductivity of the material. Considering the overall response of the sensing element as an ensemble average, the impact of defects, impurities and structural irregularities is insignificant. This is why the assumption of material homogeneity is plausible. When it comes to MEMS, however, the size of the sensing elements is comparable to individual grains that produce imperfections in the sensing material. In other words, local changes may affect the system's overall response. Therefore, it may not be plausible to assume homogeneity at this scale.

In addition, as size reduces, the surface-to-volume ratio increases and the surface characteristic of the sensing element becomes more important. To highlight this point, suppose we have a bulk sensing element that can be considered as a cube with linear dimension l. We wish to scale this element by s where $s >> l$. The surface-to-volume ratio of this element is $6l^2/l^3$, which equals $6/l$. The surface-to-volume ratio of the scaled element, on the the other hand, is $(6(l/s)^2/(l/s)^3$, which equals $(s(6/l))$. As can be seen, the surface-to-volume ratio of the scaled element increases by s. The relative importance of the surface characteristics also means there can be performance variations among MEMS that are mass produced; with one batch, or from batch to batch.

On the other hand, as the size of the sensing element approaches the size of the impurities, the total defect count decreases. When the sensing element has a simple mechanical construction, this means that the sensor is more reliable than its macro counterpart. This is, for example, true for cantilevers, which are essential components of cantilever-based accelerometers. Additional physical properties that can be affected by scaling are elastic modulus, Poisson's ratio, fracture stress, yield stress, residual in-plane stress, vertical stress gradient, and conductivity (Judy 2001).

9.1.2 Mechanical Properties

Consider Figure 9.1, in which a cantilever is fixed to a solid object. The inertial force it exerts when accelerated equals its mass multiplied by the acceleration. Since the mass is proportional to the volume, the inertial force is proportional to the volume of the cantilever. The stiffness (k) of the cantilever determines its elasticity and is a function of its dimensions:

$$k = \frac{wt^3E}{4l^3} \tag{9.1}$$

where w is the width, t is the thickness, l is the length, and E is the elastic modulus. If the linear dimensions are scaled down by a factor of s, the stiffness of the cantilever reduces by s as well, because :

$$k' = \frac{(w/s)(t/s)^3E}{4(L/s)^3} = \frac{1}{s}k \tag{9.2}$$

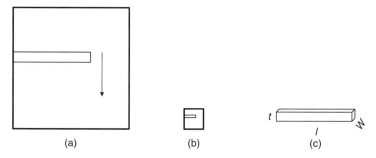

Figure 9.1 Scaling of macro structures to micro structures: (a) the macro structure; (b) the micro or miniaturised structure; (c) the three dimensional view of the cantilever which is the subject of scaling.

From Eq. (9.2), it can be seen that the mechanical strength of the cantilever reduces significantly more slowly (s) than the inertial force it generates (s^3). One of the merits of this feature is that MEMS can withstand considerably greater acceleration than their macro counterparts. Another way of looking at the same point is to consider the volume-to-surface ratio (which is proportional to s). Suppose we scale down a macro sensor from 1 cm to 10 μm; in this case surface forces will play a role 1000 times more important than inertial forces (recall that mass is proportional to volume). This means, for example, that gravity does not play an important role in MEMS except when the sensors are particularly designed to measure these effects.

9.1.3 Thermal Properties

For a thermal sensor or actuator, two of the most interesting properties are its thermal mass and the heat transfer rate (Huang et al. 2006; Janata 2010; Judy 2001; Pletcher et al. 2012). The thermal mass of an object is its ability to store heat. Mathematically, it is the product of its thermal capacity and volume. The thermal capacity of the object is the amount of heat added to or removed from it in response to a change in the resulting temperature. Both the thermal mass and the thermal capacity of an object are proportional to its volume. For temperature sensors, a large thermal mass is not desirable, as the temperature of the sensor increases or decreases more slowly than the surrounding temperature, resulting in a high response time (the thermal mass produces inertia against temperature fluctuations). In this regard, physical scaling has a desirable effect because as the linear dimensions of the sensor are reduced by s, its volume reduces by s^3. For other sensors, however, such as flow sensors, physical scaling has a negative effect.

Flow sensors typically rely on the ability of a flowing fluid to produce a thermal phenomenon on the sensing element by way of heat transfer. As the heat transfer rate is a function of the surface area of the sensor, reducing the physical dimensions by a factor of s reduces the surface area by a factor of s^2. Even so, the thermal mass reduces much faster than the heat transfer rate. Another important aspect of MEMS is that the sensing elements have very thin membranes, or they are deposited on a very thin membrane, which means that it is possible to permit heat conduction along paths of very high thermal resistance thereby achieving a very good thermal isolation.

9.1.4 Electrical and Magnetic Properties

The integrated circuit (IC) revolutions in the past four decades or so have shown that complex electrical circuits, particularly those consisting of resistors, capacitors, diodes, and transistors, scale down very well, maintaining predictable characteristics. The same can be said of their magnetic properties. But when it comes to actuation, there is a significant difference between macro and micro actuators. At the macro level all actuators relying on relays and motors have magnetic components. For motors, for instance, both the rotor and stator have magnetic components. The reason is that at the macro level it is possible to store a large amount of recoverable energy in the air gap between the stator and the motor, which is the prime cause of movement. The energy density in the air gap is a function of the applied magnetic field (correspondingly, the applied current in the case of electromagnetic energy) and the permeability of the material (Judy 2001):

$$U_\mu = \frac{B^2}{2\mu} \tag{9.3}$$

where B is the flux density. The maximum energy density is limited by the saturation magnetic flux density, which is typically in the order of 1 T (the corresponding energy density being $400\,000\ \mathrm{Jm^{-3}}$). Scaling magnetic actuators, however, has some challenging aspects including:

- a high resistive power loss
- the difficulty of integrating mechanical windings
- the potential of the magnetic field producing interfere and the need to carefully isolate it
- (for most practical magnetic materials) obtaining a uniform magnetisation region (homogeneity in magnetic domains).

At the macro level, the maximum energy density that can be stored in electrostatic actuators can be expressed as:

$$U_\epsilon = \frac{1}{2}\epsilon E^2 \tag{9.4}$$

where ϵ is the permittivity and E is the electrostatic field. This term is appreciably smaller as it is limited by the maximum electrostatic field that can be applied before an electrostatic breakdown occurs. This is approximately $3\ \mathrm{MVm^{-1}}$ and the corresponding energy density is $40\ \mathrm{Jm^{-3}}$, which is approximately $10\,000$ times smaller than the maximum energy density of magnetic actuators. At the micro level, however, electrostatic breakdown does not take place easily because as the air gap becomes smaller, fewer ionisation collisions take place, which means a large amount of voltage can be applied between the air gap, resulting in a large amount of electrostatic field.

9.1.5 Fluid Properties

In fluid mechanics, turbulence is a measure of a chaotic change in momentum diffusion, momentum convection, pressure, and velocity in a flow (Dixon and Hall 2013; Happel and Brenner 2012). A laminar flow, on the other hand, is one in which the kinetic energy of the flow diminishes smoothly as a result of molecular viscosity. The Reynolds number (Re) is used to quantify turbulence in a fluid. Mathematically, it is given as:

$$\mathrm{Re} = \frac{\rho V D}{\mu} \tag{9.5}$$

where ρ is the characteristic density, V is the characteristic velocity, and μ is the viscosity of the fluid; D is the characteristic diameter of the system through which the flow takes place (White 1999). Re < 2000 represents a laminar flow whereas Re > 4000 represents a turbulent flow. Turbulence is generally interspersed with laminar flow for $2000 \leq$ Re ≤ 4000. In macro systems, turbulence is easy to observe in a fluid flow whereas in micro systems, laminar flow dominates. This property can easily be seen in Eq. (9.5); as the linear dimensions of the system are scaled down by s, Re scales down by the same proportion, reducing it significantly. Consequently, mixing fluids and modelling fluid characteristics at a micro scale are difficult tasks.

9.1.6 Chemical Properties

There is a trade-off between the magnitude of scaling and the detection resolution of chemical sensors. As the linear dimensions of a chemical system scale down by a factor of s, the amount of samples that can be collected and hence the total number of molecules that can be detected in a fixed concentration reduce proportionally.

9.1.7 Optical Properties

Optical systems scale down without a considerable penalty on system performance. Indeed there are a substantial number of MEMS optical devices containing LEDs, laser diodes, photodiodes, phototransistors, mirrors, integrated waveguides, diffraction gratings, optical splitters, and so on. Understandably, for most practical purposes the scaling factor is limited by the wavelength of the light one is interested in. For visible light, this is, for example, between 650 and 475 nm.

9.2 Technology

The appropriate materials and technology with which MEMS are developed depend on the specific tasks of the sensors and actuators. For those systems that have mechanical parts, two of the vital features are:

- excellent elasticity or the ability of the material to deform under a constant load
- low hysteresis: the history of the applied input on the material should not influence its present response to a given input.

Other properties such as stable operations at a large range of temperature variations, small thermal expansion coefficients, high thermal conductivity, high electrical conductivity, mass production, and low prices, are also important properties and should be taken into consideration when choosing materials.

Even though no single material fulfils all the requirements equally well, silicon is by far the most extensively used material for producing MEMS. It fulfils most of the requirements listed above and simplifies the microfabrication process due to the vast amount of experience learned and the readily available micromachining technologies that can be adopted from the semiconductor industry, as well as the ease with which mechanical and electrical components can be integrated into a single, monolithic chip.

9.2.1 Growth and Deposition

Pure silicon crystals are rarely used to develop sensing and actuation elements. In fact, silicon is often used as a substrate upon which the desired material is developed. Additional materials inside or on the substrate are deposited or grown to make the desired structures or to enhance or pattern the electromechanical, magnetic, thermal, or other physical properties of the substrate. One of the simplest methods is depositing a thin silicon dioxide layer on the surface of a silicon wafer (Chabal 2012; Liu 2010; Tilli et al. 2015). When a silicon wafer is placed inside a furnace at temperatures between 800 and 1200°C, its surface interacts with the ambient oxygen (dry oxidation) producing a thin silicon dioxide layer on the wafer:

$$Si + O_2 \longrightarrow SiO_2$$

Alternatively, a water vapour (wet oxidation) can be used to produce a similar effect:

$$Si + 2\, H_2O \longrightarrow SiO_2 + 2\, H_2(g)$$

But materials can also be implanted into the silicon wafer. One of the techniques used to make room for implantation is sputtering. It is a technique that uses energetic particles to bombard the wafer and eject atoms from it. The kinetic energy of the incoming particles is much greater than 1 eV. Doping is one of the mechanisms used to implant ions into the substrate. Doping can be achieved in various ways. For example, ions with a high energy can hit the silicon wafer and penetrate in it to a certain depth, precisely controlled by the energy of the ions. Implantation energies in the order of thousands of electronvolts are required to reach depths of a few hundred nanometres. A further annealing process may take place to spread the doping more homogeneously over a thicker layer. Another method of doping is indiffusion at elevated temperatures. This is typically accomplished by putting the silicon wafers side by side with boron or phosphorous wafers in a tube at a high temperature.

9.2.2 Photolithography

Once the silicon substrate is pre-processed, the next stage is developing the desired structure. There are several ways of fabricating or replicating structures using elastomeric stamps, molds, and conformable photomask, but the most important technique is photolithography. It is a process whereby ultraviolet light is applied on either a positive or a negative photoresist film to develop the desired structure. A positive photoresist is soluble when exposed to light (the unexposed part is insoluble) whereas a negative photoresist becomes insoluble when exposed to light (the unexposed part is soluble). The most widely used photoresist for MEMS design is an organosilicon polymer known as polydimethylsiloxane (PDMS), due to its many desirable features; for example, it is optically clear, chemically inert, non-toxic, and non-flammable (Judy 2001). A mask containing the desired structure is placed on or near the photoresist or at a projecting position and a light is applied to it, after which a wet-etching or a dry-etching takes place. Figure 9.2 illustrates the application of positive and negative photoresist.

Figure 9.3 provides a more detailed illustration of the photolithography process. In (a) the silicon wafer is displayed. In (b), a thermal oxidation takes places to wrap the silicon wafer with silicon oxide. In the final analysis, it is this layer we wish to shape or mold in order to produce the sensing or actuation element. In (c) a thin photoresist material is

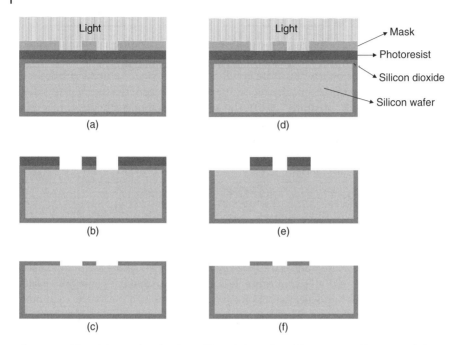

Figure 9.2 Photolithography. The desired layer (silicon dioxide) is patterned by using a light, a mask, a photoresist, and an etching process. On the left-hand side, a positive photoresist is used; on the right, a negative photoresist is used. In (a) and (d), an ultraviolet light shines on a mask, which blocks the light from reaching the photoresist in certain areas whilst permitting it to reach others. In (b) and (e) the photoresist and the oxidised layer are patterned. In (c) and (f), a wet-etching process removes the photoresist.

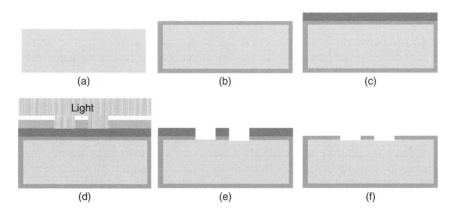

Figure 9.3 Illustration of photolithography in micromachining: (a) a silicon substrate (wafer); (b) the surface of the substrate is coated with silicon dioxide; (c) the photoresist is deposited on top of the silicon dioxide layer. In (d) and (e) ultraviolet light is applied to the photoresist. A mask is used to pattern the layer we wish to micromachine. (f) After the silicon dioxide is patterned, the photoresist is removed by wet-etching.

deposited above the silicon oxide layer. When this layer is exposed to light, it becomes soluble (assuming it is a positive photoresist). In (d), a mask bearing the structure we wish to impress upon the silicon oxide layer is placed on the photoresist and a light is illuminated on the mask. In (e), the mask is removed and the portion of the photoresist which is exposed to light is chemically removed (wet-etching). Finally, in (f), a further etching process takes place to remove the part of the silicon oxide which is exposed to light. Apparently, the silicon oxide layer now bears the shape of the mask.

9.2.3 Etching

Etching is a process by which layers from the surface of a substrate are chemically removed. There are several etching techniques, but the most frequently used are wet-etching and dry-etching. In wet-etching, liquid chemicals are applied to remove a layer; in dry-etching, the layer is bombarded with highly accelerated ions or a chemical reaction takes place with the layer followed by the removal of the reaction products. Etching can be isotropic (etch-speed being equal in all directions) or anisotropic (etching speed depends on the crystallographic directions). The former produces rounded structures whereas the latter produces sharp edges and corners. Figurer 9.4 illustrates the use of wet- and dry-etching processes.

9.3 Micromachining

Micromachining is the process of developing multiple-patterned materials as a unified sensing or actuation structure. Broadly speaking, micromachining can be categorised

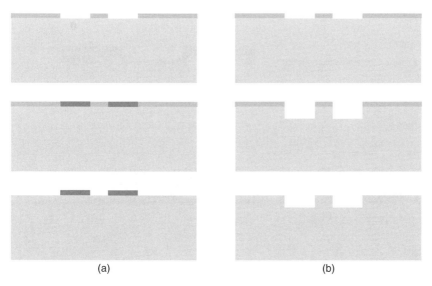

(a) (b)

Figure 9.4 In micromachining, photolithography is often followed by deposition and etching. In (a), first the wafer is patterned using photolithography (top), the desired material is deposited (middle), and the photoresist is removed using wet-etching (bottom). In (b), after the wafer is patterned (top), dry-etching takes place to remove a part of the wafer (middle) in order to make the desired shape, followed by the removal of the photoresist using wet-etching (bottom).

into surface micromachining and bulk micromachining. In surface micromachining the structure is developed on the substrate and the substrate itself is, by and large, intact. In bulk micromachining, the structure is developed in the substrate itself. The appropriate choice of micromachining depends on the complexity of the desired structure. Surface micromachining is a relatively simple process, but the structure is relatively simple as well. On the other hand, complex structures can be developed with bulk micromachining but the process is elaborate.

9.3.1 Surface Micromachining

Surface micromachining consists of depositing, patterning, and etching thin films on a substrate. The thickness of these films ranges from a few micro meters to 100 μm. To facilitate the process, two important layers on top of the substrate are deposited and processed. The first is the structural layer, which develops into the desired structure (mostly mechanical). The second layer, which facilitates the structure but in the end is undesired in itself, is called the sacrificial layer. Figure 9.5 illustrates the surface micromachining process, which consists of four essential steps.

a) First, the sacrificial layer is deposited on the substrate.
b) This layer is shaped using photolithography and wet-etching to overlay the structured layer.
c) The structure layer is deposited on top of the sacrificial layer and, possibly, on the substrate. Once again photolithography and wet-etching take place to develop and shape the desired structure.
d) Finally, the sacrificial layer is removed using wet-etching, thereby freeing the structural layer.

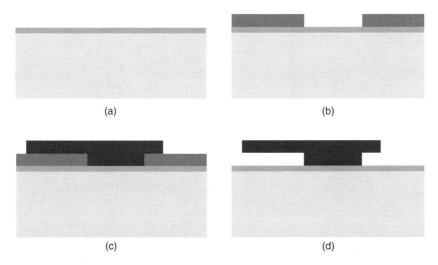

Figure 9.5 Surface micromachining: (a) and (b) the substrate is coated with silicon dioxide, a sacrificial layer is deposited on top of the silicon dioxide, and the sacrificial layer is patterned with photolithography; (c) the structural layer is deposited on top of the sacrificial layer and the silicon dioxide; (d) the sacrificial layer is removed by etching.

Figure 9.6 A one-dimensional metal ring gyroscope produced by surface micromachining. Courtesy of M.W. Putty and K. Najafi, (*Proc. of IEEE Solid State Sensor and Actuator Workshop,* 1994).

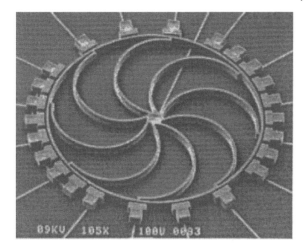

Different combinations of structural and sacrificial materials are possible, for example silicon nitride and polysilicon, gold and titanium, nickel and titanium, polyimide and aluminium, tungsten and silicon-dioxide, and aluminum and polymer (Dornfeld et al. 2006; Gattass and Mazur 2008; Judy 2001). Nevertheless, the most widely applied surface micromachining technology employs polysilicon as the structural material and either silicon dioxide or phosphor silicate glass as the sacrificial layer. Figure 9.6 shows a surface-micromachined gyroscope.

9.3.2 Bulk Micromachining

Bulk micromachining is a technique for producing three-dimensional structures by selectively removing layers from the silicon substrate. The final aim of bulk micromachining is to release or undercut the structures, so that they can freely move with respect to the substrate (which serves as a reference frame). The complexity of bulk micromachining depends on the number or layer of masks required. The structures most frequently produced with bulk micromachining are beams and suspensions, membranes, cavities, nozzles, and mesas. Figures 9.7 and 9.8 illustrate different types of structure that have been fabricated with bulk micromachining.

The most important steps in bulk micromachining are similar to those in surface micromachining, their only difference being the type and magnitude of etching required on the substrate. Bulk micromachining often uses deep and reactive etching. Figure 9.9 summarises some of the important steps.

a) Silicon dioxide is deposited and the photoresist is spin-coated on the substrate.
b) A mask is placed on the photoresist and the photoresist is exposed to an ultraviolet light.
c) The photoresist is patterned by chemically removing the part of the photoresist which is not exposed to light (wet-etching). The photoresist is used as a mask to pattern the silicon dioxide with a high frequency signal followed by doping the surface of the silicon substrate with boron (using the silicon dioxide as a mask).
d) The silicon dioxide is removed from the substrate and a deep reactive etching on the silicon substrate takes place to develop a free-moving boron beam.

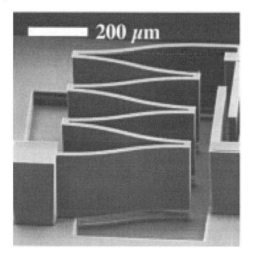

Figure 9.7 A single-crystal silicon leaf spring is etched through a top wafer which is bonded to an underlying silicon substrate with pre-etched pits. Courtesy of Lucas NovaSensor, Fremont, CA.

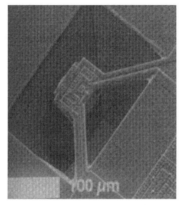

Figure 9.8 Suspension and bridge microstructure fabricated using bulk micromachining. Courtesy of E.H. Klaassen, R.J. Reay, and G.T.A. Kovacs, (*Eurosensors IX*, 1995).

Two of the most important technologies for bulk micromachining are reactive ion etching (RIE) and micromolding.

9.3.2.1 Reactive Ion Etching

Reactive ion etching (RIE) is a dry-etching process employed to cut deep structures into a substrate. Often it involves both physical and chemical etching. The physical etching is useful for anisotropic etching whereas the chemical etching is used for an isotropic etching. The physical etching involves the bombardment of the substrate with accelerated ions, which eject a chain of atoms from the substrate. This step is then followed by a chemical etching to cut deep into the substrate.

Figure 9.10 illustrates the essential components of RIE. Two electrodes forming an anode and a cathode are connected to an RF signal generator. The silicon wafer is a part of the cathode. Between the two electrodes is diffused a plasma. Some of the free electrons leaving the cathode en route to the anode will be absorbed by the plasma (which generates ions and activated neutrals), as a result of which various chemical reactions

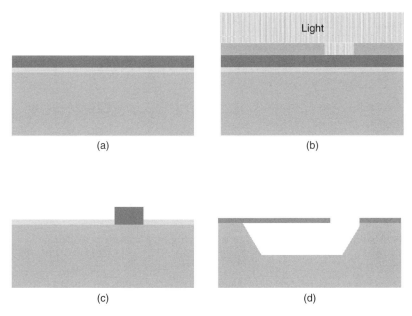

Figure 9.9 The essential steps of bulk micromachining. (a) and (b) Similar to surface micromachining, the substrate is coated with silicon dioxide, a photoresist layer is deposited on top of the silicon dioxide, and the photoresist is patterned with photolithography. (c) The silicon dioxide layer is patterned with high frequency signal and afterwards the photoresist is removed with wet-etching. (d) The substrate is dry-etched and a thin cantilever is released.

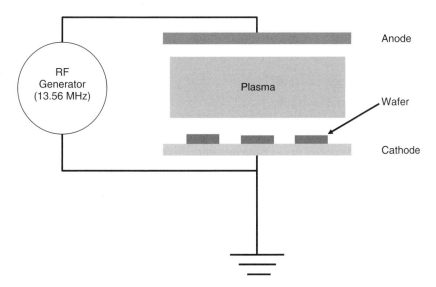

Figure 9.10 Plasma etching as a form of reactive ion etching. The RF generator produces a high-frequency signal (13.56 MHz) to produce plasma (ions) from a gaseous material, which is then absorbed by the substrate. As a result, a chemical reaction takes place, removing a part of the substrate.

take place. If, for example, the plasma is tetrafluormethan (CF_4), the following chemical reactions take place:

Dissociation

$$CF_4 + e^- \longrightarrow CF_3 + F + e^-$$

Dissociative ionisation

$$CF_4 + e^- \longrightarrow CF_3^+ + F + 2e^-$$

Excitation

$$CF_4 + e^- \longrightarrow CF_4^* + e^-$$

where F^* represents fluorine atoms with electrons having high vapour pressure.

Ionisation

$$CF_4 + e^- \longrightarrow CF_4^+ + 2\ e^-$$

Recombination

$$CF_3^+ + F + e^- \longrightarrow CF_4$$
$$F + F \longrightarrow F_2$$

Furthermore, the by-products chemically interact with the silicon substrate and chemical desorption takes place, thereby removing a part of the substrate:

$$CF_4 \longrightarrow F^* + CF_3$$
$$CF_4 + e^- \rightleftharpoons CF_3^+ + F^* + 2e^-$$
$$Si + 4F^* \longrightarrow SiF_4 \uparrow$$

By controlling the diffusion, absorption, acceleration, pressure, and desorption of ions, it is possible to guide and control the etching process. There are different approaches to ensure that etching progresses to the desired depth and in the desired direction (selectivity and anisotropy). One of them is an iteration process involving directional bombardment of the substrate atoms followed by the deposition of side-wall inhibitors followed by isotropic chemical etching. Figure 9.11 illustrates the deposition of chemically inert species as side-wall inhibitors to protect a side-wall from chemical etching (reaction).

As an alternative to the use of RF, photons can be used to facilitate chemical etching, in which case, a laser beam is used to create and diffuse into the silicon substrate highly reactive radicals, which then chemically interact with the substrate. Because of the narrowness of the laser beam, it is possible to undertake precise anisotropic chemical etching. In Figure 9.12, highly reactive chlorine radicals are generated by a laser beam and then interact with the silicon atoms, thereby selectively and directionally removing a part of the substrate.

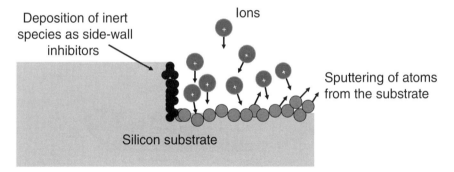

Figure 9.11 Inhibitors are deposited on a side-wall to prevent isotropic etching.

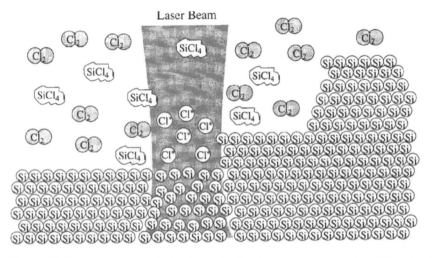

Figure 9.12 Laser-assisted chemical etching with Cl_2. Here, photons are used to diffuse chlorine ions into the silicon substrate and to facilitate a chemical reaction. Courtesy of Kovacs et al. (1998); reconstructed after Bloomstein and Ehrlich (1991).

9.3.2.2 Micromolding

Micromolding is a fabrication process that combines RIE and conformal depositions to fabricate micro structures. The process begins by cutting deep structures into the substrate (the sacrificial layer) using bulk micromachining. Then, a sequential conformal deposition takes place inside the structure. Lastly, the sacrificial layer is removed from the micromold. These steps are illustrated in Figure 9.13.

The most widely used deposition methods are low pressure chemical vapour deposition (LPCVD) and plasma-enhanced chemical vapour deposition (PECVD). Essentially, both approaches are similar to plasma etching, except a film is deposited rather than being removed. The essential steps are the following:

1) Gaseous (plasma) species are generated and transported into the substrate.
2) The active species are absorbed by the substrate.
3) Chemical reaction takes place between the species and the surface of the substrate.
4) Unwanted by-products and leftover reactants are removed.

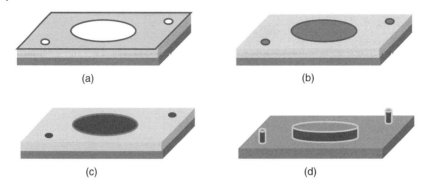

Figure 9.13 Microfabrication with LIGA: (a) and (b) a deep plating mold is structured using X-ray lithography; (c) material is deposited into the mold by electroplating; (d) the photoresist is detached from the micromold structure.

LPCVD relies on thermal decomposition for the diffusion and absorption of active species whereas PECVD relies on an RF signal. Hence for LPCVD a wafer surface temperature ranging from 600 to 800°C is required, while the temperature requirement for PECVD is much less. LPCVD requires a simpler set up, the procedure is reliable (reproducible), and a more uniform deposition of thickness can be achieved. On the other hand, PECVD is much faster, cleaner, and deposition can achieve higher film densities.

9.3.2.3 Non-silicon Micromolding

As MEMS find applications in a wide range of areas, including the development of biomedical sensors, non-silicon microfabrication is becoming ubiquitous. The driving motivations are, among others, enhanced aspect ratios, inexpensiveness of the fabrication process, reproducibility, fast production times due to process parallelism, and bio-compatibility (Bhushan 2007; Judy 2001; Liu 2010). The most significant developments in this regard are LIGA and plastic molding with PDMS.

LIGA stands for three German words: Lithographie (lithography), Galvanoformung (electroplating), and Abformung (molding). Needless to say, the process involves these three steps. However, unlike silicon-based photolithography, LIGA relies on intensive X-ray lithography instead of UV light to produce deep side-walls. Hence the photoresist material is both different and considerably thicker. The primary advantage of LIGA is its ability to make large aspect ratio structures (a thickness of up to 1000 μm and a width of only several microns is achievable).

The most important step during the lithography process is the preparation of the mask, which consists of three parts:

- absorber
- carrier membrane
- supporting frame.

The absorber, which is usually a thick material with a high atomic weight, prevents the X-ray from reaching the photoresist. The carrier membrane supports the absorber but also permits the X-rays to pass through. Therefore, it should be of material with a low atomic weight and as thin as possible. Moreover, the residual stress of the absorber should be as small as possible to prevent the mask pattern from distortion. Beryllium

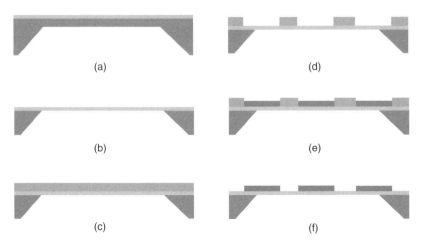

Figure 9.14 Fabrication of an X-ray mask for LIGA: (a) sputtering is used to prepare the membrane (it has to be thin); (b) the back of the silicon supporting frame is etched to form the membrane; (c) and (d) a photoresist is deposited on the membrane and patterned using photolithography; (e) and (f) the absorber layer is formed using electroplating and the photoresist is removed.

and titanium are the most frequently used carrier membranes, while gold is the most frequently used absorber. Figure 9.14 shows the most important steps in developing a LIGA mask.

Apart from the choice of the photoresist and the use of X-rays for etching, the essential steps in X-ray lithography are similar to that of photolithography. The most popular LIGA photoresist is polymethyl methacrylate (PMMA), a plastic polymer coated on the substrate by casting or glueing rather than spinning, since the photoresist is thick. The next step after X-ray lithography depends on the nature of the structures to be developed. If the structures require micromolding, then micromolding takes place. This is typically done by injecting the structure to be developed into a mold. First the material is melted and then injected into the mold by applying pressure to it. Alternatively, a mold is applied to a molten polymer in hot embossing. Then the material is cooled down and solidified. Whether or not micromolding is carried out, a LIGA process is completed with electroplating, since the structures fabricated by X-ray lithography are plastic in nature. With electroplating, their surface is made metallic. Nickel is the most commonly used material for electroplating. Figures 9.15 and 9.16 illustrate the essential steps of LIGA.

With the inclusion of a sacrificial layer and photolithography in the process, movable parts can be realised with LIGA as well. Hence the process begins by depositing a metallic layer on the substrate and coating the sacrificial layer on the metallic layer. Then the sacrificial layer is patterned with photolithography. Afterwards, a thick layer of PMMA is cast on top of both the metallic and the sacrificial layers. After this, X-ray lithography takes place, with the sacrificial layer finally removed by etching. Figure 9.17 illustrates the LIGA technique combined with surface micromachining.

9.3.2.4 Plastic Micromolding

Biomedical sensors have some specific requirements. For example, they have to be inexpensive and mass-produced because they are, more often than not, use-and-throw.

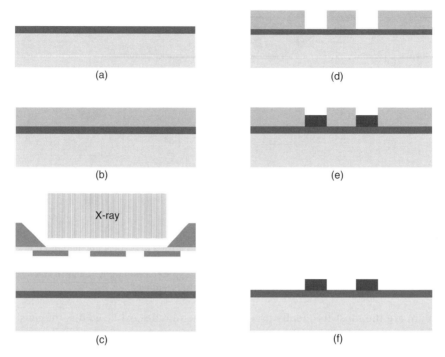

Figure 9.15 Microfabrication with LIGA: (a) The substrate is sputtered and a carrier membrane is deposited on top; (b) a thick photoresist is cast upon the carrier membrane; (c) and (d) the photoresist is patterned using X-ray lithography; (e) the microstructure is created by electroplating; (f) the photoresist is removed. Notice that, unlike silicon-based micromachining (lithography), a LIGA photoresist is considerably thicker than in silicon-based micromachining, where it is a thin film. Consequently, LIGA can rely on X-rays, which generate a higher kinetic energy than UV light, and it can pattern the thicker photoresist, producing deeper and more complex structures than silicon-based surface micromachining.

Besides this, the sensors should also be biocompatible and have to fit into fine biochemical milieus without disturbing or inhibiting biochemical processes (Cheng et al. 2010; Giboz et al. 2007; Mabeck and Malliaras 2006; Madou et al. 2006). PDMS, a transparent elastomer that can be poured over a mold, fulfils many of these requirements and it is the most frequently used material for plastic molding. Firstly, many inexpensive parts can be fabricated from a single PDMS mold. Secondly, it faithfully reproduces sub-micron features in a mold. Thirdly, owing to its transparent nature, cells and tissues can easily be imaged through it (Abedinpour et al. 2007; Harrison et al. 2007; Hunt and Armani 2010; Judy 2001; Nikolou and Malliaras 2008; Patolsky et al. 2006; Tian et al. 2007; Waggoner and Craighead 2007).

9.4 System Integration

Like any other sensors, MEMS require conditioning circuits to amplify, filter, and pre-process the output signals. Due to MEMS' size as well as the size of the overall device (sensor or actuator) we wish to produce, ICs are the most feasible way of achieving this.

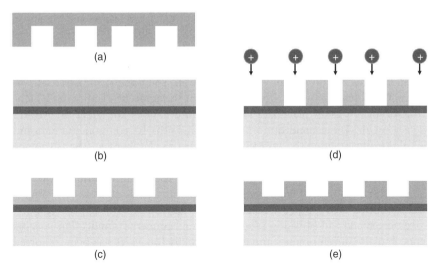

Figure 9.16 Hot embossing with LIGA: (a) a mold; (b) a plastic polymer is softened by raising its temperature just above its glass transition temperature; (c) the mold is impressed on the plastic polymer; (d) the undesirable part of the polymer is removed by etching; (e) the desired structure is formed by electroplating.

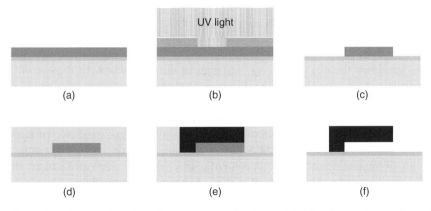

Figure 9.17 The creation of moving structures using S-Liga: (a)–(c) a photoresist is patterned using surface micromachining; (d) a thick photoresist is deposited on top of the sacrificial layer; (e) the thick photoresist is patterned using X-ray lithography and material is deposited into the mold by electroplating; (f) the photoresist as well as the sacrificial layer is removed and the moveable structure is freed.

There are, however, two challenges. The first is the difference in the temperature ceilings that are imposed by the materials used for processing. The temperature ceiling can be high (well above 800°C) for MEMS materials, whereas it is typically low for ICs (below 400°C). The second challenge is the different types of substrates MEMS and ICs require. MEMS typically require a substrate with a large surface topology, while ICs require a highly planarised (smooth) substrate.

The easiest way to integrate MEMS with ICs is to produce them separately and assemble them at a later stage. This, however, is not optimal. One of the shortcomings of this approach is the presence of parasitic (undesirable capacitive) effects at connecting points, which reduces performance. Monolithic integration eliminates parasitic effects but the process cost is high. There can be three different monolithic integrations. The first option is to produce the MEMS structures first and then monolithically integrate IC fabrication. But this requires an intermediate step in which the IC substrate is planarised before IC fabrication begins. The advantage of this approach is that the IC fabrication, which typically takes place at a low temperature, does not affect the MEMS structures very much. The second option is to produce the IC components first and then to monolithically integrate the MEMS components at a latter stage. The advantage of this approach is that the integration process is simpler, because IC fabrication, which requires a more careful dopant distribution, can be carried out in isolation. However, the latter stage, which requires a high processing temperature, may affect the IC components. The third option is to interleave IC and MEMS fabrication. This approach shortens the processing speed and can address specific design concerns (application-specific requirements) but the development time is considerably longer. It has been reported that Analog Devices, Inc. uses this approach to produce inertial sensors (Brand 2006; Fedder et al. 2008; Judy 2001).

9.5 Micromechanical Sensors

Sensing with MEMS can take place electrically, optically, or magnetically. The most frequent electrical sensing principles are piezoresistive, capacitive, and piezoelectric. In the former a physical phenomenon, such as pressure or force, changes the resistance of a MEMS structure thereby producing a change in output voltage. In the second, the mechanical phenomenon changes the distance of separation between the capacitive plates of a MEMS structure. The effect of this displacement can be detected by observing the change in output capacitance, voltage, or resonance frequency. In piezoelectric sensing, a mechanical stress on a MEMS structure results in a change in the electric current flowing inside a circuit involving the MEMS structure. Similarly, in optical sensing, a physical phenomenon (such as a displacement) can change the time of flight of an optical signal inside a micro fibre (Peng et al. 2007; Scheuer and Yariv 2006).

9.5.1 Pressure and Force Sensors

In Section 4.1 we saw that the resistance of a material changes when there is a change in its:

- physical dimensions or
- resistivity.

This was expressed by Eqs. (4.7) and (4.8). When these changes are induced by an applied pressure or force, the material is said to be a piezoresistor. When the change is mainly in the resistivity of the material and little or no emphasis is given to the change in dimension, then the material is specifically said to be piezoresistive. Whereas the resistance of a strain gauge mainly changes as a result of a change in its physical dimensions, the resistance of a piezoresistive material mainly changes due to a change in its resistivity. Alternatively, pressure or force can induce a piezoelectric effect by displacing charge

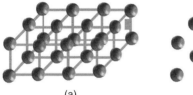 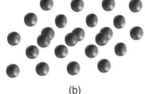

(a)   (b)

Figure 9.18 The mean atomic spacing in a crystal lattice influences the mobility of charge carrier thereby affecting its resistance. The atomic spacing is subject to change when the crystal lattice experiences a physical strain: (a) the crystal lattice with lines depicting the covalent bonding between atoms; (b) lines depicting the covalent bonding are removed to illustrate atomic spacing.

carriers when a structure undergoes deformation. This displacement in turn induces a flow of current in a predetermined direction. A silicon crystal is an excellent piezoresistive and piezoelectric material. When a force or pressure is applied to it, the crystal undergoes a deformation, both physical and chemical, and in consequence it is possible to change its resistivity or to liberate charge carriers and produce an electric current. The magnitude of the cross-coupling of mechanical and electrical energy depends on many factors, such as the type and orientation of the silicon crystal, the type and concentration of dopant, and the temperature of the substrate.

At the micro level, the deformation of a silicon crystal lattice due to an applied force or pressure affects the mobility of charge carriers. The exact amount of mobility depends, however, on the mean free time between carrier collision events and the effective mass of a carrier. These quantities, in turn, are related to the mean atomic spacing in the crystal lattice (as shown in Figure 9.18), which is subject to change when the crystal lattice experiences strain.

In order to understand the macro-level phenomena of piezoelectric and piezoresistive effects, it is useful to first examine the effects of force on electromechanical structures. When a force is applied to a three-dimensional structure, the structure undergoes deformation in three dimensions. This is true even if the force is one-dimensional. Hence the deformation results in a three-dimensional strain. The strain normal to the applied force is called the normal strain whereas the strains lateral to the applied force are called shear strains. Similarly, a stress produced at a surface as a result of a normal force is denoted by σ, whilst a stress produced at a surface as a result of a lateral force is depicted by τ. Recall from Section 4.1 that stress (σ) and strain (ϵ), measures of deformation, are related to one other. Figure 9.19 illustrates the three components of the strain of a hexagonal structure as a result of an applied force (not depicted in the figure). Figure 9.20 provides a more complete summary relating force, cross-sectional area, and stress in a semiconductor crystal. For example, the three components of a stress that can be measured at the surface area of the cube perpendicular to the x-axis are the following:

$$\sigma_{xx} = \frac{\Delta F_x}{\Delta A}$$

$$\tau_{xz} = \frac{\Delta F_z}{\Delta A}$$

$$\tau_{xy} = \frac{\Delta F_y}{\Delta A}$$

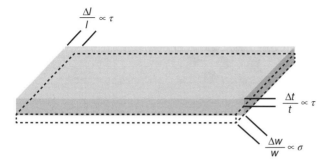

Figure 9.19 A force applied to one of the six faces of the hexagonal structure produces a three-dimensional deformation. For example, a force applied to the surface area which is perpendicular to the z-axis produces a normal strain in the z-direction and shear strain in the x- and y-directions.

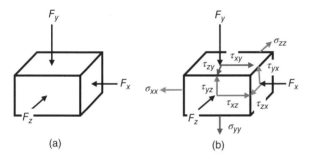

(a)   (b)

Figure 9.20 Normal and lateral forces applied to a semiconductor crystal produce normal and shear stress: (a) forces applied normal to a cross-sectional area; (b) resulting normal and lateral stress components.

9.5.1.1 Piezoelectric Effect

Consider Figure 9.21, in which a piezoelectric material is sandwiched between two capacitive plates. When an external load is applied to the structure (sensor), the piezoelectric material generates charges in response. The equation that quantifies the cross-coupling from mechanical to electrical energy is given as:

$$D_i = e_{ij}S_j + \epsilon_0 \epsilon_{ik}^S E_k \tag{9.6}$$

where D is the electric charge displacement, S is the mechanical strain, and E is the electric field vector. A charge displacement can take place in all directions. If we designate the x, y, z directions as $i = 1, 2, 3$, the possible charge displacements in these directions are results of the six strain components: two in the x-direction (positive and negative), two in the y and two in the z-direction. We designate these components as $j = 1, 2, 3, 4, 5, 6$. Thus the displacement in a single direction is a result of six strain components. The contribution of each component depends on the cross-coupling coefficient between direction i and strain component j. The coefficient is denoted by e_{ij}. Likewise, ϵ_0 is the permittivity of free space and ϵ_{ik}^S is the relative permittivity of the piezoelectric material under constant strain.

Equation (9.6) has two essential components, one coming from the mechanical strain and another coming from an external electric field that can be confined between the

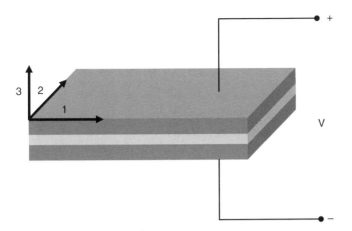

Figure 9.21 A piezoelectric sensor sandwiched between two electrodes. If the sensor is biased by a voltage V, then the current flowing through the capacitive structure has two components. The first component is a result of the electric field set up by the potential different introduced by the biasing voltage. The second component is a piezoelectric current generated by the applied pressure or force. The magnitude of the resulting current is subject to the magnitude and direction of the electric field as well as the applied stress.

capacitive plates. These two components are shown in Figure 9.22. In the absence of an external electrical field, the charges collected by the capacitive plates (in the direction of 3) – due to in-plane stress or strain induced by the substrate to which one of the plates is attached – can be simplified by the following equation:

$$D_3 = \frac{Q_3}{A_3} = e_{31}(S_1 + S_2) \tag{9.7}$$

where Q_3 is the electric charge collected by the capacitive plate and A_3 is the cross-sectional area of the capacitive plates (assuming the surface areas of both plates are equal).

Example 9.1 In order to relate the voltage that can be measured across the capacitive plates and the strain applied to the piezoelectric material, it is useful to approximate Figure 9.21 with an electrical equivalent circuit. Determine the structure of the circuit and derive an expression for the terminal voltage.

Figure 9.23 shows an example equivalent circuit for the piezoelectric sensor. The resistor in parallel with the capacitor represents the internal resistance of the piezoelectric material while the voltage source V_n represents the voltage due to white thermal noise, which equals:

$$V_n = \sqrt{4k_B TR} \tag{9.8}$$

where k_B is Boltzmann's constant and T is the temperature of the piezoelectric material in Kelvin. To calculate the output voltage in frequency domain, we can apply the superposition theorem, as shown in Figure 9.24. In (a), we short circuit V_n and in (b), we open circuit the charge source. In (a), the output voltage is the voltage drop across the parallel

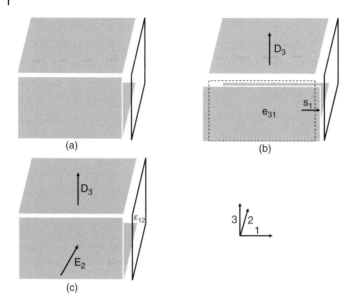

Figure 9.22 A three-dimensional representation of the relationship between charge displacement due to piezoelectric effect, strain, and electric field: (a) the three-dimensional view of the planes of interest; (b) and (c) the electric charge displacement is perpendicular to both the external electric field and the strain.

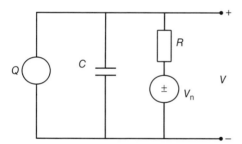

Figure 9.23 An equivalent electrical circuit for the piezoelectric sensor sandwiched between two electrodes. In the figure, Q represents the piezoelectric charge generated by the applied pressure or force (the piezoelectric effect) whilst V represents the applied biasing voltage. R and C represent the capacitance and internal resistance of the piezoelectric structure: C is in parallel with both the biasing voltage and the piezoelectric charge generator and R is in series with the biasing voltage.

RC circuit, which is the same as the current multiplied by the parallel impedance:

$$V' = \frac{R \cdot (1/sC)}{R + (1/sC)} \cdot I = \frac{IR}{1 + sRC} \tag{9.9}$$

To calculate the output voltage in (b), we should first calculate the current that flows in the circuit. The output voltage is the same as the voltage across the capacitive element. With the current source open circuited, the resistor and the capacitor are connected in

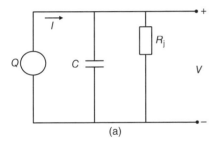 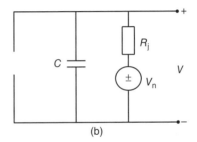

Figure 9.24 Using the superposition principle to calculate the output voltage of a piezoelectric sensor: (a) the white thermal noise voltage source is short circuited; (b) the current source is open circuited.

series. The current that flows in the circuit is:

$$I_c = \frac{V_n}{R + 1/sC} = \frac{sCV_n}{1 + sRC} \tag{9.10}$$

Likewise, the voltage drop across the capacitor is:

$$V'' = \frac{sCV_n}{1 + sRC} \cdot \frac{1}{sC} = \frac{V_n}{1 + sRC} \tag{9.11}$$

Now using the superposition theorem, the output voltage is the summation of the two superposition voltages: $V = V' + V''$, which is:

$$V = \frac{IR + V_n}{sRC + 1} \tag{9.12}$$

The current generated as a result of the charges collected by the capacitive plates equals $I = sQ_3 = j\omega Q_3$. Hence, if we substitute Eq. (9.7) into Eq. (9.12), we get:

$$V = \frac{sRA_3 e_{31}}{sRC + 1}(S_1 + S_2) + \frac{1}{sRC + 1}V_n \tag{9.13}$$

Note that, at high frequency, $\omega \gg 0$, the second term in Eq. (9.13) vanishes, suggesting that the piezoelectric sensor acts as a high-pass filter. Therefore, it is desirable to use piezoelectric sensors for dynamic signal sensing as opposed to static sensing. The capacitance of the capacitor in Figure 9.23 can be determined by taking ϵ_{33} (the relative permittivity of the piezoelectric material in the y-direction) and the dimensions of the piezoelectric material into consideration:

$$C = \epsilon_0 \epsilon_{33} \frac{A_3}{t}$$

Therefore, at a high frequency, the voltage across the two capacitive plates can be approximated by:

$$V = \frac{e_{31}t}{\epsilon_0 \epsilon_{33}^S}(S_1 + S_2) \tag{9.14}$$

From Eq. (9.14) it is possible to conclude that the output voltage is a sum of the in-plane strain times a constant. The constant, in turn, is predetermined by the selection and dimensions of the piezoelectric material. Note that the sensitivity of the piezoelectric sensor can be improved by selecting a thicker material.

Example 9.2 We wish to establish a quantitative relationship between a mechanical force that exerts on the piezoelectric sensor of Figure 9.21 (the measurand) and the capacitance of the sensor.

The electrostatic energy that is stored by the capacitor when a biasing DC voltage is applied across its plates (the electrodes) is given as:

$$E = \frac{1}{2}CV^2 \tag{9.15}$$

But the capacitance of the capacitor is given as:

$$\epsilon_0 \epsilon_r \frac{A}{d} \tag{9.16}$$

If we assume that the permittivity of the capacitor does not change for a small change in d, then the electrostatic energy changes when the capacitor is compressed by the applied force. The change in the electrostatic energy with respect to the change in the separating distance of the capacitor describes the applied force:

$$F = \frac{\partial E}{\partial d} = \frac{\partial}{\partial d}\left(\frac{1}{2}\epsilon_0 \epsilon_r \frac{A}{d}V^2\right) = -\frac{1}{2}\frac{CV^2}{d} \tag{9.17}$$

9.5.1.2 Piezoresistance

In piezoresistance sensing, the main goal is to change the resistance of a structure as a result of an applied pressure or force. Conventionally, the resistance of a resistor is measured longitudinally; that is, along the direction of the current flow. In a piezoresistor, particularly in a silicon crystal, the change in resistance depends on:

- the direction of current flow
- the orientation of the crystal
- the magnitude and direction of the applied force.

If, for example, the crystal is oriented along the x-axis (in crystallography this is depicted as [100]), the change in resistance along one of the axes (say the x-axis) is a result of six stress components (similar to the electric displacement we considered above): two due to forces normal to the x-axis (normal stress) and four due to forces lateral to it (shear stress). Refer once again to Figure 9.20 to visualise the phenomenon. The change in resistivity in the crystal lattice along the x-axis is the sum total of changes in resistivity due to the six stress components. Mathematically, the percentage change in resistivity of a semiconductor crystal having an arbitrary orientation is expressed as:

$$\frac{\Delta \rho_i}{\rho_0} = \sum_{j=1}^{6} \pi_{ij}\epsilon_j \tag{9.18}$$

where i and j take values from 1 to 6, denoting the parameters in the corresponding strain directions, ϵ_j is the component of the stress tensor in six-component vector notation, and π_{ij} is the component of the piezoresistive tensor (electroresistance coefficient). For a semiconductor crystal having a cubic symmetry and oriented along the x-axis, the stress tensor can be described using only three constants ($\pi_{11}, \pi_{12}, \pi_{44}$):

$$[\pi_{ij}] = \begin{bmatrix} \pi_{11} & \pi_{12} & \pi_{12} & 0 & 0 & 0 \\ \pi_{12} & \pi_{11} & \pi_{12} & 0 & 0 & 0 \\ \pi_{12} & \pi_{12} & \pi_{11} & 0 & 0 & 0 \\ \pi_0 & \pi_0 & \pi_0 & \pi_{44} & 0 & 0 \\ \pi_0 & \pi_0 & \pi_0 & 0 & \pi_{44} & 0 \\ \pi_0 & \pi_0 & \pi_0 & 0 & 0 & \pi_{44} \end{bmatrix} \tag{9.19}$$

Hence,

$$\begin{bmatrix} \Delta\rho_1/\rho_0 \\ \Delta\rho_2/\rho_0 \\ \Delta\rho_3/\rho_0 \\ \Delta\rho_4/\rho_0 \\ \Delta\rho_5/\rho_0 \\ \Delta\rho_6/\rho_0 \end{bmatrix} = \begin{bmatrix} \pi_{11} & \pi_{12} & \pi_{12} & 0 & 0 & 0 \\ \pi_{12} & \pi_{11} & \pi_{12} & 0 & 0 & 0 \\ \pi_{12} & \pi_{12} & \pi_{11} & 0 & 0 & 0 \\ \pi_0 & \pi_0 & \pi_0 & \pi_{44} & 0 & 0 \\ \pi_0 & \pi_0 & \pi_0 & 0 & \pi_{44} & 0 \\ \pi_0 & \pi_0 & \pi_0 & 0 & 0 & \pi_{44} \end{bmatrix} \begin{bmatrix} T_1 \\ T_2 \\ T_3 \\ T_4 \\ T_5 \\ T_6 \end{bmatrix} \tag{9.20}$$

where $T_1 = \sigma_{xx}, T_2 = \sigma_{yy}, T_3 = \sigma_{zz}, T_4 = \tau_{yz}, T_5 = \tau_{xz}, T_6 = \tau_{xy}$.

9.5.1.3 Fabrication of a Piezoresistive Sensor

MEMS piezoelectric pressure/force sensors can be fabricated using either surface or bulk microfabrication. Figure 9.25 shows the surface microfabrication of a strain gauge using polysilicon as a piezoelectric material (Kon and Horowitz 2008). In (1), a 1 μm phosphosilicate glass is deposited on the substrate as a sacrificial layer using LPCVD, and a 1 μm undoped polysilicon film is deposited on top of the phosphosilicate layer, again using LPCVD. The sacrificial layer serves also as a source of dopant to the polysilicon layer. In (2), silicon nitride is deposited using a low-stress LPCVD and patterned. After patterning, the dopants are diffused into the silicon substrate using an annealing process at 1000°C. The aim is to achieve a uniform dopant concentration in the silicon substrate. In (4), a seed layer is sputtered onto the wafer. In (5), thick gold bonding bumps are formed by electroplating of the wafer, which is patterned with a thick photoresist (recall the LIGA process). Afterwards the photoresist is removed and the substrate is submerged in 5 : 1 buffered hydrofluoric acid to wet-etch the sacrificial layer. At this stage the photoresist sensors are supported by silicon tethers through anchor points on the silicon substrate. This is illustrated in (6).

The next two steps (7) and (8) are application specific. The authors of this work intend to mount the piezoresistive sensors on a disk drive to monitor strain of the order of hundreds of nanostrain. Step (7) and (8) are required to transfer the fabricated silicon piezoresistive sensor onto steel substrates through metal-to-metal bonding. In this process, a sturdy landing layer is produced by evaporating gold onto the silicon substrate (after gold patterning, the exposed silicon is removed) and by face-to-face bonding the fabricated silicon wafer onto the steel substrate at the gold bumps.

9.5.2 Flow Sensors

Measuring fluid viscosity, fluid shear stress, and fluid type is useful for many reasons. In medicine, measuring the shear stress on the atrial and ventricular walls of the heart as well as on the walls of arteries and veins enables inference of myocardial infarction

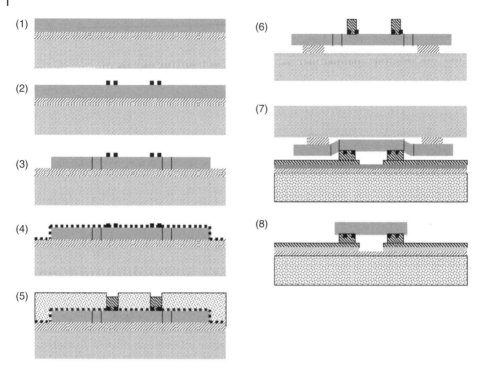

Figure 9.25 The fabrication process of a piezoresistive strain sensor. See main text for explanation of the parts of the figure. Courtesy of Kon and Horowitz (2008), (*IEEE Sensors*).

and potential ruptures (Aarnoudse et al. 2007; Kim et al. 2011; Kubli et al. 2008; Pijls et al. 1995; Steffen et al. 2007). In industry processes, measuring flow properties is vital during extrusion, coating, calendering, and molding (Jelali and Kroll 2012; Lipták 2013; Shajii et al. 1992). In oil, gas, and water pipelines, measuring shear stress is useful to determine fluid quality and to localise potential leakages during transportation (Ismail et al. 2005).

There are different approaches for measuring flow properties; some of them use a direct method while some of them use an indirect approach, such as ultrasound or infrared absorption and scattering (Lynnworth 2013). For direct sensing, piezoresistive sensors can be implanted into the flow, some aspect of it changing the resistivity of the sensors. This is typically done by imparting shear stress to the sensors as the flow moves parallel to them and the wall of the structure. Even though the approach is intrusive, with careful design and deployment, the mechanism can provide continuous and long-term sensing.

9.5.2.1 Floating Plate

Figure 9.26 demonstrates the construction of a simple flow sensor consisting of a floating plate and four tethers. The tethers function as mechanical support for the plate and as piezoresistive materials to convert mechanical stress to a change in electrical resistivity. When a flow passes over the floating plate and parallel to the length of the tether, it generates shear stress on the plate, in the direction of the flow. At the same time the tethers restrain the plate from making a significant movement. The force of the floating

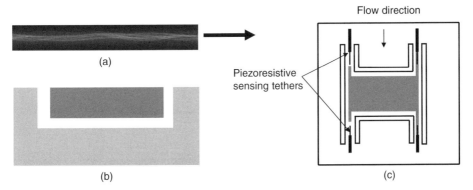

Figure 9.26 Flow sensing using MEMS: (a) flow through a structure produces shear stress in its walls; (b) a floating element between tethers fabricated with surface micromachining; (c) the basic structure of the flow sensor consisting of four tethers and a floating plate: two of the tethers are photoresistive materials and form a half-bridge electrical circuit to measure strain due to a compression and tensile stress.

plate will be equally divided between the four tethers. Two of them (top) experience compression whereas the other two (bottom), experience a tensile stress. These two type of stresses in turn produce axial strain fields through the tether structure producing a change in resistivity on the photoresistive tethers.

If the floating plate does not deform due to the shear stress coming from the flow, the output of the piezoresistive tethers can be calculated as follows:

1) The portion of stress received by each tether (σ_t) is a function of its relative cross-sectional area:

$$\sigma_t = \frac{\tau_p A_p}{4A_t} \tag{9.21}$$

where A_p is the cross-sectional area of the floating plate, A_t is the cross-sectional area of the tether, and τ is the shear stress on the floating plate.

2) The change in the resistance of the tether is:

$$\frac{\Delta R}{R} = GF\epsilon_t \tag{9.22}$$

where GF is the gauge factor and ϵ_t is the average strain experienced by the tether. For tethers fabricated with single silicon crystals, $\sigma_t = E_{Si}\epsilon$.

3) Combining Eqs. (9.21) and (9.22) yields:

$$\frac{\Delta R}{R} = \left(\frac{GFA_p}{4A_t E_{Si}}\right)\tau_p \tag{9.23}$$

As can be seen, the change in resistance is linearly related to the applied shear stress. Moreover, the sensitivity of the piezoresistive sensors can be controlled by the geometry of the floating plate.

Figure 9.27 displays a partial view of the surface microfabrication process of the floating plate of a flow sensor proposed by Shajii et al. (1992). Two silicon wafers are required to produce the floating plate. These wafers are joined using silicon wafer-bonding. The

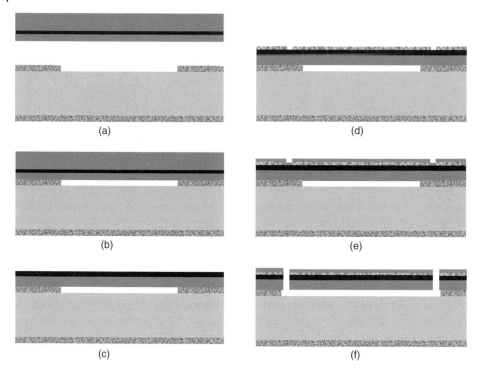

Figure 9.27 The cross-sectional view of a surface microfabrication of the floating plate of a flow sensor; two wafers (handle and device wafer) are required to produce the floating element: (a) the handle layer is coated with silicon dioxide and wet-etched to produce a bridge; (b) the two wafers are bonded together using a silicon wafer-bonding; (c) the top wafer is wet-etched leaving only a 5 µm lightly doped n-type epitaxial; (d) aluminium is deposited on top of the silicon epitaxial layer and patterned and sintered; (e) a thin layer of silicon is deposited using PECVD to protect the aluminium layer from subsequent stages, thereby acting as a mask for patterning the floating plate; (f) the silicon layer is etched to produce trenches and release the floating plate. Redrawn after Shajii et al. (1992), *IEEE Journal of Microelectromechanical Systems.* A similar approach is adopted by Barlian et al. (2007).

first wafer (the handle) is essentially used for confining the floating plate while the second (the device wafer) constitutes the plate itself. In the beginning, the handle, which is a 4-inch, $10 - 20\ \Omega$ cm n-type silicon oriented in the [100] direction (*x*-axis), is coated with 1.4 mm of silicon dioxide and patterned using wet etching at 950°C. The silicon dioxide is removed later using plasma etching. Next, the front of this wafer is bonded with the device wafer. The device wafer has a lightly doped, 5 µm, n-type epitaxial silicon layer grown on top of a densely doped p^+ boron region. Next, the device wafer is thinned using wet-etching until approximately 40 µm of silicon remains. The p^+ layer is subsequently removed with wet-etching, leaving a 5 µm thick silicon epitaxial layer bonded to a patterned oxide layer. Next an aluminium layer is deposited using electron-beam deposition, patterned and sintered in a nitrogen atmosphere. Finally, a thin layer of amorphous silicon is deposited on the aluminium layer using plasma etching to protect it from subsequent etching. In other words, this layer functions as a mask for patterning the structure to produce the floating plate. A final etching is carried out, via which trenches are formed and the floating plate is then released from the handle wafer.

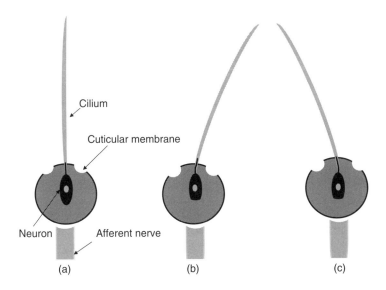

Figure 9.28 A hair cell and its mechano-transduction principle: (a) the basic constituents of a hair cell; (b) and (c) the direction and magnitude of the bending of a hair cell produces a corresponding potential variation inside the hair cell, changing a mechanical stress (shear stress) to an electrical energy.

9.5.2.2 Artificial Hair Cell

Natural mechanoreceptive hair cells can be found in many animals and in many parts of the body. These cells transform mechanical energy (bending) into electrical energy. They have simple construction and transduction mechanisms and yet perform remarkable tasks. Biological hair cells have different functions in different animals, but the most significant functions are sensing mechanical and acoustic vibrations, fluid and air flows, gravitation, pressure, and orientation. A very good summary of natural and man-made hair cells in the context of microelectromechanical sensing and actuation can be found in Engel et al. (2006) and the references therein.

Figure 9.28a shows the basic constituents of a hair cell, which are a cilium, a cuticular membrane, a neuron, and a neurotransmitter (Bermingham et al. 1999; Brownell et al. 1985; Hudspeth 2014; Nagata et al. 2005). In some cases, a single cell may contain a bundle of cilia. The cilium protrudes from the apical surface of the cell into a fluid containing rich concentration of ions, or connects to an external neuron. The other tip of the cilium is connected to an inner neuron, which in turn is connected to a nerve fibre.

The size of a cilium varies from organism to organism and depends on the specific purpose of the hair cell. In mammals and vertebrates, hair cells inside the ear transform acoustic vibrations and radial motion to electrical energy. The transduction process enables hearing and balancing. The hair cells in the inner ear (stereocilia) have sizes ranging between 2 and 8 μm and a diameter ranging from 0.1 to 0.3 μm. Fish use hair cells in their lateral line to sense water flow and vibrations in their surroundings, so that they can perceive nearby predators, prey, or obstacles. The same hair cells enable them to stabilise and balance in turbulent water. The array of hair cells in a lateral line are significantly larger than those in the inner ear in mammals, with a cupula structure enclosing the hairs reaching a length of up to 400 μm. In some insects (such as cockroaches and crickets), hair cells have multiple functions, such as sensing flow, vibration, and joint flexion. The cercal wind-receptor hairs of a cricket can have a length widely varying from 30 to 1500 μm and a diameter ranging from 1 to 9 μm.

When a mechanical shear force deflects the cilium of a hair cell, the deflection changes the permeability (the openings) of the cell membrane; the permeability either increases or decreases or the membrane closes altogether. If the permeability increases, it opens ion channels for ions to migrate from the external fluid into the hair cell or vice versa. The migration changes the ion concentration in the cell and thereby the resting electrical potential of the cell and triggers the neuron to fire a pulse, which is then transported by neurotransmitters to the nerve fibre. Thus the input mechanical energy is transformed into a neural signal.

For example, in humans, the inner hair cells inside the cochlea (a part of the inner ear) change the mechanical vibration of the basilar membrane to electrical (or neuronal) signals. When there is no mechanical vibration, the electrical resting potential of an unstimulated cell is approximately −50 mV. In this setting, about 10 % of the ion channels in the plasma membrane of the cell are open. In the presence of vibration, a positive stimulus that displaces the stereocilia towards the tall edge opens additional channels. This results in an influx of positive ions, which depolarises the cell. Likewise, a negative stimulus displacing the stereocilia towards the short edge shuts the channels that are open at rest and hyperpolarises the cell. The receptor potential of a hair cell is graded; as the stimulus amplitude increases, the receptor potential grows increasingly larger, up to a maximal point of saturation. The relationship between a bundle's deflection and the resulting electrical response is S-shaped. Figure 9.29 displays the relationship between the deflection of a hair cell bundle and the change in the electrical potential within the cell. As can be seen, a displacement of the cell by approximately 100 nm represents 90 % of the response range of the hair cell.

Figure 9.30 shows an air flow sensor emulating an artificial hair cell and fabricated at the Micro and Nanotechnology Laboratory, University of Illinois, Urbana (Engel et al. 2006). The sensor consists of a vertical beam (artificial cilium) rigidly attached to a silicon substrate. An external flow parallel to the sensor substrate imparts a drag force (due to friction or pressure) upon the vertical cilium. Due to a rigid connection between the substrate and the vertical cilium, a mechanical bending moment is transferred to the substrate, inducing a longitudinal strain at the base of the substrate. The magnitude of the induced strain is captured by a strain gauge attached at the joint of the substrate and the artificial cilium. The strain gauge is made up of a thin film nichrome (NiCr) resistor. The vertical part of the artificial hair cell is surface micromachined and deflected out of plane using magnetic three-dimensional assembly. The vertical portion remains in a deflected position due to plastic deformation at the joint. An external force applied to the cilium, either through direct contact with another object or by the drag force from fluid flow, deflects the beam and causes the strain gauge to stretch or compress. The strain gauge region itself is rigidly attached to the substrate, while the cilium is free to deflect. The magnitude of the induced strain (ϵ) is largest at the base, where the strain gauge is located, and can be determined by:

$$\epsilon = \frac{Mt}{2EI} \tag{9.24}$$

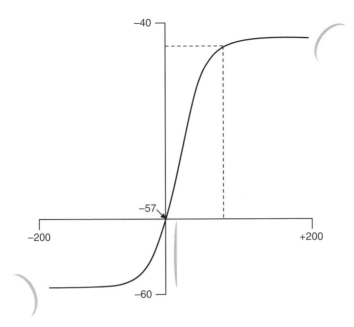

Figure 9.29 Relationship between the deflection of a hair cell bundle and the change in the corresponding electrical potential of the receptor. The *x*-axis represents the deflection distance (in nm) and the *y*-axis represents the associated membrane potential (in mV).

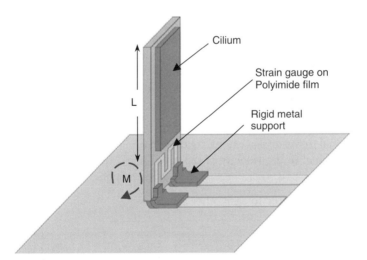

Figure 9.30 The schematic diagram of a polymer-based artificial hair cell sensor for measuring flow. Courtesy of Engel et al. (2006), (*IEEE Journal of Microelectromechanical Systems*, 2006).

where M is the moment experienced at the base, t is the thickness of the cilium fabricated with polyimide, and E and I are the modulus of elasticity and the moment of inertia of the polyimide, respectively. Figure 9.31 displays an array of such artificial hair cells fabricated on a single silicon wafer.

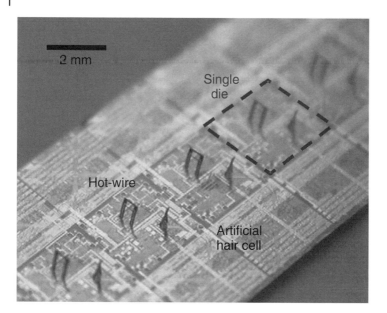

Figure 9.31 The microphotograph detail of the fabrication of an array of hair cell sensors and their integration with conditioning circuits. Courtesy of Engel et al. (2006), (*IEEE Journal of Microelectromechanical Systems*, 2006).

9.5.3 Accelerometers

Consider Figure 9.32, in which a simple cantilever with a proof mass at one of its ends is attached to a ridged frame. Suppose at standstill the proof mass is at a distance zero with respect to the internal frame of reference. If the ridged body is accelerated to move forward a distance of *y* with respect to the external frame of reference, the proof mass will also be accelerated and begins to oscillate. Three forces will be acting upon the proof mass as a result:

- an inertial force proportional to the proof mass
- a spring force proportional to the displacement, tending to oscillate the proof mass
- a damping force proportional to the velocity of the proof mass, tending to stop its oscillation.

Mathematically, these forces are related to each other as follows:

$$m\frac{d^2x}{dt^2} + c\frac{dx}{dt} + kx = -m\frac{d^2y}{dt^2} \tag{9.25}$$

Factoring out the mass term results in:

$$\ddot{x} + \frac{c}{m}\dot{x} + \frac{k}{m}x = -\ddot{y} = a(t) \tag{9.26}$$

It suffices to determine $x(t)$ in order to determine $a(t)$. In the frequency domain, the transfer function of Eq. (9.26) is expressed as:

$$H(s) = -\frac{1}{s^2 + \frac{c}{m}s + \frac{k}{m}} \tag{9.27}$$

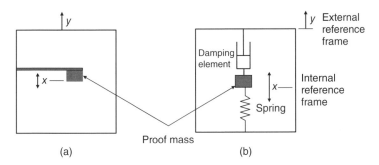

Figure 9.32 The basic principle of a cantilever accelerometer: (a) a cantilever suspended at one end; (b) a model of acceleration using a single spring and a damper.

For frequencies approaching to zero, the magnitude of the transfer function becomes independent of frequency and can be expressed as:

$$H(s) = -\frac{m}{k} \tag{9.28}$$

Hence it is possible to determine the acceleration of the ridged frame by:

$$a(t) = -\frac{k}{m}x(t) \tag{9.29}$$

Equation (9.29) is called the fundamental equation of accelerometers.

9.5.3.1 Fabrication of an Accelerometer

The most frequently used mechanism to sense $x(t)$ is capacitive sensing, although there are also accelerometers that employ piezoresistive and piezoelectric sensing. In capacitive sensing, the proof mass is a part of one of the electrodes of a capacitor. As the proof mass is accelerated, the distance between the moving electrode(s) and the fixed electrode(s) changes there by changing the overall capacitance of the microstructure. Figure 9.33 illustrates the basic principle of capacitive sensing in MEMS and Figure 9.34 shows the basic capacitive components of the ADXL 250, two-dimensional accelerometer fabricated by Analog Devices Inc. Similarly, Figure 9.35 displays a highly sensitive MEMS accelerometer realised with capacitive sensing. The accelerometer is capable of measuring acceleration below 1 g, where 1 g = 9.8 ms^{-2}. The fixed electrode

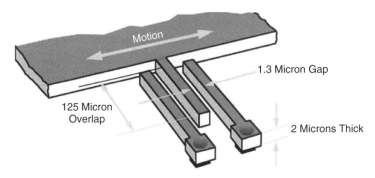

Figure 9.33 The basic principle of capacitive sensing to measure acceleration. Courtesy of Paschal Meehan and Keith Moloney, Limerick Institute of Technology.

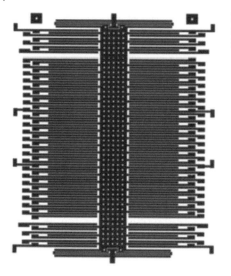

Figure 9.34 The fixed and moveable fingers of the ADXL 250 capacitive accelerometer sensor produced by Analog Devices Inc.

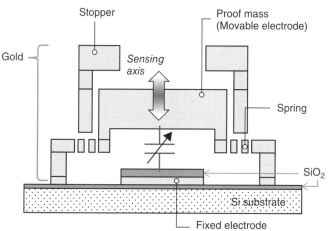

Figure 9.35 A combination of surface and bulk micromachining to realise an accelerometer with capacitive sensing. Courtesy of Yamane et al. (2014), (*Applied Physics Letters*).

of the capacitor is realised with SiO_2 film deposited on top of a silicon substrate. The moving electrode, which is the proof mass, is realised with the deposition and surface micromachining of a gold film. The spring constant is designed to have 0.3 Nm^{-1} in the z-direction. The stoppers on both sides of the proof mass are intended to protect the structure from excessive acceleration. Figure 9.36 displays the micrographs of the MEMS accelerometer with the squared proof mass.

9.5.4 Gyroscopes

A gyroscope is a device used to measure or maintain the orientation of a mechanical system such as a moving aircraft, ship, spacecraft, or robot. It has a simple construction, consisting of a spinning disk (rotor) that rotates around a fixed spin axis. The spin may have circular and rotating frames (gimbals), as can be seen in Figure 9.37. The gimbals,

Figure 9.36 A microscopic image of the MEMS accelerometer sensor of Figure 9.35: (a) chip view and (b) close-up image. Courtesy of Yamane et al. (2014), (*Applied Physics Letters*).

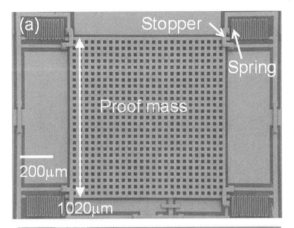

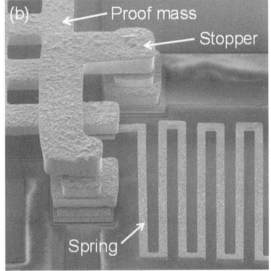

however, are not essential components of a gyroscope but they are useful to manage its movement. One of the gimbals (the outer) can be free to rotate in an axis determined by the position of the system the orientation of which we wish to maintain or determine. When the rotor is not spinning, the gyroscope is as good as a useless device; it does not support itself nor does it exhibit any interesting characteristics. But when the rotor is spinning, a number of interesting phenomena can be observed, all of which have something to do with conservation of angular momentum. The most important physical phenomenon is that a spinning gyroscope tends to resist any force that tries to change its orientation. The resistance force of the gyroscope is manifested by a new kind of motion the gyroscope undergoes to maintain its original orientation.

To explain this phenomenon, consider Figure 9.38 where we have a simple spinning gyroscope horizontally suspended at one of its ends, which is attached to the frictionless and rotating top of a supporting pylon. Suppose the gyroscope is initially spinning

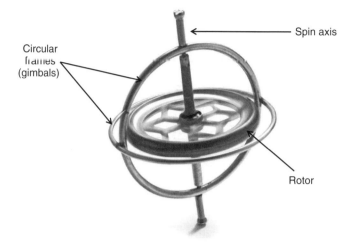

Figure 9.37 A gyroscope consisting of a spinning rotor and gimbals oriented perpendicular to each other.

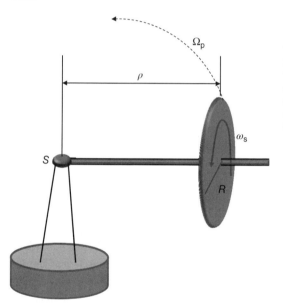

Figure 9.38 A spinning gyroscope horizontally suspended by one of its ends which is attached to the rotating top of a supporting pylon. The axis of rotation of the gyroscope undergoes precession (a change in the orientation of the rotational axis) when the centre of mass of the gyroscope is subject to a force that tends to disturb its original orientation.

around a fixed position in a counter-clockwise direction with a constant angular velocity of ω rad/s and no other force is exerting on the gyroscope. If, after a while, we apply a force such that this force tends to disturb the position of its spinning axis, the gyroscope will react to this force by displacing the axis of rotation in a direction which is perpendicular to both the original position of the axis and the direction of the applied force. In so doing, the gyroscope preserves its total angular momentum. The change in the orientation of the rotational axis is known as precession. If we know the angular velocity of the spinning rotor around its axis and the magnitude and direction of the applied force, then its possible to determine the displacement vector or the new orientation of the gyroscope's spin axis. Alternatively, if we know the new position of the axis, it is possible to determine the magnitude and direction of the applied force.

Example 9.3   We shall study the precession phenomenon more closely. In many practical applications, a gyroscope can be used to determine orientation, rotation, or acceleration. One of the most important measurable quantities providing sufficient information about these quantities is the precessional angular velocity of the gyroscope. Assume that the gyroscope in Figure 9.38 initially spins at a constant angular velocity around its axis in a counter-clockwise direction and the gyroscope is not subject to a gravitational force. However, as soon as a gravitational force is applied to the centre of mass of the gyroscope, it precesses (the axis rotates) in a counter-clockwise direction, which is perpendicular to both the original position of the axis of spinning and the gravitational force (namely, around the pylon). Since the gravitational force is constant at any of the gyroscope's positions and we are assuming that the spinning angular velocity of the gyroscope is constant, the rotation of the gyroscope around the pylon will be a uniform circular motion. Determine the precessional angular frequency.

The derivation of the precessional angular velocity may appear lengthy and confusing, but if you understand the different steps that should be taken, then it is comprehensible. The idea is the following: By the time we configure the gyroscope, we know the following parameters:

- the spinning angular velocity (ω_s) – both its direction and magnitude – because we give the gyroscope a deliberate spin with a known magnitude and direction;
- the centre of mass of the gyroscope (m);
- the distance of the centre of mass of the gyroscope from S (ρ);
- the moment of inertia (I_s) of the spinning gyroscope, which is a function of the centre of mass and the radius of the gyroscope, both of which are known.

The entire analysis deals with the motion of objects around two axes, namely, around the gyroscope's axis (the motion due to the spinning of the gyroscope's rotor) and around the supporting pylon (the motion due to precession). We know all the parameters pertaining to the first motion and wish to relate some of the parameters of the second motion in terms of the parameters of the first motion.

A brief summary of elementary mechanics and vector multiplication is useful. Consider Figure 9.39a, where we have a rotating disk. The relationship between the angular and linear displacements of the disk is a function of the radius of the disk R. Suppose we have a fixed point at the edge of the disk (the distance between this point and the centre of the disk is the radius of the disk). Suppose also at time $t = 0$, this point touches the ground. If we rotate the disk forward towards the right, our reference point recedes backward towards the left. The angle between the point touching the ground at any given time t and our reference point, $\theta(t)$, is a function of R and the arc between the two edges of the disk, $s(t)$:

$$\theta(t) = \frac{s(t)}{R} \tag{9.30}$$

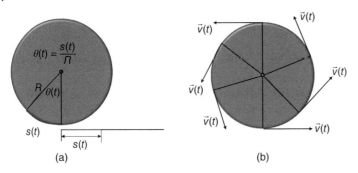

Figure 9.39 (a). The relationship between angular and linear displacements (from which it is possible to derive the relationship between linear and angular velocity and momentum). (b) The velocity of a spinning gyroscope (disk) is always tangent to the axis of rotation.

$\theta(t)$ is called the angular displacement and $s(t)$ corresponds to the linear displacement of the disk. From Eq. (9.30), different angular and linear relationships can be easily established:

$$v = \frac{ds}{dt} = R\frac{d\theta}{dt} = R\omega \tag{9.31}$$

where v and ω are the linear and angular velocities of the disk, respectively. An interesting phenomenon to observe is that the direction of the linear velocity of the disk is always tangent to its axis of rotation, as shown in Figure 9.39b. Likewise,

$$a = \frac{dv}{dt} = \frac{d^2s}{dt^2} = R\frac{d^2\theta}{dt^2} = R\alpha \tag{9.32}$$

where a and α are the linear and angular accelerations, respectively.

For an object undergoing a uniform circular motion around a central reference point S (refer to Figure 9.40), the acceleration due to a centripetal force is of interest (it is the centripetal force that keeps the object rotating in a circle; otherwise, the object would move away from a centre in a rectilinear motion). To derive an expression for this quantity, consider Figure 9.40a, where we have an object moving at a uniform velocity in a counter-clockwise direction. The magnitude of the tangential velocity is always constant, but the object changes its direction continuously. In order to calculate the centripetal acceleration, consider the velocity of the object at two time instances t_1 and t_2 and assume that $\Delta t = t_2 - t_1$ is very small. Hence, the centripetal acceleration of the object is expressed as:

$$a_c(t) = \frac{v(t_2) - v(t_1)}{\Delta t} = \frac{\Delta v}{\Delta t} \tag{9.33}$$

The change in velocity is depicted in Figure 9.40b using a vector representation. We have reversed the direction of $v(t_1)$ to obtain $-v(t_1)$. As can be seen, the change in velocity is directed inward. For a very small change in time (Δt), the distance travelled by the object (the arc **AB**) is expressed as $v\Delta t$ and can be approximated by a straight line. Hence, the triangle formed due to the linear displacement of the object (**SAB**) and the triangle formed by the changes in velocity vector, namely, $v(t_1)$, $v(t_2)$, and Δv, have two aspects in common:

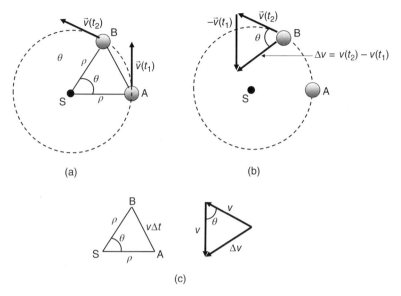

Figure 9.40 The precession of a gyroscope modelled as a uniform circular motion around S: (a) the velocity of the gyroscope at two different time instances; (b) the change in velocity between the two time instances – since the gyroscope is moving with a constant speed, the change is only a change in direction; (c) the displacement of the gyroscope during $\Delta t = t_2 - t_1$ is the arc **AB** which equals $v\Delta t$ (for a very small Δt, the arc can be approximated by a straight line). Hence, the two triangles (the one depicting the distance travelled by the gyroscope and the other depicting the change in the velocity vector) form two isosceles triangles having the same apex angle θ.

1) Both triangles are isosceles triangles. The triangle **SAB** has two equal sides, the dimension of each of which is equal to the radius ρ whilst the triangle formed by the velocity vector has two equal sides the dimension of each of which equals v due to the assumption that the magnitude of the velocity vector is constant.
2) Both triangles have the same apex angle θ. This is because the change in Δv is due to the angular displacement θ.

Consequently, using the two triangles, we can derive the centripetal acceleration:

$$\frac{\Delta v}{v} = \frac{v\Delta t}{\rho} \tag{9.34}$$

Rearranging terms yields:

$$a_c = \lim_{\Delta t \to 0} \left(\frac{\Delta v}{\Delta t} \right) = \frac{v^2}{\rho} \tag{9.35}$$

Furthermore, since angular velocity and linear velocity are related to each other (Eq. (9.31)), we have:

$$a_c = \rho\omega^2 \tag{9.36}$$

Going back to the rolling disk of Figure 9.39, what will happen to the angular velocity of the disk, if the disk maintains a constant linear velocity? From Eq. (9.30), we may be tempted to think that the angular velocity, too, should be constant, since R is a constant and $\omega = v/r$. This is, however, only a partial truth. The magnitude of the angular velocity

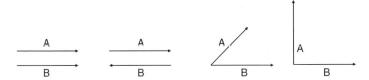

Figure 9.41 A summary of the cross product of two vectors.

will, indeed, be constant, but because the disk rotates, the angular velocity, which is always tangent to the direction of rotation, changes direction. As a result, the angular velocity is not constant. The same is true to the angular acceleration.

In order to examine the interaction between the different forces acting upon the gyroscope, it is worth to remember how we should be dealing with the cross product of two vectors. The cross product of two vector quantities that are parallel to each other is zero. In other words, the two vector quantities do not act upon each other. The resultant of two vector quantities acting upon each other becomes a maximum when they are orthogonal to each other (when the angle between them is 90°). This relationship is summarised in Figure 9.41. Consequently, the magnitude of the resultant vector is given as:

$$\vec{A} \times \vec{B} = |A||B| \sin \beta \tag{9.37}$$

where β is the angle between the two vectors. The direction of the resultant vector is determined by the following rule (see Figure 9.42):

$$\vec{i} = \vec{j} \times \vec{k}$$
$$\vec{j} = \vec{k} \times \vec{i}$$
$$\vec{k} = \vec{i} \times \vec{j}$$

Notice that the labelling of the axes in Figure 9.42 is a convention, whilst the direction of the resultant vector is not. In general, as you multiply vectors in a counter-clockwise direction, the direction of the resultant vector is the axis that comes next. For example, the multiplication of a vector directed towards $\vec{i}$ with a vector directed towards $\vec{j}$ (notice that we are multiplying in a counter-clockwise direction $(\vec{i} \times \vec{j})$), yields a resultant vector having a direction towards $\vec{k}$. Recall also, the cross product is anti-commutative; that is, $\vec{j} \times \vec{i} = -(\vec{i} \times \vec{j}) = -\vec{k}$, and so on.

Going back to the analysis of the different forces acting upon the gyroscope of Figure 9.38, the rotor of the gyroscope spins around its axis with an angular velocity:

$$\vec{\omega}_s = \omega_s \hat{r} \tag{9.38}$$

where we choose $\hat{r}$ as a reference frame aligning with the direction of the gyroscope's spinning axis. Remember our assumption that the rotor of the gyroscope rotates around its axis at a constant angular velocity. When a gravitational force (or any other force, for that matter) is applied to the gyroscope with the tendency to flip it downwards, the gyroscope precesses due to the torque exerted on the axis by the gravitational force at the contact point S. The orientation of this torque ($\hat{\theta}$) is orthogonal to both the orientation of the gravitational force ($\hat{k}$) and the orientation of the spin axis ($\hat{r}$); that is, $\hat{\theta} = \hat{k} \times \hat{r}$. As

Figure 9.42 The components of a vector **a** in a three-dimensional space and the three basis vectors $\hat{\mathbf{i}}, \hat{\mathbf{j}},$ and $\hat{\mathbf{k}}.$

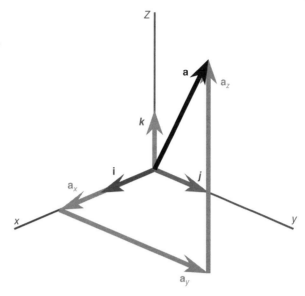

a result, the axis begins to rotate around the vertical (z) axis with an angular velocity:

$$\vec{\Omega}_{\mathrm{p}} = \Omega_{\mathrm{p}}\hat{\mathbf{k}} \tag{9.39}$$

$\hat{\mathbf{k}}$ indicating that the rotation is around the pylon, which is the vertical axis. The overall angular momentum of the gyroscope with respect to the contact point between the gyroscope's spin axis and the tip of the pylon (S) is given as:

$$\vec{\omega} = \vec{\Omega}_{\mathrm{p}} + \vec{\omega}_{\mathrm{s}} \tag{9.40}$$

If $\Omega_p << \omega_s$, which is the case with typical gyroscopes, then the overall angular velocity of the system is almost equal to the angular velocity due to the spin of the gyroscope:

$$\omega \approx \omega_{\mathrm{s}} \tag{9.41}$$

where ω is the magnitude of $\vec{\omega}$. To sum up our description of the precessing gyroscope, we have three forces acting upon the contact point: the gravitational force, an upward reaction force keeping the gyroscope from falling, and the inwardly directed centripetal force that keeps the gyroscope rotating around the vertical axis. Figure 9.43 illustrates the relationship between the forces acting upon a precessing gyroscope, the resulting precessional angular velocity, and the configuration of the gyroscope.
    Hence,

$$F_{\mathrm{K}} - mg = 0 \tag{9.42}$$

$$F_{\mathrm{c}} = ma_{\mathrm{c}} \tag{9.43}$$

Taking into account the relationship between the centripetal acceleration and the angular velocity (Eq. (9.36)), the last term in the above equation can be expressed as:

$$F_{\mathrm{c}} = m\rho\Omega_{\mathrm{p}}^2 \tag{9.44}$$

The torque exerted upon the spinning gyroscope when it is subject to the gravitational force is a function of both the gravitational force and the position of the centre of mass

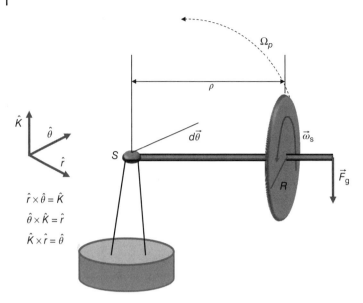

Figure 9.43 Description of the three forces acting upon a spinning gyroscope subject to gravitational force.

of the gyroscope from S. Since the position of the gyroscope continuously changes due to the precession, the position of the gyroscope's centre of mass with respect to S should be regarded as a position vector:

$$\vec{\tau}_s = \vec{\mathbf{r}} \times \vec{\mathbf{F}}_g = \rho \hat{\mathbf{r}} \times mg \ (-\hat{\mathbf{k}}) = \rho mg \hat{\mathbf{\theta}} \tag{9.45}$$

where ρ is the distance of the centre of mass of the gyroscope from S and $\hat{\mathbf{r}}$ depicts the orientation of the spin axis. As the gyroscope precesses around the pylon, the spin axis also continuously changes its position at a velocity v, which is equal to:

$$\vec{\mathbf{v}}_p = \frac{d\vec{\mathbf{r}}}{dt} \tag{9.46}$$

This velocity is tangent to the circle around the pylon and its magnitude is a function of ρ and the precessional angular velocity:

$$v_p = \rho \frac{d\theta}{dt} \tag{9.47}$$

where θ is the angular displacement (refer also to Eqs. (9.30) and (9.31)). When the gyroscope spins, the spin angular momentum about the centre of mass of the gyroscope points along the axis and is directed radially outward:

$$\vec{\mathbf{L}}_s = I_s \omega_s \vec{\mathbf{r}} \tag{9.48}$$

where I_s is the moment of inertia (angular mass) of the gyroscope's rotor with respect to the gyroscope's axis of spinning. The magnitude of the angular momentum is constant because the angular velocity is constant, but as the position of the spin axis changes, so does the direction of the momentum change, by the same rate as the change in the position vector of the axis. Hence,

$$\frac{d\vec{L}_s}{dt} = \left|\vec{L}_s\right| \frac{d\theta}{dt}\hat{\theta} \tag{9.49}$$

but,

$$\frac{d\theta}{dt} = \Omega_p \tag{9.50}$$

This is because the precessional angular velocity is the same as the rate of change of the angular displacement of the gyroscope around the pylon. Thus the rate of change of the gyroscope's angular momentum around its spinning axis can be expressed as:

$$\frac{d\vec{L}_s}{dt} = \left|\vec{L}_s\right| \Omega_p \hat{\theta} = I_s \omega_s \Omega_p \hat{\theta} \tag{9.51}$$

The angular momentum of the gyroscope with respect to S has two components due to:

- the spinning of the gyroscope around its own axis
- the rotation (precession) of the gyroscope around the pylon.

$$\vec{L}_{tot} = \vec{L}_s + \vec{L}_p \tag{9.52}$$

The angular and linear momentums of the gyroscope due to the above components are related with each other as follows:

$$\vec{L}_p = \vec{r} \times \vec{p}_p \tag{9.53}$$

where $\vec{r}$ is the precession position vector and $\vec{p}_p$ is the precession linear momentum. Notice also that $\vec{p}_p$ is always tangential to S, the axis of rotation around the pylon. Consequently,

$$\vec{P}_p = mv_p\hat{\theta} = m\rho\Omega_p\hat{\theta} \tag{9.54}$$

where we have made use of Eqs (9.47) and (9.50). Furthermore, bearing in mind that the position vector from S to the centre of mass of the gyroscope is $\vec{r} = \rho\hat{r}$ and $\hat{r} \times \hat{\theta} = \hat{k}$:

$$\vec{L}_p = (\rho\hat{r}) \times (\rho m\Omega_p\hat{\theta}) = m\rho^2\Omega_p\hat{k} \tag{9.55}$$

All the parameters in Eq. (9.55) are constant and therefore the angular momentum vector with respect to S is constant. Thus:

$$\frac{\vec{L}_p}{dt} = 0 \tag{9.56}$$

As a result, the angular momentum about S is due only to the angular momentum of the spinning of the gyroscope. So, if we fix the total angular momentum, we can determine the torque that generates the precession of the gyroscope because we know that only the angular momentum due to the spinning of the gyroscope changes. Hence,

$$\vec{\tau}_s = \frac{d\vec{L}_s}{dt} \tag{9.57}$$

Substituting Eqs (9.45) and (9.51) into Eq. (9.57) yields:

$$\rho mg = I_s \omega_s \Omega_p \tag{9.58}$$

From which we finally have:

$$\Omega_p = \frac{\rho mg}{I_s \omega_s} \qquad (9.59)$$

9.5.4.1 Fabrication of a Gyroscope

A MEMS gyroscope can be realised by exploiting the Coriolis effect (Prikhodko et al. 2013; Trusov et al. 2013; Xie and Fedder 2003). The concept is as follows. An object moving at a velocity $\vec{v}$ inside a rotating frame (rotating at $\vec{\Omega}$ angular velocity) experiences a Coriolis force that is perpendicular both to the direction of rotation of the frame $(\vec{\Omega})$ and the velocity of the object $(\vec{v})$ and proportional to its mass (Liu et al. 2009; McLusky 2013). This force is expressed as:

$$\vec{F}_c = m\vec{a}_c \qquad (9.60)$$

where m is the mass of the moving object and $\vec{a}$ is the Coriolis acceleration, which in turn can be expressed as:

$$\vec{a}_c = 2\vec{v} \times \vec{\Omega} \qquad (9.61)$$

This simple principle can be modelled by suspending a proof mass with springs within a rotary frame as shown in Figure 9.44b (Xie and Fedder 2003). If the reference frame rotates around the z-axis, the proof mass will undergo oscillation (displacement) in the x-axis (the sensing mode) inducing a Coriolis acceleration in the y-axis (the drive mode). The displacement $\vec{x}$ due to the Coriolis force (or the drive force) is given by:

$$\vec{x} = \frac{\vec{F}_c}{k_s} = \frac{m\vec{a}_c}{k_s} = \frac{\vec{a}_c}{\omega_{s,r}^2} \qquad (9.62)$$

where $\omega_{s,r}$ is the resonant frequency of oscillation of the springs to which the proof mass is fixed ($\omega_{r,s} = \sqrt{k_s/m}$) and k_s is the stiffness of the springs. From Eq. (9.62) it is clear

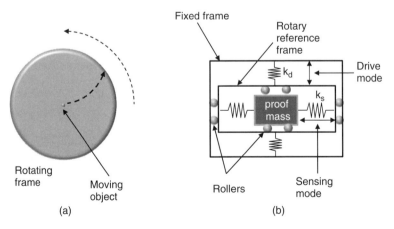

Figure 9.44 Applying the Coriolis effect to realise a gyroscope: (a) basic principle; (b) a two-dimensional conceptual model of a Coriolis gyroscope. The inner suspension ensures that the proof mass oscillates only on the x-axis relative to the frame of reference whilst the outer suspension ensures that the reference frame moves relative to the fixed frame only in the y-direction.

that a gyroscope can be considered as an accelerometer, but whereas an accelerometer measures translational acceleration or displacement, the accelerometer in a gyroscope must vibrate with a known velocity. Suppose the springs attached to the inner gyroscope frame have a stiffness of k_d and the outer frame oscillates (vibrates) sinusoidally due to the Coriolis acceleration (in other words, due to a change in the gyroscope's orientation). Hence,

$$y(t) = Q_d \frac{F_d}{k_d} \sin(\omega_{r,d} t) \tag{9.63}$$

where F_d is the amplitude of the drive force, Q_d is the mechanical quality factor of the drive mode and the springs are oscillating at resonance. Hence, the velocity of the drive mode is:

$$v_y(t) = \frac{dy(t)}{dt} = Q_d \frac{\omega_{r,d} F_d}{k_d} \cos(\omega_{r,d} t) \tag{9.64}$$

If we define K_s as:

$$K_s = \left[\left(1 - \left(\frac{\omega_{r,d}}{\omega_{r,s}} \right)^2 \right)^2 + \left(\frac{\omega_{r,d}}{Q_s \omega_{r,s}} \right)^2 \right]^{-1/2} \tag{9.65}$$

then, taking the second-order mechanical transfer function (in the frequency domain) into consideration, the magnitude and phase of the displacement vector can be expressed as:

$$|x| \angle \phi_s = K_s \frac{|a_c|}{\omega_{r,s}^2} \angle \phi_s = 2K_s \frac{|v_y||\Omega|}{\omega_{r,s}^2} \angle \phi_s \tag{9.66}$$

where Ω is the rotation rate of the inner gyroscope frame around the z-axis and ϕ_s is given as:

$$\phi_s = \tan^{-1} \left[\frac{\omega_{r,d}}{Q_s \omega_{r,s}} \left(1 - \left(\frac{\omega_{r,d}}{\omega_{r,s}} \right)^2 \right)^{-1} \right] \tag{9.67}$$

If the rotation signal is sinusoidal, $\Omega(t) = \Omega_m \cos(\omega_\Omega t)$, and $\omega_\Omega \ll \omega_{r,d}$, then substituting Eq. (9.64) into Eq. (9.66) yields:

$$x(t) = K_s Q_d \frac{F_d}{k_d} \frac{\omega_{r,d} \Omega_m}{\omega_{r,s}^2} \{ \cos[(w_{r,d} - \omega_\Omega)t + \phi_s] + \cos[(w_{r,d} + \omega_\Omega)t + \phi_s] \} \tag{9.68}$$

From Eq. (9.68) it is apparent that the mechanical sensitivity is determined by the actuation amplitude, the drive and sense resonant frequencies, and the sensing element quality factor. Increasing the proof mass reduces the resonance of the embedded accelerometer, thereby increasing the sensitivity of the gyroscope. The amplitude and resonance of the drive mode must also be kept large to attain high sensitivity. Moreover, Eq. (9.65) reveals that the sensitivity is maximised by matching the resonance frequencies of the two modes.

The fabrication of a microelectromechanical vibrating gyroscope follows similar steps as the fabrication of cantilever accelerometer sensors. Hence a change in the orientation of the gyroscope produces a translational displacement, which in turn produces either

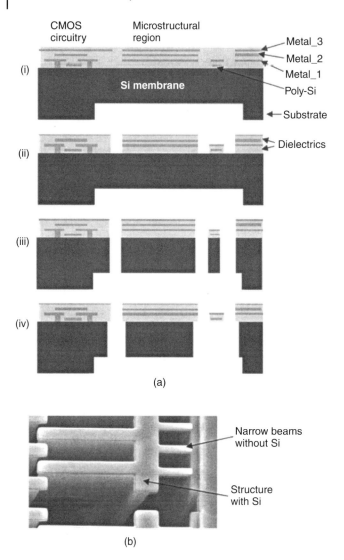

Figure 9.45 A deep reactive ion etching CMOS-MEMS process in which the metal interconnect layers are used as etching mask. (a) Cross-sectional view of the process steps: (i) backside deep silicon etching; (ii) anisotropic oxide etching; (iii) deep silicon etching for release; (iv) silicon undercut. (b) Micrograph of a fabricated comb drive, which has electrically isolated silicon electrodes. Courtesy of Xie and Fedder (2003), (*IEEE Sensors*).

a capacitive, piezoresistive, or piezoelectric effect. As a result, the capacitor, resistance, or voltage of the underlying structure changes. The most frequently used mechanism is capacitive sensing, in which a comb-drive actuator is used to transform a change in orientation to a change in output voltage. Figure 9.45 shows the application of a deep reactive ion etching process and the specific steps that can be taken to produce the stationary and moveable fingers of the comb drive. Figure 9.46 displays a Coriolis gyroscope fabricated with bulk microfabrication (Xie and Fedder 2003).

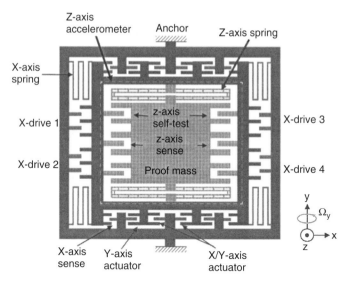

Figure 9.46 The topology of a vibrating MEMS gyroscope fabricated with a deep reactive ion etching process in a CMOS technology. Courtesy of Xie and Fedder (2003), (*IEEE Sensors*).

References

Aarnoudse W, van't Veer M, Pijls NH, ter Woorst J, Vercauteren S, Tonino P, Geven M, Rutten M, van Hagen E, de Bruyne B *et al.* 2007 Direct volumetric blood flow measurement in coronary arteries by thermodilution. *Journal of the American College of Cardiology*, **50**(24), 2294–2304.

Abedinpour S, Bakkaloglu B and Kiaei S 2007 A multistage interleaved synchronous buck converter with integrated output filter in 0.18 *μ*m SiGe process. *Power Electronics, IEEE Transactions on*, **22**(6), 2164–2175.

Arlett J, Myers E and Roukes M 2011 Comparative advantages of mechanical biosensors. *Nature Nanotechnology*, **6**(4), 203–215.

Bao M 2005 *Analysis and Design Principles of MEMS Devices*. Elsevier.

Barlian AA, Park SJ, Mukundan V and Pruitt BL 2007 Design and characterization of microfabricated piezoresistive floating element-based shear stress sensors. *Sensors and Actuators A: Physical*, **134**(1), 77–87.

Bermingham NA, Hassan BA, Price SD, Vollrath MA, Ben-Arie N, Eatock RA, Bellen HJ, Lysakowski A and Zoghbi HY 1999 Math1: an essential gene for the generation of inner ear hair cells. *Science*, **284**(5421), 1837–1841.

Bhushan B 2007 Nanotribology and nanomechanics of MEMS/NEMS and bioMEMS/bioNEMS materials and devices. *Microelectronic Engineering*, **84**(3), 387–412.

Bloomstein T and Ehrlich D 1991 Laser deposition and etching of three-dimensional microstructures. *Solid-State Sensors and Actuators, 1991. Digest of Technical Papers, TRANSDUCERS'91., 1991 International Conference on*, pp. 507–511.

Brand O 2006 Microsensor integration into systems-on-chip. *Proceedings of the IEEE*, **94**(6), 1160–1176.

Brownell WE, Bader CR, Bertrand D and De Ribaupierre Y 1985 Evoked mechanical responses of isolated cochlear outer hair cells. *Science*, **227**(4683), 194–196.

Chabal YJ 2012 *Fundamental Aspects of Silicon Oxidation*, vol. **46**. Springer Science & Business Media.

Challa VR, Prasad M, Shi Y and Fisher FT 2008 A vibration energy harvesting device with bidirectional resonance frequency tunability. *Smart Materials and Structures*, **17**(1), 015035.

Cheng CM, Martinez AW, Gong J, Mace CR, Phillips ST, Carrilho E, Mirica KA and Whitesides GM 2010 Paper-based ELISA. *Angewandte Chemie International Edition*, **49**(28), 4771–4774.

Cook-Chennault K, Thambi N and Sastry A 2008 Powering MEMS portable devices – a review of non-regenerative and regenerative power supply systems with special emphasis on piezoelectric energy harvesting systems. *Smart Materials and Structures*, **17**(4), 043001.

Dixon SL and Hall C 2013 *Fluid Mechanics and Thermodynamics of Turbomachinery*. Butterworth-Heinemann.

Dornfeld D, Min S and Takeuchi Y 2006 Recent advances in mechanical micromachining. *CIRP Annals-Manufacturing Technology*, **55**(2), 745–768.

Engel JM, Chen J, Liu C and Bullen D 2006 Polyurethane rubber all-polymer artificial hair cell sensor. *Microelectromechanical Systems, Journal of*, **15**(4), 729–736.

Fedder GK, Howe RT, Liu TJK and Quevy EP 2008 Technologies for cofabricating MEMS and electronics. *Proceedings of the IEEE*, **96**(2), 306–322.

Gattass RR and Mazur E 2008 Femtosecond laser micromachining in transparent materials. *Nature Photonics*, **2**(4), 219–225.

Giboz J, Copponnex T and Mélé P 2007 Microinjection molding of thermoplastic polymers: a review. *Journal of Micromechanics and Microengineering*, **17**(6), R96.

Happel J and Brenner H 2012 *Low Reynolds Number Hydrodynamics: with Special Applications to Particulate Media*, vol. **1**. Springer Science & Business Media.

Harrison RR, Watkins PT, Kier RJ, Lovejoy RO, Black DJ, Greger B and Solzbacher F 2007 A low-power integrated circuit for a wireless 100-electrode neural recording system. *Solid-State Circuits, IEEE Journal of*, **42**(1), 123–133.

Huang W, Ghosh S, Velusamy S, Sankaranarayanan K, Skadron K and Stan MR 2006 Hotspot: A compact thermal modeling methodology for early-stage VLSI design. *Very Large Scale Integration (VLSI) Systems, IEEE Transactions on*, **14**(5), 501–513.

Hudspeth A 2014 Integrating the active process of hair cells with cochlear function. *Nature Reviews Neuroscience*, **15**(9), 600–614.

Hunt HK and Armani AM 2010 Label-free biological and chemical sensors. *Nanoscale*, **2**(9), 1544–1559.

Ismail I, Gamio J, Bukhari SA and Yang W 2005 Tomography for multi-phase flow measurement in the oil industry. *Flow Measurement and Instrumentation*, **16**(2), 145–155.

Janata J 2010 *Principles of Chemical Sensors*. Springer Science & Business Media.

Jelali M and Kroll A 2012 *Hydraulic Servo-systems: Modelling, Identification and Control*. Springer Science & Business Media.

Judy JW 2001 Microelectromechanical systems (MEMS): fabrication, design and applications. *Smart Materials and Structures*, **10**(6), 1115.

Kaajakari V *et al.* 2009 *Practical MEMS: Design of Microsystems, Accelerometers, Gyroscopes, RF MEMS, Optical MEMS, and Microfluidic Systems*. Small Gear Publishing.

Kim DH, Lu N, Ghaffari R, Kim YS, Lee SP, Xu L, Wu J, Kim RH, Song J, Liu Z *et al.* 2011 Materials for multifunctional balloon catheters with capabilities in cardiac electrophysiological mapping and ablation therapy. *Nature Materials*, **10**(4), 316–323.

Kon S and Horowitz R 2008 A high-resolution MEMS piezoelectric strain sensor for structural vibration detection. *Sensors Journal, IEEE* **8**(12), 2027–2035.

Korvink J and Paul O 2010 *MEMS: A Practical Guide of Design, Analysis, and Applications*. Springer Science & Business Media.

Kovacs GT, Maluf NI and Petersen KE 1998 Bulk micromachining of silicon. *Proceedings of the IEEE*, **86**(8), 1536–1551.

Krüger J, Singh K, O'Neill A, Jackson C, Morrison A and O'Brien P 2002 Development of a microfluidic device for fluorescence activated cell sorting. *Journal of Micromechanics and Microengineering*, **12**(4), 486.

Kubli DA, Quinsay MN, Huang C, Lee Y and Gustafsson ÅB 2008 Bnip3 functions as a mitochondrial sensor of oxidative stress during myocardial ischemia and reperfusion. *American Journal of Physiology-Heart and Circulatory Physiology*, **295**(5), H2025–H2031.

Lipták BG 2013 *Process Control: Instrument Engineers' Handbook*. Butterworth-Heinemann.

Liu C 2010 *Foundations of MEMS*. Pearson Education India.

Liu K, Zhang W, Chen W, Li K, Dai F, Cui F, Wu X, Ma G and Xiao Q 2009 The development of micro-gyroscope technology. *Journal of Micromechanics and Microengineering*, **19**(11), 113001.

Lynnworth LC 2013 *Ultrasonic Measurements for Process Control: Theory, Techniques, Applications*. Academic Press.

Lyshevski SE 2002 *MEMS and NEMS: Systems, Devices, and Structures*. CRC Press.

Ma X and Kuo GS 2003 Optical switching technology comparison: optical MEMS vs. other technologies. *Communications Magazine, IEEE*, **41**(11), S16–S23.

Mabeck JT and Malliaras GG 2006 Chemical and biological sensors based on organic thin-film transistors. *Analytical and Bioanalytical Chemistry*, **384**(2), 343–353.

Madou M, Zoval J, Jia G, Kido H, Kim J and Kim N 2006 Lab on a CD. *Annual Reviews of Biomedical Engineering*, **8**, 601–628.

Maluf N and Williams K 2004 *Introduction to Microelectromechanical Systems Engineering*. Artech House.

McLusky D 2013 *The Estuarine Ecosystem*. Springer Science & Business Media.

Nagata K, Duggan A, Kumar G and García-Añoveros J 2005 Nociceptor and hair cell transducer properties of TRPA1, a channel for pain and hearing. *The Journal of Neuroscience*, **25**(16), 4052–4061.

Nikolou M and Malliaras GG 2008 Applications of poly (3, 4-ethylenedioxythiophene) doped with poly(styrene sulfonic acid) transistors in chemical and biological sensors. *The Chemical Record*, **8**(1), 13–22.

Patolsky F, Zheng G and Lieber CM 2006 Fabrication of silicon nanowire devices for ultrasensitive, label-free, real-time detection of biological and chemical species. *Nature Protocols*, **1**(4), 1711–1724.

Peng C, Li Z and Xu A 2007 Optical gyroscope based on a coupled resonator with the all-optical analogous property of electromagnetically induced transparency. *Optics Express*, **15**(7), 3864–3875.

Pijls NH, Van Gelder B, Van der Voort P, Peels K, Bracke FA, Bonnier HJ and El Gamal MI 1995 Fractional flow reserve a useful index to evaluate the influence of an epicardial coronary stenosis on myocardial blood flow. *Circulation*, **92**(11), 3183–3193.

Pletcher RH, Tannehill JC and Anderson D 2012 *Computational Fluid Mechanics and Heat Transfer*. CRC Press.

Prikhodko IP, Trusov AA and Shkel AM 2013 Compensation of drifts in high-Q MEMS gyroscopes using temperature self-sensing. *Sensors and Actuators A: Physical*, **201**, 517–524.

Reina A, Jia X, Ho J, Nezich D, Son H, Bulovic V, Dresselhaus MS and Kong J 2008 Large area, few-layer graphene films on arbitrary substrates by chemical vapor deposition. *Nano Letters*, **9**(1), 30–35.

Scheuer J and Yariv A 2006 Sagnac effect in coupled-resonator slow-light waveguide structures. *Physical Review Letters*, **96**(5), 053901.

Shajii J, Ng KY and Schmidt MA 1992 A microfabricated floating-element shear stress sensor using wafer-bonding technology. *Journal of Microelectromechanical Systems*, **1**(2), 89–94.

Staples M, Daniel K, Cima MJ and Langer R 2006 Application of micro-and nano-electromechanical devices to drug delivery. *Pharmaceutical Research*, **23**(5), 847–863.

Steffen M, Aleksandrowicz A and Leonhardt S 2007 Mobile noncontact monitoring of heart and lung activity. *Biomedical Circuits and Systems, IEEE Transactions on*, **1**(4), 250–257.

Tian B, Zheng X, Kempa TJ, Fang Y, Yu N, Yu G, Huang J and Lieber CM 2007 Coaxial silicon nanowires as solar cells and nanoelectronic power sources. *Nature*, **449**(7164), 885–889.

Tilli M, Motooka T, Airaksinen VM, Franssila S, Paulasto-Krockel M and Lindroos V 2015 *Handbook of Silicon Based MEMS Materials and Technologies*. William Andrew.

Trusov AA, Prikhodko IP, Rozelle D, Meyer A and Shkel AM 2013 1 ppm precision self-calibration of scale factor in MEMS coriolis vibratory gyroscopes. *Solid-State Sensors, Actuators and Microsystems (TRANSDUCERS & EUROSENSORS XXVII), 2013 Transducers & Eurosensors XXVII: The 17th International Conference on*, pp. 2531–2534.

Waggoner PS and Craighead HG 2007 Micro-and nanomechanical sensors for environmental, chemical, and biological detection. *Lab on a Chip*, **7**(10), 1238–1255.

White FM 1999 *Fluid Mechanics*. McGraw-Hill.

Xie H and Fedder GK 2003 Fabrication, characterization, and analysis of a DRIE CMOS-MEMS gyroscope. *Sensors Journal, IEEE*, **3**(5), 622–631.

Yamane D, Konishi T, Matsushima T, Machida K, Toshiyoshi H and Masu K 2014 Design of sub-1g microelectromechanical systems accelerometers. *Applied Physics Letters*, **104**(7), 074102.

10

Energy Harvesting

The sensing and design principles we considered so far can have another and crucial application, which is the harvesting of energy from the environment to power the sensors themselves. You may recall that in Chapter 1 I defined sensors as devices that transform one form of energy to another (often to electrical energy). You can see that I have already acknowledged that the environment in which sensors are deployed produces energy that can be used to operate sensors by charging their batteries or energy storage devices. Some environments, such as the human body, produce heat and movement. Researchers have identified several mechanisms by which electrical energy can be harvested from a human body including: use of glucose oxidation, electric potentials of the inner ear, mechanical movements of limbs, and natural vibrations of internal organs (Mitcheson et al. 2008; Ramadass and Chandrakasan 2010; Saha et al. 2008; Vullers et al. 2009). Bridges and water produce continuous ambient movement (vibration) (Beeby et al. 2006, 2007; Stephen 2006); other environments such as open fields, give the opportunity to use solar and wind energy (Kansal et al. 2007; Raghunathan et al. 2005; Sudevalayam and Kulkarni 2011; Taneja et al. 2008). Thermocouples can be used to generate energy from heat (Tan and Panda 2011), piezoelectric devices can be used to generate energy from vibration (Erturk and Inman 2011; Khaligh et al. 2010; Sirohi and Mahadik 2011; Stanton et al. 2010), inductive motors can be used in conjunction with mechanical ambient vibrations to generate energy from the environment, and photovoltaic cells (PN-junction crystalline silicon solar cells) can be used to generate energy from light or solar radiation (López-Lapeña et al. 2010; Qiu et al. 2011, Tan and Panda 2011). Energy can even plausibly be harvested from ambient radio-frequency (RF) radiation (or noise) (Nishimoto et al. 2010; Shi et al. 2011), although this is a very optimistic scenario, as the amount of energy that can be generated from RF sources is extremely small.

10.1 Factors Affecting the Choice of an Energy Source

Whereas the principles of energy harvesting are straightforward, the selection of the appropriate source, nevertheless, should take several factors into account. These preconditions, or the factors affecting the choice of the appropriate energy source, are the following:

- the lifetime of the sensor deployment
- the amount of energy the sensor consumes as a load

Principles and Applications of Ubiquitous Sensing, First Edition. Waltenegus Dargie.
© 2017 John Wiley & Sons, Ltd. Published 2017 by John Wiley & Sons, Ltd.
Companion Website: www.wiley.com/go/dargie2017

- the amount, the predictability, and controllability of the energy that can be harvested
- the capacity and quality of the energy storage device.
- the complexity of the regulation circuit interfacing the energy harvesting, the energy storage, and the sensing systems.

We shall discuss these factors in the following subsections.

10.1.1 Sensing Lifetime

Energy harvesting requires harvesting and regulation components. It also requires harvesting time. Whether or not the effort of designing and deploying energy-harvesting components is worthwhile depends, first and foremost, on how long the components will be used for. The longer the lifetime, the more worthwhile is the cost. Other factors, such as the weight of the harvesting system, should also be taken into account when lifetime is considered. For example, the heat dissipation produced by a harvesting system during excessive charging can cause irritation to a human body and its long-term impact can be detrimental. Likewise, the movement of objects can be constrained when they are loaded with additional components. Therefore, both aspects of deployment lifetime should be taken into account when designing and deploying energy-harvesting systems.

10.1.2 Sensor Load

The sensing system may consist of many components, including sensor(s), conditioning circuits, analogue-to-digital-converter(s), a processor (microcontroller), and, perhaps, a wireless transceiver. The overall power consumption of the sensing system is the sum total of the power consumed by each component. In most practical cases, as the measurand, for which the sensing system is deployed, changes slowly over time, the sensing system can operate in active and inactive states to save energy. The duty cycle of the sensing system can be tuned in accordance with the average amount of power that can be harvested from the environment. Consider, for example, that the average power that can be steadily delivered by the combined effort of the harvesting and storage systems is p_{avg} watts at v_{dd} volts. Hence, the average current that should be drawn by the sensing system should be:

$$i_{avg} = \frac{p_{avg}}{v_{dd}} \tag{10.1}$$

The duty cycle of the sensing system (the alternation between active and inactive states) is defined as:

$$D = \frac{\tau_a}{\tau_a + \tau_{in}} \tag{10.2}$$

Then, the average power consumption of the sensing system can be expressed as:

$$p_{avg} = p_a \tau_a + p_{in} \tau_{in} \propto i_a \tau_a + i_{in} \tau_{in} \tag{10.3}$$

where i_a and i_{in} are the currents drawn by the sensing system when active and inactive, respectively. Assuming v_{dd} remains constant when the current drawn by the sensing system varies, we have,

$$i_{avg} = i_a \tau_a + i_{in} \tau_{in} \tag{10.4}$$

The duration of the active time τ_a can be expressed as a fraction of the duration of the inactive time,

$$\tau_a = \alpha \tau_{in}, \alpha << 1 \tag{10.5}$$

Setting the idle time to unity, we have,

$$i_{avg} = \alpha i_{in} + i_{in} \tag{10.6}$$

α can be fixed according to the application context for which the sensor and the energy harvesting system are deployed:

$$\alpha = \frac{i_{avg} - i_{in}}{i_a} \tag{10.7}$$

Finally, the duty cycle can be expressed in terms of the average current that should be drawn from the energy-harvesting system:

$$D = \frac{\alpha \tau_{in}}{\alpha \tau_{in} + \tau_{in}} = \frac{\alpha}{\alpha + 1} = \frac{i_{avg} - i_{in}}{i_{avg} - i_{in} + i_a} \tag{10.8}$$

10.1.3 Energy Source

There are several vital aspects of the energy source that should be taken into account to decide which source to tap and which harvesting component to design.

The first is the expected amount of energy that can be harvested in a unit time (this refers to the amount of charge that can be accumulated in a unit time at a given voltage) and the rate of fluctuation of the available energy. The latter not only affects the amount of charge that can be collected but also:

- the rate at which the energy storage or the battery should be charged
- the complexity of the regulation component that controls and regulates the charge storage process, which in turn affects the design cost of the energy-harvesting system.

The second aspect is the predictability and controllability of the energy source. Controllability refers to the ability to control both the amount and the timing of energy generation, and predictability refers to the ability to foresee the fluctuation in the amount of energy that can be harvested. Almost all ambient sources (solar, wind, thermal, vibration, and so on) are uncontrollable by nature, but for some specific settings, they can be predictable. Some ambient sources (for example, heat or vibration from a human body) can be considered so deterministic that we can take them for controllable sources. Controllability and predictability intimately influence the choice of the energy-storage system as well as the complexity of the regulation system.

The third aspect is the efficiency with which the energy of the source can be converted to useful energy (electrical energy). A significant amount of energy is lost during conversion: at the interface between the source and the sensing element (the converting element), between the sensing element and the regulator, and in the storage system itself. Table 10.1 provides a compact summary of some of the available energy sources, the sensing elements they use to convert one form of energy to electrical energy, the conversion efficiency, and the obtainable energy.

Table 10.1 A summary of energy harvesting sources, mechanisms, and efficiency

Source	Sensing	Conversion efficiency	Obtainable power
Solar	Photovoltaic cells	16 – 17%	$12\,\mathrm{mW/cm^2}$
Ambient indoor light	Photovoltaic cells	16 – 17%	$100\,\mathrm{mW/cm^2}$
Thermoelectric	Thermocouple	$\leq 1\%$ for $\Delta T < 40°C$	$60\,\mu\mathrm{W/cm^2}$ at $\Delta T = 5°C$
Ambient air flow	MEMS turbine	—	$1\,\mathrm{mW/cm^2}$ at $30\,l/min$
Wind	Anemometer	59% (theoretical limit)	up to $1200\,\mathrm{mW/day}$
Footfalls	Piezoelectric	7.5%	$5\,\mathrm{W}$
Finger motion	Piezoelectric	11%	$2.1\,\mathrm{mW}$
Exhalation	Breath mask	40%	$0.4\,\mathrm{W}$
Breathing	Ratchet-flywheel	50%	$0.42\,\mathrm{W}$
Blood pressure	Microgenerator	40%	$0.37\,\mathrm{W}$

A portion of the data is gathered from Paradiso and Starner (2005) and Kansal et al. (2007).

10.1.4 Storage

The storage system in the design and application of an energy-harvesting system plays a vital role. There are two essential aspects that necessitate energy storage. First of all, the amount of energy that can be harvested from the environment fluctuates considerably, and therefore cannot be directly fed to the load (the sensing system). Secondly, there is a discrepancy between the supply of and demand for energy in the sensing system: at times, the load may not consume all the energy a source can provide and at other times the source may not provide the amount of energy the load demands. To deal with these two concerns, the viable solution is to accumulate and store energy, whenever it is available, for later use.

There are various options from which a suitable storage system can be chosen (Winter and Brodd 2004). Broadly speaking, these options can be classified into two groups: special-purpose capacitors (supercapacitors) and rechargeable batteries. Special-purpose capacitors are double-layer electrolytic capacitors that employ a highly porous, ultralight dielectric material having a large surface area (a carbon aerogel) to store charges (Conway 2013, Pandolfo and Hollenkamp 2006, Zhu et al. 2011). These capacitors have much faster charging and discharging rates, much higher power densities, and are more environmental-friendly than their chemical counterparts. However, their energy densities and capacities are significantly smaller. Rechargeable batteries, on the other hand, have relatively high energy density, high capacity, and charge-discharge cycles typically of the order of 200–2000 cycles. But they are more expensive and adverse to the environment. The difference between power and energy densities should be clear. Energy is the capacity to do work. Power is the rate at which energy is consumed or stored. Power can be better understood as the amount of charge that can be accumulated or drawn per unit time. Figure 10.1 compares the energy and power density of rechargeable batteries and capacitors.

There is a wide selection of commercially available rechargeable batteries. The selection criteria include mass–energy-density, volume–energy-density, mass, volume, capacity, self-discharge (leakage), and charging method, among others. The density criteria refer to the amount of energy that can be stored in a unit volume or mass.

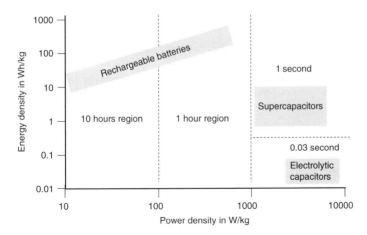

Figure 10.1 A comparison of power and energy densities of rechargeable batteries and capacitors as energy storage systems.

The self-discharging criterion is a measure of the tightness of the storage system to contain the charge it accumulates for a given period of time. Table 10.2 compares the parameters of some commercially available storage systems.

10.1.5 Regulation

The regulator of an energy-harvesting system controls the speed at which the energy storage is charged. It is responsible for protecting the storage system from damage, by controlling the amount of charge with which the storage system is supplied. If desired, a regulator detects when the storage system is fully charged and isolates the storage system from the source to prevent further charging. The complexity of the regulator (and, therefore, its cost) mainly depends on the type of the storage or battery and the charging rate.

The rate at which an energy storage system can be charged is broadly categorised as fast and slow (trickle). The duration of fast charging is typically less than two hours, whereas the duration of slow charging can be indefinite (in other word, the charging current can be applied to the storage indefinitely without damaging it). The maximum amount of slow charging depends on both the chemistry and construction of the internal elements (electrodes and electrolytes). One of the merits of slow charging is that it does not require an extra circuit to detect the completion of charging, because the charging rate is such that the input energy can always be balanced by the output energy supplied to the load. In other words, regulators of trickle charging are relatively cheap. However, care must be taken to ensure that the input and output energies are balanced, otherwise the load (the sensing system) can overload and damage the storage or the regulator if it draws excessive energy.

The amount of current that can be discharged is often expressed as a C-rate, which is a measure of the rate at which a storage system can be discharged (the amount of current that can be drawn from it) relative to its maximum capacity, which is expressed in terms of milliampere hours. A 1C rate corresponds to the discharge of the entire current the storage system can supply in 1 h. For a storage system that has a capacity of 100 mA h, 1C

Table 10.2 A summary of the essential parameters by which commercially available storage systems can be evaluated

Type	Lead acid	NiCd	NiMH	Li-ion	Li-polymer	Supercap
Manufacturer	Panasonic	Sanyo	Energizer	Ultralife	Ultralife	Maxwell
Model number	LC-R061R3P	KR-1100AAU	NH15-2500	UBP053048	UBC433475	BACP0350
Nominal voltage (V)	6.0	1.2	1.2	3.7	3.7	2.5
Energy (Wh)	7.8	1.32	3.0	2.8	3.4	0.0304
Capacity (mAh)	1300	1100	2500	740	930	350 (F)
Mass energy density (Wh/kg)	26	42	100	165	156	5.06
Volume energy density (Wh/L)	67	102	282	389	296	5.73
Mass (g)	300	24	30	17	22	60
Volume (cm^3)	116.4	8.1	8.3	9.3	12.8	53.0
Self-discharge (in % per month)	3–20	10	30	< 10	< 10	5.9 (per day)
Charge-discharge efficiency (%)	70–92	70–90	66	99.9	99.8	97-98
Memory effect	no	yes	no	no	no	no
Charging method	trickle	trickle/pulse	trickle/pulse	pulse[a]	pulse	trickle

a) Pulse charging is a mechanism by which a series of voltage or current pulses is supplied to a storage system. The rise time, width, frequency, and amplitude of the pulses can be controlled to make pulse charging suitable for a variety of size, voltage, capacity, and chemistry. The table is adapted from Taneja et al. (2008).

equates to the discharge of 100 mA in 1 h. A 2C rate of this storage system corresponds to 200 mA h (and, therefore, that the charge will be exhausted in 30 min), and a C/2 rate corresponds to 50 mA h (and, therefore, the charge will be exhausted in 2 h). Most Ni-Cd batteries tolerate sustained charging at a C/10 rate (110 mA h) with no additional protection circuit. In contrast, the slow charging rate for Ni-MH cells is between C/40 (62.5 mA h) and C/10 (250 mA h).

Fast charging has a significantly shorter charge duration, but requires protection circuitry. Moreover, its safety is temperature dependent, with the safest charging temperature for most commercially available batteries often lying between 10 and 40°C and 25 °C considered optimal. Without protection circuitry, an excessively charged storage system can internally produce gas, which leads to the recombination of ions, reducing the efficiency of the storage system. The gas in turn builds up pressure within the storage system. Most storage systems are equipped with a venting mechanism that automatically opens to release gas. The opening of the internal vent, however, reduces the lifetime of the battery cell. The type of gas that can be released is chemistry dependent. In Ni-Cd cells, for example, it is oxygen, which is harmless, but for Ni-MH it is hydrogen, which is explosive.

Fast charging may also result in a rapid build up of pressure within a cold storage system. Interestingly, the charging reaction of different storage systems can be different.

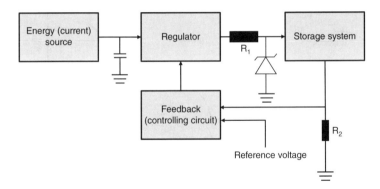

Figure 10.2 A feedback regulator charges an energy storage with a constant current. The feedback system establishes a voltage drop at one of its terminals by tapping a portion of the current entering into the storage system and comparing the voltage drop with a reference voltage to detect overcharging.

For instance, in Ni-Cd cells, the charging reaction is endothermic, which means that charging makes the battery cell colder. In Ni-MH cells, however, the charging reaction is exothermic, which means it increases the temperature of the battery cell. Consequently, Ni-Cd cells tolerate higher rate charging than Ni-MH. Another factor that affects fast charging is the internal resistance of the storage system, because the internal dissipation of power in the form of heat is a function of the internal resistance: $P = I^2 R$.

Figure 10.2 shows the basic components of a fast-charging regulator. The main purpose of the regulator is ensuring the constant flow of current into the energy storage. The capacitor between the energy source and the regulator prevents a sudden change of voltage at the source whereas the Zener diode (which is reverse biased) between the regulator and the storage system keeps the charging voltage at a fixed level. When the charging current exceeds its specified limit, a reverse-biased current begins to flow through the Zener diode, thereby reducing the voltage across the storage system. A portion of the current entering into the storage system is tapped to create a voltage drop across one of the terminals of the feedback system (R_2), which is then compared with a reference voltage. If the voltage drop exceeds the reference voltage, the feedback system generates an output voltage to turn off the regulator.

The protection circuit in Figure 10.2 does not take the temperature of the storage system into account. The temperature of the storage system can be a direct indication of the extent to which the storage system is charged. Figure 10.3 displays the temperature-vs-voltage curve for two technologies (Ni-Cd and Ni-MH). As can be seen, the temperature of the two battery technologies steadily increases as the charging nears completion, although obviously the two technologies have different temperature footprints (Ni-Cd does not show any appreciable temperature increment until the battery is fully charged – recall that the internal charge reaction in Ni-Cd is endothermic – whilst a steady increase can be observed from the outset for the Ni-MH). Common to both technologies, however, is a rise in temperature by approximately 10°C as the battery cells are fully charged.

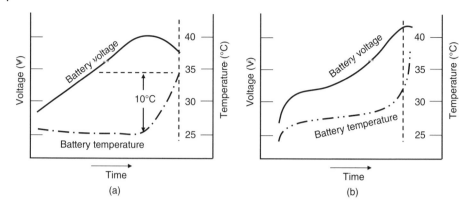

Figure 10.3 Voltage-versus-temperature curves for two different battery cells: (a) Ni-Cd; (b) Ni-MH. The curves were drawn on the basis of measurements taken at a 1C charging rate at 25 °C. As can be seen, the temperature of the batteries significantly rises when the batteries are nearly fully charged. Courtesy of Chester Simpson, National Semiconductor.

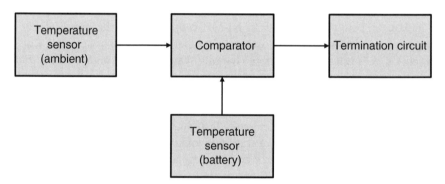

Figure 10.4 ΔT detector for a fast-charging regulator. The temperature of the battery is compared with the ambient temperature to extract the change in temperature due to charging. Often a timer is used in conjunction with the ΔT detector to make sure that the termination signal is generated after the rapid rise in temperature is detected.

Figure 10.4 shows the schematic diagram of a temperature-sensitive end-of-charge detector. Two temperature sensors are used: one to sense the ambient temperature surrounding the storage system as a reference, and the other to directly sense the temperature of the storage system. When the comparator senses a difference of $\Delta T = 10\,°C$, it generates a termination signal that turns off the regulator. The temperature sensors output a voltage that is proportional to the rise in temperature. For example, the Texas Instruments LM35 temperature sensor generates $10\,mV\,°C^{-1}$. Consequently, a voltage of $100\,mV$ should correspond to $\Delta T = 10\,°C$.

Example 10.1 One of the tasks of a voltage regulator is to supply the load with the appropriate DC voltage. Suppose the voltage that can be supplied by the storage system is 3.7 V but the voltage the sensing system requires is 2.3 V. Propose a voltage regulator that converts the 3.7 V supply voltage to 2.3 V load voltage.

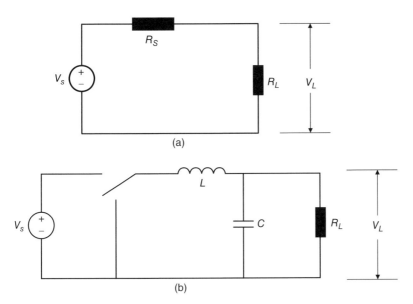

Figure 10.5 A DC-DC converter: (a) a simple but lossy resistive DC-DC converter; (b) a buck converter consisting of a switching device and a low-pass filter. The low-pass filter suppresses the AC component from reaching the load whereas the switching device fixes the duty cycle of the source voltage.

Obviously, what we need is a DC-DC converter. The simplest DC-DC converter would be a single-stage voltage divider circuit, as shown in Figure 10.5a, for

$$V_L = \frac{R_L}{R_L + R_S} V_S \tag{10.9}$$

Since the load resistance is known, it is possible to adjust the source resistance such that the desired voltage can be obtained. However, when a load current flows through the source resistance, there will be a power loss the magnitude of which equals: $P_S = I^2 R_S$. This is rather a waste. The more efficient DC-DC converter is the one shown in Figure 10.5b, which is called a buck converter. Ideally it should be lossless because the capacitor and the inductor are energy-storage components, but they have internal resistance as a result of which there is a small amount of power dissipation in the form of heat. Moreover, because they are imperfect energy storage, they leak, which is why we should expect some inefficiency when we employ voltage regulators. First we shall analyse the circuit logically in order to understand how it works, and then we shall derive the mathematical equations to establish a relationship between the source and load voltages.

The switching component acts as a single-pole, double-throw switch. When the switch is "on" (when it is connected to the source), a DC current flows through the load. Because the inductor acts as a short circuit to DC, ideally there is no voltage drop across the inductor. Likewise, since the capacitor acts like an open circuit for DC, there will be no current branching off into the capacitor, as a result of which all the current generated in the circuit passes through the load, which is what we want. When, however, the switch is "off", the source is decoupled from the load and, as a result, there will be no current flowing into the load coming from the source. However, since the capacitor has been

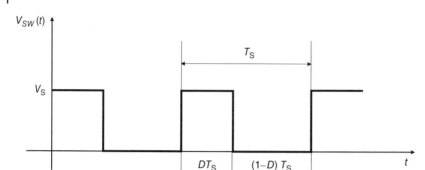

Figure 10.6 The duty cycle of a DC-DC converter.

charging in the previous cycle, now it will discharge through the load and it acts as a current source. Subsequently, the average voltage, consisting of the voltage due to the source from the previous cycle and the voltage due to the discharging capacitor in the present cycle, should provide the desired load voltage.

Suppose we periodically switch "on" and "off" the switch. Furthermore, suppose the duration of the switch in the "on" state is DT_s and in the "off" state is $(1 - D)T_s$ where T_s is the switching period and $\{D : 0 < D < 1\}$ is defined as the duty cycle of the regulator, as can be seen in Figure 10.6. Thus the switching frequency of the DC-DC converter is:

$$f_s = \frac{1}{T_s} \tag{10.10}$$

One of the consequences of the periodic switching is that the voltage across the switch, $V_{SW}(t)$ contains high-frequency components. However, if we choose the cut-off frequency (f_c) of the low-pass filter carefully (by making it significantly smaller than the switching frequency), then it is possible to suppress the high-frequency components:

$$f_c = \frac{1}{\sqrt{LC}} << f_s \tag{10.11}$$

The DC component of V_{SW} can be determined by using the inverse Fourier transform:

$$V_{SW}(t) = \frac{1}{T_s} \int_0^{T_s} V_{SW}(j\omega) e^{j\omega t} dt \tag{10.12}$$

Since we are interested in the DC voltage ($\omega = 0$), we can reduce the above expression as follows:

$$V_{SW} = \frac{1}{T_s} \int_0^{DT_s} V_S dt = DV_S \tag{10.13}$$

For our case, $V_{SW} = 2.3$ V and $V_S = 3.7$ V. Hence,

$$D = \frac{2.3}{3.7} \approx 0.62$$

10.2 Architecture

So far, we have seen how different factors can affect the design of an energy-harvesting system. The architecture of this system, both the choice of building blocks and how the building blocks should be interconnected, is likewise influenced by those factors. Conceptually speaking, however, it is possible to identify the most significant components and to determine the input-output relationships between these components, as shown in Figure 10.7. Since we are interested in supplying energy to a sensing system, there are understandably two types of measurand: the first is the main physical process or object for which the sensing system is designed whilst the second is our energy source. The two may or may not be similar. Even when they are similar, the sensing techniques we employ are essentially different, as the outputs of the two systems are different.

The energy sensor (or converter) transforms the physical quantity (radiation, wind, heat, vibration, and so on) to electrical power. The sensor is typically characterised by its conversion efficiency and suitability with respect to the sensing system in terms of its size, weight, and the amount of heat it produces. Since the sensing system is typically small, the energy harvester, when compared to its macro counterparts, is typically small as well, which means the issues we raised and discussed in Chapter 9 as regards miniaturisation concern us here. One of the crucial problems with miniaturised energy harvesting is that the efficiency of the converter decreases as its size decreases.

In general there are two regulators. One of them (the input regulator) conditions the output of the energy converter to meet the operational requirements of both the storage and the sensing systems. For example, the output voltage and current of the converter may not match the input voltage and current requirements of the storage system, in which case the regulator adjusts the output voltage and current to meet the requirement. As we have already seen, the input regulator also manages the charging process of the storage system. Similarly, the output regulator conditions the output of the storage system and the input regulator to meet the operational requirements of the sensing system. The essential difference between the input and the output regulators is that whereas the charging current of the storage system is constant, the operational current of the load varies, depending on its activity level. The activity level, in turn, depends on the complexity of the sensing system. If the sensing system integrates a processor, then, the amount of current it draws from the output regulator significantly varies with the variation of the processor's current state. For example, Table 10.3 displays the nominal

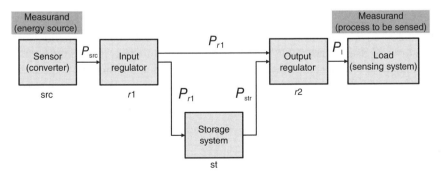

Figure 10.7 The main components constituting the architecture of an energy-harvesting system.

Table 10.3 Nominal current draw of the
ATmega128L Microcontroller

Power mode	Current (mA)
Active	8.0
Idle	3.2
ADC noise reduction	1.0
Power-down	0.103
Power-save	0.110
Standby	0.216
Extended standby	0.223

current an ATmega128L microcontroller draws in different power modes (Dargie 2012). Consequently, the output regulator should be able to provide different amounts of current to the sensing system and protect the storage system from overload.

When the available energy source is inadequate to directly supply power to the sensing system (load), the storage system has to charge first. However the energy source may have a surplus of energy available to directly supply the load with power whilst the storage system is charged. In this case, the input regulator should meet the demands of the storage system as well as of the output regulator. The overall efficiency of the energy-harvesting system depends on the efficiency of the input and output regulators as well as of the storage system (itself dependent on the self-discharge rate of the storage system). Hence,

$$\eta = \eta_{\text{src}} \times \eta_{\text{r1}} \times \eta_{\text{str}} \times \eta_{\text{r2}} \tag{10.14}$$

Equation (10.14) plays a useful role in the design of an energy-harvesting system because it easily relates the power that is required by the load (P_l) to the power that can be supplied by the source (P_{src}):

$$P_l = \eta \times P_{\text{src}} \tag{10.15}$$

Alternatively,

$$P_{\text{src}} = \frac{1}{\eta} \times P_l \tag{10.16}$$

Example 10.2   We wish to design a power supply system that harvests energy from blood pressure and uses 300 mg of lithium polymer as its storage system. Suppose the storage system has a self-discharge rate of 2 %, both the output and the input regulators have 50 % efficiency, and the sensing system requires 20 mW of power, how much power should the energy source deliver?

According to Table 10.2, up to 0.37 W can be harvested from the blood pressure with a conversion efficiency of 40 %. Hence we have:

$$\eta_{\text{src}} = 0.4$$
$$\eta_{\text{r1}} = 0.5$$

$$\eta_{st} = 0.98$$
$$\eta_{r2} = 0.5$$

And,

$$\eta = 0.4 \times 0.5 \times 0.98 \times 0.5 = 0.098$$

Therefore, the amount of power the energy source should deliver is:

$$P_{src} = \frac{1}{\eta} \times P_l = \frac{1}{0.098} \times 20 \text{ mW} = 204 \text{ mW}$$

If we assume a 10 % error in our modelling, the maximum power consumed by the sensing system would still be 224.4 W, which is below 370 mW, which can be obtained from the energy source.

10.3 Prototypes

At the macro scale, the use of renewable energy is becoming to be a ubiquitous phenomenon around the world. As of July 2015, statistics from the European Wind Energy Association (EWEA) reveal that:

> In the first six months of 2015, Europe fully grid connected 584 commercial offshore wind turbines, with a combined capacity totalling 2342.9 MW. Additionally, 15 commercial wind farms were under construction. Once completed, these wind farms will have a total capacity of over 4268.5 MW.[1]

At the macro scale, however, harvesting energy is not as ubiquitous. In this section, we shall consider some representative prototypes.

10.3.1 Microsolar Panel

A simple crystalline PN-junction crystalline silicon is the basis for constructing a solar panel to transform light energy into electrical energy. Silicon is a semiconductor and therefore has four electrons in its outer orbit. Atoms in a silicon crystal form a covalent bond to get eight electrons in their outer orbit, which makes the crystal stable. When considered separately, each silicon atom is electrically neutral, as the number of electrons (which are negatively charged) and protons (which are positively charged) are always equal. So crystalline silicon is also electrically neutral. Incidentally, metals also have equal numbers of electrons and protons, but the electrons in the outer orbits of metals, which are always less than three, are loosely held or attracted by the positive nucleus, as a result of which they can be easily excited out of their orbits.

A non-metal atom, such as phosphorus, which has five electrons in its outer orbit, can be injected into a silicon crystal by a process known as doping. The process does not immediately change the electrical neutrality of the now doped crystalline silicon, as the number of electrons and protons are still equal, but it creates the necessary condition to make it unstable. Similarly, a metal atom, such as boron, with three electrons in its

---

1 Source: European Wind Energy Association (2016).

outer shell, can also be injected into the silicon crystal to create the necessary condition for an electrical imbalance. The quantity of dopant (impurities) in the crystalline silicon significantly influences its electrical properties. When the crystal is doped lightly, it is often referred to as extrinsic semiconductor because its semiconductor property dominates; in contrast, when the crystal is highly doped, it behaves more like a conductor rather than a semiconductor and is referred to as a degenerate semiconductor.

Now consider Figure 10.8. Suppose we first have an undoped crystalline silicon as shown in (a). On the left-hand side of part (b), an N-type crystal is doped with atoms which have five outer electrons. Each silicon atom contributes four outer electrons. So in the neighbourhood of the dopant, there are altogether nine outer electrons, eight tightly held by a covalent bond, but one free to move. The more impurities there are, the more free electrons there will be. Similarly, on the right-hand side of part (b), an N-type crystal is doped with atoms having three electrons. Therefore, in the neighbourhood of this atom, there are only seven electrons, with a deficiency of one electron to make that region stable (in a sense, there is a positively charged hole in this region into which a single electron can fall).

When an N-type and a P-type crystal are physically brought together, an interesting phenomenon occurs. The free electrons in the N-type crystal will be accelerated towards

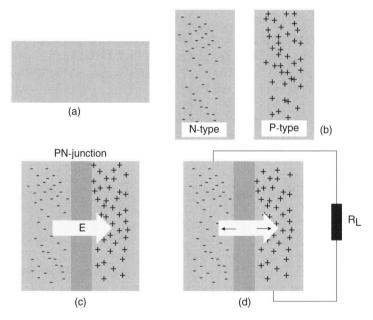

Figure 10.8 The formation of a PN-junction is the basis for the construction of a crystalline silicon solar panel: (a) a pure crystalline silicon in which silicon atoms form covalent bonding to form the crystal; (b) negatively and positively doped crystalline silicon. In both cases, the doped silicon, when separately considered, is electrically neutral, as the number of positively charged and negatively charged particles are equal, effectively cancelling each others' charges; (c) the combination of N- and P-type crystals enables some of the free electrons in the N-type crystal to migrate to the P-type crystal in order to combine with some of the positive charges – this way, the net charge density in the N-type crystal becomes positive and the net charge density in the P-type crystal becomes negative and the charge imbalance creates an electric field; (d) a closed circuit formed by an external load resistance enables the flow of current due to the charge concentration imbalance in the PN-junction solar panel.

the P-type crystal to fill up the deficiency there. In doing so, however, now that they have left their original crystal, there will be an imbalance between the electrons and protons in the N-type crystal (the net charge of the crystal will be positive). Similarly, there will be an imbalance in the P-type crystal with the P-type crystal becoming negative. As can be seen in part (c), the charge imbalance in these regions creates an electric field, the strength of which depends on the rate at which free electrons initially migrate and the concentration of dopant in both regions. The tendency of this field is to oppose the further migration of electrons and holes across the PN-junction.

Meanwhile, what happens when energy of some kind frees some of the electrons in the PN junction? They will be accelerated by the electric field towards the N-type crystal, while the free holes will be accelerated towards the P-type crystal. If the circuit is closed, as shown in Figure 10.8d, then the accelerated electrons will repel some of the free electrons in the N-type crystal whilst their counterparts, the accelerated holes, will repel some of the holes in the P-type crystal. The repelled electrons and holes in their turn repel other electrons and holes, and so the propagation of electrons and holes will continue towards the edges and through the connecting wires, and this way, a flow of electrons (a current) will be created.

A crystalline silicon (c-Si) solar cell is essentially a PN-junction semiconductor that releases electrons when the PN-junction is exposed to radiation. A moderately doped P-type square wafer is used as an absorber material. Its typical dimensions are 10×10 cm and a thickness of approximately 300 μm. Both sides of the wafer are highly doped: the top N-type and the bottom P-type. The solar panel is organised into cells, modules, and arrays (as shown in Figure 10.9). The power that can be generated by a single cell is essentially small, but by connecting multiple cells in series modules can be formed. These are usually a sealed or encapsulated unit of convenient size. Similarly, when multiple modules are connected, they form arrays. The power generated by solar panels is environmentally friendly, because the panels produce no air, thermal, or water pollution. The only disadvantage of solar panels at present is their high cost and the fact that the production process has a negative environmental impact, primarily from the energy required. A small amount of heavy metals (typically lead and cobalt) are also produced in the purification of crystalline silicon.

Taneja et al. (2008) at UC Berkeley implemented the architecture by employing a micro-solar panel (see Figure 10.7). They began their design by first analysing the power requirement of their sensing system, which consists of sensing, processing, and radio subsystems. The purpose of the sensor node was "to collect widespread, high-frequency, and automated observations of the life cycle of water as it progresses through a forest ecosystem". The sensor node can be configured to operate with different duty cycles. For their purposes, it was enough to operate the processor with a 0.4% and the radio with a 1.2% duty cycle. The peak active current the node draws when the processor is on and the radio is in receiving mode is 0.53 mA at 3.3 V. The node draws 15 μA when sleeping.

Their selection of the input and output regulators as well as the storage system was primarily influenced by cost and the efficiency with which power could be transferred from the solar panel to the storage system. The solar panel matching the size of their sensor node was a 4 V, 100 mA panel from Silicon Solar Inc. This panel delivers a maximum output power of 276 mW h at 3.11 V. The authors argue that experimenting with an input regulator showed that the panel was forced to operate at a point far from its

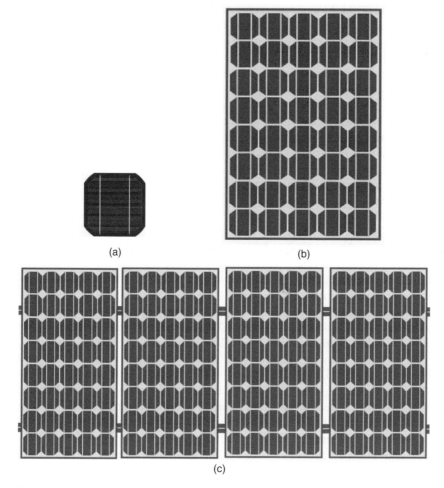

Figure 10.9 The components of a solar (photovoltaic) panel: (a) a single cell; (b) a single module; (c) an array of modules.

maximum power point. Therefore, they decided to directly supply the output of the solar panel to the storage system.

Their choice of storage system took into consideration the simplicity with which it could be integrated into their overall system. For this reason, they decided on Ni-MH cells, with $\eta_{st} = 0.66$. Similarly, in their decision regarding the output regulator, the researchers took into account the operating ranges of the storage system and the sensing system (the load). With two Ni-MH AA batteries, the nominal operating voltage of the storage system should be 2.4 V. However, the operating voltage of the sensing system varied between 2.7 and 3.6 V. Therefore, the output regulator had to be able to boost the voltage of the storage system. The output regulator was set up to provide a stable supply voltage. Finally, the authors chose an LTC1751 regulator which, unfortunately, has an efficiency of only 50 % ($\eta_{r2} = 0.5$).

Going backward, with the sensing system having an average daily energy requirement of approximately 45.3 mW h and the output regulator having a 50 % efficiency, the storage system had to deliver 79.2 mW h. Thus the input power to the storage system had to be:

$$P_{src} = \frac{1}{\eta_{str}} \times P_{str} = (1/0.66) \times 79.2 \text{ mW h} = 120 \text{ mW h}$$

According to the researchers, the solar panel generates 139 mW h per day if it receives 30 min sunlight per day. Figure 10.10 shows the prototype they designed and deployed at the Angelo Reserve in Northern California. Altogether, the authors deployed 19 nodes over a 220 m × 260 m area stretching across a deep ravine.

10.3.2 Microgenerator

The principle of an electric generator is straightforward and obeys Faraday's electromagnetic induction law (more precisely, Maxwell-Faraday's equation). If two magnets are placed near one another, as in Figure 10.11, a magnetic field will be set up between

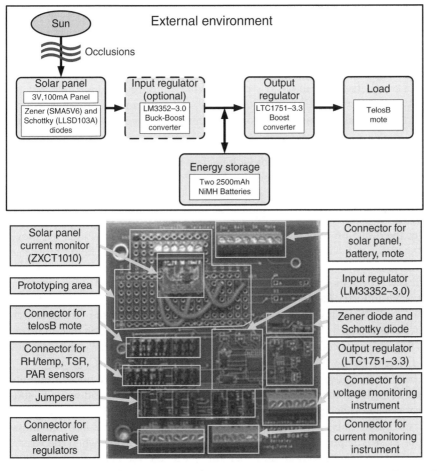

Figure 10.10 A wireless sensor node prototype using a solar panel to harvest energy from sunlight. Courtesy of Taneja et al. (2008) (*Proceedings of the 7th International Conference on Information Processing in Sensor Networks*, IEEE 2008).

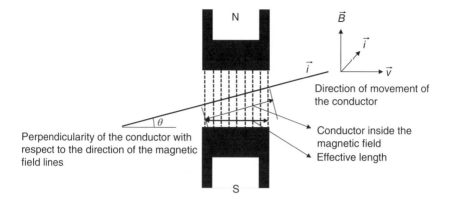

Figure 10.11 The basic principle of an electric generator. A conductor moving inside a magnetic field induces an electric current within itself. The magnitude and direction of the induced current depend on the effective length of the conductor inside the magnetic field, the relative direction of the movement of the conductor with respect to the direction of the field, and the strength of the magnetic field.

them. The magnetic lines of force are always directed from a north pole to a south pole. If a simple conductor of length l moves inside the magnetic field (assuming the conductor makes a complete loop), an electric current will be induced in the conductor and begins to flow. Alternatively, the conductor can be stationary and the two magnets can be moving. In either case, the magnitude of the current depends on the following factors:

- the effective length of the conductor
- the relative position of the conductor with respect to the magnetic field (the component of the conductor which is perpendicular to the magnetic field)
- the speed of the conductor
- the magnetic field strength.

Similarly, the direction of the flow of current depends on the direction of the movement of the conductor with respect to the direction of the magnetic field. In order to produce an appreciable amount of current, the effective length of the conductor must be long. This can be achieved by winding a long wire around a ferromagnetic material. The wristwatch industry has been effectively exploiting this simple principle for more than half a decade. It employs the movement of wrists to move the rotor of the microgenerator to which permanent magnets are attached. The moving magnetic field induces a voltage in the stationary coils of a stator. The AC voltage thus generated is then supplied to an AC-DC converter, which produces a DC current that drives a DC motor.

As can be seen in Figure 10.12a, the ETA Autoquartz watch uses a micromotor and an electronic system to manage the movements of the hour, minute, and second hands. The storage capacitor (accumulator) provides the motor with DC electrical power. It is charged itself by the microgenerator, which produces current as a result of the movement of the wrist. The proof-mass on top of the generator oscillates when the wrist is moved, thereby moving the rotor of the generator. The Seiko system has similar building blocks to generate electrical power using an oscillating proof-mass and to temporarily store the power for regulatory purpose. The proof-mass is conditioned to move in one dimension only (for example, along the x-axis), but the direction of movement along

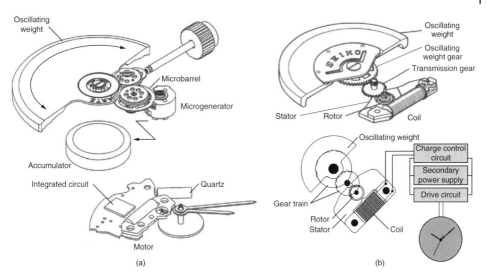

Figure 10.12 Commercially available self-winding mechanisms for micro-generators. (a) Self-winding system for the ETA Autoquartz. The proof-mass winds a spring that pulses the micro-generator at an optimal rate of 15,000 RPM for 50 ms, producing 6 mA at a voltage greater than 16 V. The electric power produced by the microgenerator charges a capacitor, which serves as the energy storage of the Autoquartz. (b) The self-winding mechanism of Seiko's automatic generating system. The mechanism omits the intermediate spring and produces 5 μW on average from an ambient movement (when the wristwatch is worn) and 1 mW when the wristwatch is forcibly shaken. Courtesy of Paradiso and Starner (2005), *IEEE Pervasive Computing* 2005.

that axis (left or right) does not matter, because full-wave rectifiers can be used to convert the AC voltage produced as a result of the left-right movement of the proof-mass into a DC voltage.

The sensor and MEMS research community is attempting to take advantage of similar approaches to fabricate microgenerators to supply microelectromechanical sensors with power. Beeby et al. (2007) designed and fabricated a microgenerator with a stationary coil and a moving cantilever, on the top and bottom of which high-density magnets were bonded. The magnets are produced from sintered neodymium iron boron (NdFeB). Cyanoacrylate is used to bond the magnets to the cantilever. Each magnet has dimensions of $1 \times 1 \times 1.5$ mm, the longer dimension (1.5 mm) being normal to the magnetic field. The way the magnets are arranged produces a concentrated flux gradient through the stationary coil as they vibrate. The free end of the cantilever is connected with tungsten wires to provide it with additional mass. The density of the magnet is 7.6 g cm^{-3}. Figure 10.13 shows the dimensions of the cantilever.

The researchers introduced slots and holes into the beam to accommodate the coil and bolt. Moreover, the corners of the beam are shaped so as to reduce the effect of concentrated stress. The cantilever was fabricated from double-polished, single-crystal silicon wafers. The cantilever produced a resonance (vibration) frequency of 50 and 60 Hz when its thickness was 40 μm. The cantilever beam was fabricated with deep reactive ion etching through the 50 μm thickness, and the wafers were resist bonded to a host wafer. The cantilever beam assembly was clamped onto the base using an M1-sized nut and bolt and a square washer, the square washer giving a straight clamped edge

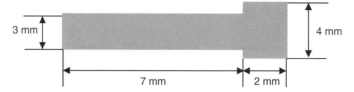

Figure 10.13 The dimensions of the cantilever (the rotor of the microgenerator) to which the high energy density magnets are attached.

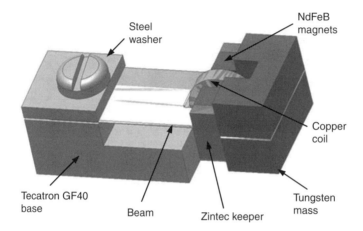

Figure 10.14 The architecture of the cantilever-based microgenerator, which produces an electric power from an ambient vibration. Courtesy of Beeby et al. (2007), *Journal of Micromechanics and Microengineering* (2007).

perpendicular to the beam length. Similarly, the stationary coil was manually bonded to a semi-circular recess machined in the base. It had an outside radius of 1.2 mm, an inside radius of 0.3 mm and a thickness of 0.5 mm. It was wound from 25 μm diameter enamelled copper wire and had 600 turns. Figure 10.14 shows the architecture of the micro-generator.

The researchers report that the micro-generator produced 46 μW in a resistive load of 4 kΩ when the cantilever received an acceleration of 0.59 m s^{-2} at a resonance frequency of 52 Hz. A voltage of 428 mV (RMS) was obtained from the generator with a 2300-turn coil. The efficiency of the generator was reported to be 30 % (that is, 30 % of the power supplied from the environment was transformed to useful electrical power in the load).

10.3.3 Piezoelectricity

In piezoelectric energy harvesting, a mechanical vibration can be directly converted to electrical energy, with no need for an intermediate stage. This is because when the atomic structure of piezoelectric materials (such as crystals or certain ceramics) is compressed (strained), an electric charge accumulates in the material producing a potential difference. The amount of energy that can be harvested this way is typically small (in the microwatt range), but it can be sufficient to power MEMS sensors.

Scientists at universities and research centres in the US and China have identified several medical applications for which the energy harvested from piezoelectric materials can be useful. These are heart rate monitors, pacemakers, implantable cardioverter defibrillator (ICD), and neural stimulators (Dagdeviren et al. 2014). The scientists indicate that existing technologies for these devices (bioelectronic devices) require batteries to provide continuous diagnostics and therapy, but the operational lifetime of these batteries is inherently short (a few days for wearable devices and a few years for implants) due to practical constraints. Consequently, surgical procedures are required to replace the batteries of implantable devices, which may expose patients to health risks, heightened morbidity, and even potential mortality; moreover it is needless to state that surgical procedures are costly.

The researchers proposed and developed a piezoelectric harvester to produce energy from the normal rhythms of the heart and the lungs. The structure of the basic element of the harvester consists of a lead zirconate titanate (PZT) ribbon, which is sandwiched between two electrodes, thus forming a capacitor-like structure. The PZT ribbon has a thickness of 500 nm and width and length of 100 μm and 2.02 mm respectively. It is produced by wet chemical etching on a silicon wafer. The top electrode, which has dimensions of 50 μm × 2 mm, was formed by the deposition of Au/Cr (200 nm/10 nm) with an electron beam evaporator on the surface of a multilayer stack of $Pb(Zr_{0.52}Ti_{0.48})O_3/Pt/Ti/SiO_2$ (500 nm/300 nm/20 nm/600 nm) on a silicon wafer. The bottom electrode (Pt/Ti) has dimensions of 140 μm × 2.02 mm and was patterned by wet chemical etching.

A further encapsulation process of the assembly using polyimide (PI), a biocompatible material, isolates the harvester from bodily fluids and tissues and minimises the risks of failure or an immune response. The researchers report that the entire structure is highly flexible, with computed bending stiffnesses (per unit width) of 0.22 N.mm and 0.10 N.mm for regions coincident with and away from the PZT structures, respectively. For a bending radius of 2.5 cm, the researchers reported a maximum strain in the PZT of 0.1 %. Figure 10.15 shows the structure of the piezoelectric harvester, an optical

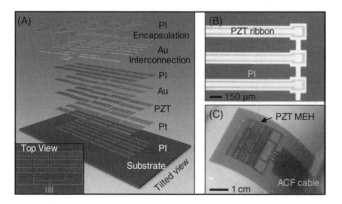

Figure 10.15 A flexible piezoelectric energy harvester: (A) a schematic illustration of the components consisting of the piezoelectric harvester, namely, a piezoelectric ribbon sandwiched between two electrodes and encapsulated within polyimide (PI); (B) an optical microscopic image of PZT ribbons fabricated onto a thin film of PI; (C) image of the entire harvester. MEH, mechanical enegy harvester. Courtesy of Dagdeviren et al. (2014), *PNAS* (2014).

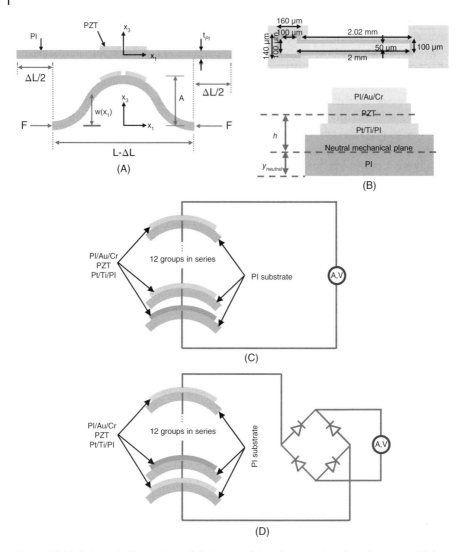

Figure 10.16 Schematic illustrations of the layout of the piezoelectric energy harvester: (A) the theoretical shape for buckling the PZT harvester under compression; (B) a top view of a single PZT ribbon capacitor structure (top), and a cross-section showing the position of the neutral mechanical plane of the device (bottom). A buckled array of PZT ribbon capacitor structures on a PI substrate: (C) without a rectification circuit and (D) with a rectification circuit. Courtesy of Dagdeviren et al. (2014), *PNAS* (2014).

microscope image of the PZT ribbons printed onto a thin film of PI, and the PZT image together with connecting cables. The researchers reported that the harvester can generate a peak current ranging from 0.06 to 0.15 μA with a conversion efficiency of 1.7 %, which is sufficient to power the medical devices that motivated the work.

A single assembly essentially makes up a single PZT capacitor that charges when it is vibrated. The researchers organised the elements into 12 groups, a single group consisting of ten PZT capacitors connected in parallel and each group connected in series to increase the output voltage. Figure 10.16 displays the way the PZT capacitors

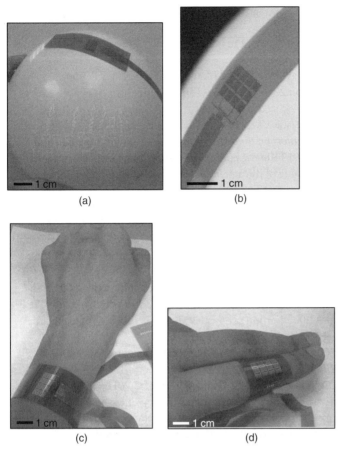

Figure 10.17 The flexibility of the PZT harvester and its conformity to various surfaces: (A) deployment on a balloon; (B) a magnified view of the harvester; (C) deployment on a human arm; (D) deployment of the harvester on a finger. Courtesy of Dagdeviren et al. (2014), *PNAS* (2014).

are organised, along with the electronic components that are required to condition the output voltage. Figure 10.17 displays the flexible deployment of the energy harvester on different objects.

References

Beeby SP, Torah R, Tudor M, Glynne-Jones P, O'Donnell T, Saha C and Roy S 2007 A micro electromagnetic generator for vibration energy harvesting. *Journal of Micromechanics and Microengineering*, **17**(7), 1257.

Beeby SP, Tudor MJ and White N 2006 Energy harvesting vibration sources for microsystems applications. *Measurement Science and Technology*, **17**(12), R175.

Conway BE 2013 *Electrochemical Supercapacitors: Scientific Fundamentals and Technological Applications*. Springer Science & Business Media.

Dagdeviren C, Yang BD, Su Y, Tran PL, Joe P, Anderson E, Xia J, Doraiswamy V, Dehdashti B, Feng X *et al.* 2014 Conformal piezoelectric energy harvesting and storage from motions of the heart, lung, and diaphragm. *Proceedings of the National Academy of Sciences,* **111**(5), 1927–1932.

Dargie W 2012 Dynamic power management in wireless sensor networks: State-of-the-art. *Sensors Journal, IEEE,* **12**(5), 1518–1528.

Erturk A and Inman DJ 2011 *Piezoelectric Energy Harvesting.* John Wiley & Sons.

European Wind Energy Association 2016 Statistics webpage. URL: http://www.ewea.org/library/statistics/offshore/.

Kansal A, Hsu J, Zahedi S and Srivastava MB 2007 Power management in energy harvesting sensor networks. *ACM Transactions on Embedded Computing Systems (TECS),* **6**(4), 32.

Khaligh A, Zeng P and Zheng C 2010 Kinetic energy harvesting using piezoelectric and electromagnetic technologies: state of the art. *Industrial Electronics, IEEE Transactions on,* **57**(3), 850–860.

López-Lapeña O, Penella MT and Gasulla M 2010 A new MPPT method for low-power solar energy harvesting. *Industrial Electronics, IEEE Transactions on,* **57**(9), 3129–3138.

Mitcheson PD, Yeatman EM, Rao GK, Holmes AS and Green TC 2008 Energy harvesting from human and machine motion for wireless electronic devices. *Proceedings of the IEEE,* **96**(9), 1457–1486.

Nishimoto H, Kawahara Y and Asami T 2010 Prototype implementation of ambient RF energy harvesting wireless sensor networks. *Sensors, 2010 IEEE,* pp. 1282–1287–IEEE.

Pandolfo A and Hollenkamp A 2006 Carbon properties and their role in supercapacitors. *Journal of Power Sources,* **157**(1), 11–27.

Paradiso JA and Starner T 2005 Energy scavenging for mobile and wireless electronics. *Pervasive Computing, IEEE* **4**(1), 18–27.

Qiu Y, Van Liempd C, Op het Veld B, Blanken PG and Van Hoof C 2011 5μw-to-10mw input power range inductive boost converter for indoor photovoltaic energy harvesting with integrated maximum power point tracking algorithm *Solid-State Circuits Conference Digest of Technical Papers (ISSCC), 2011 IEEE International,* pp. 118–120IEEE.

Raghunathan V, Kansal A, Hsu J, Friedman J and Srivastava M 2005 Design considerations for solar energy harvesting wireless embedded systems *Proceedings of the 4th International Symposium on Information Processing in Sensor Networks,* p. 64.

Ramadass YK and Chandrakasan AP 2010 A battery-less thermoelectric energy-harvesting interface circuit with 35 mV startup voltage. *IEEE Journal of Solid-State Circuits.* **46**(1), 333–341.

Saha C, O'Donnell T, Wang N and McCloskey P 2008 Electromagnetic generator for harvesting energy from human motion. *Sensors and Actuators A: Physical,* **147**(1), 248–253.

Shi Y, Xie L, Hou YT and Sherali HD 2011 On renewable sensor networks with wireless energy transfer *INFOCOM, 2011 Proceedings IEEE,* pp. 1350–1358.

Sirohi J and Mahadik R 2011 Piezoelectric wind energy harvester for low-power sensors. *Journal of Intelligent Material Systems and Structures,* **22**(18), 2215–2228.

Stanton SC, McGehee CC and Mann BP 2010 Nonlinear dynamics for broadband energy harvesting: investigation of a bistable piezoelectric inertial generator. *Physica D: Nonlinear Phenomena,* **239**(10), 640–653.

Stephen N 2006 On energy harvesting from ambient vibration. *Journal of Sound and Vibration,* **293**(1), 409–425.

Sudevalayam S and Kulkarni P 2011 Energy harvesting sensor nodes: Survey and implications. *Communications Surveys & Tutorials, IEEE,* **13**(3), 443–461.

Tan YK and Panda SK 2011 Energy harvesting from hybrid indoor ambient light and thermal energy sources for enhanced performance of wireless sensor nodes. *Industrial Electronics, IEEE Transactions on,* **58**(9), 4424–4435.

Taneja J, Jeong J and Culler D 2008 Design, modeling, and capacity planning for micro-solar power sensor networks *Proceedings of the 7th international Conference on Information Processing in Sensor Networks*, pp. 407–418.

Vullers R, van Schaijk R, Doms I, Van Hoof C and Mertens R 2009 Micropower energy harvesting. *Solid-State Electronics,* **53**(7), 684–693.

Winter M and Brodd RJ 2004 What are batteries, fuel cells, and supercapacitors? *Chemical Reviews,* **104**(10), 4245–4270.

Zhu Y, Murali S, Stoller MD, Ganesh K, Cai W, Ferreira PJ, Pirkle A, Wallace RM, Cychosz KA, Thommes M *et al.* 2011 Carbon-based supercapacitors produced by activation of graphene. *Science,* **332**(6037), 1537–1541.

11

Sensor Selection and Integration

In the previous chapters we considered different fundamental sensing techniques (electrical, thermocouple, ultrasonic, optical, and magnetic sensing). These techniques can be employed in different ways to produce thermoelectric sensors (Dürig 2005), photoelectric sensors (Carotenuto et al. 2007), photomagnetic sensors (Giri et al. 2002), magnetoelectric sensors (Fiebig 2005; Nan et al. 2013), thermomagnetic sensors (Chen et al. 2014), thermooptic sensors (Berruti et al. 2013; Choi et al. 2008; Watts et al. 2013), elastomagnetic sensors (Jiles 1995), elastoelectric sensors (Dong et al. 2004), and thermoelastic sensors (Duwel et al. 2003, 2002; Roszhart 1990). Table 11.1 provides a brief summary of the different possibilities for input-output relationships between a measurand and the electrical quantity we wish to process by the subsequent conditioning circuit. For example, a temperature sensor can be realised using thermocouples, resistance temperature detectors (RTDs), thermistors, infrared detectors, or acoustic sensors. The decision to pick one of these technologies mainly depends on two essential criteria: the quality trade-off and the ease with which we can integrate the sensor into the rest of the system we wish to develop or monitor. In this chapter we shall discuss in some detail the sensor quality parameters and their integration aspects.

11.1 Sensor Selection

The quality of a sensor is judged by many parameters. Strictly speaking, the list of parameters required to specify a sensor can be formidably long; most of them are interdependent too. Furthermore, the significance of a particular parameter depends on the requirements of the overall system. While some of the parameters solely depend on the sensing element, some depend on the sensor as a whole (sensing and conditioning) as well as on the overall system. In this section, we shall discuss the parameters that apply to most existing sensors.

11.1.1 Accuracy

The accuracy of a sensor is a measure of the nearness of its output to the true value. The most significant challenge in measuring the accuracy of a sensor is obtaining the true value. Typically, standard references in a laboratory setting are used to determine a true value with which the sensor's output can be compared. The steps are as follows

Principles and Applications of Ubiquitous Sensing, First Edition. Waltenegus Dargie.
© 2017 John Wiley & Sons, Ltd. Published 2017 by John Wiley & Sons, Ltd.
Companion Website: www.wiley.com/go/dargie2017

Table 11.1 The selection of a sensing technology depends on the conditioning and subsequent electronic circuits

Measurand	Sensing element	Output
Magnetic	Hall effect	Voltage
	Magneto-resistive	Resistance
Temperature	Thermocouple	Voltage
	RTD	Resistance
	Thermistor	Resistance
	Infrared	Current
Humidity	Capacitive	Capacitance
	Infrared	Current
Force, weight,	Strain gauge	Resistance/voltage
Pressure, vibration	Piezo-electric	Voltage or charge
	LVDT	AC voltage
	Microphone	Voltage
	Accelerometer	Voltage
Flow	Magnetic flowmeter	Voltage
	Mass flowmeter	Resistance, voltage
	Ultrasound/Doppler	Frequency
Fluid level, volume	Ultrasound	Time delay
	Potentiometer	Resistance, voltage
	Capacitor	Capacitance
	Switch	On/off
Light	Photodiode	Current
Chemical	pH	Electrode voltage
Solution	Conductivity	Resistance/current

$m(t)$ $v_{out}(t)$

Figure 11.1 The accuracy of a sensor is usually determined in a lab setting by exposing the sensor to a measurand of known values (which can be measured by a standard device) and by comparing its response to reference values.

(refer also to Figure 11.1). In a controlled environment, the sensor is exposed to a measurand of known magnitude (here the term "known" refers to the use of a standard reference system to determine the value of the measurand) and the output of the sensor is observed. After a large set of measurements are taken (refer to Figure 11.2), the relationship between the input and the output is established in a probabilistic sense.

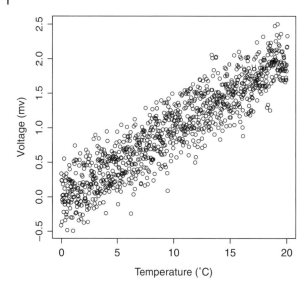

Figure 11.2 An example of the input output relationship between a measurand (temperature in °C) and the corresponding output (in mV) of a temperature sensor that can be established in a lab setting using a standard reference (by which we have measured the temperature of the measurand). This measurement can be used to establish the joint probability density function between the input temperature and the output voltage as well as the expected value between the input temperature and output voltage which will serve as the basis to establish the accuracy of the sensor.

A more detailed analysis of estimation will be given in Chapter 12. Generally, a sensor may not produce the same accuracy for all expected magnitudes of a measurand: often, the accuracy decreases towards the two extremes in the measurand's span. Furthermore, the accuracy varies when the measurand's frequency varies. Therefore, the accuracy of a sensor is calculated as the expected error (in percentage terms) of the sensor, which is obtained by taking both the magnitude and the frequency variations of the measurand into consideration.

11.1.2 Sensitivity

The sensitivity, when it refers to the sensing element, is the minimum magnitude of a measurand that can be picked up by the sensor to produce a corresponding output. Sensitivity, when it refers to the entire sensor, is not necessarily the quality of the sensing element alone but also the quality of the conditioning circuit and, most importantly, the quality of the preamplifier.

11.1.3 Zero-offset

The zero-offset of a sensor is the magnitude of the output when the measurand is zero. It can be expressed in different ways, depending on the sensor's output. For most electrical sensors, it is expressed in millivolts or milliamps. It can also be expressed as a unitless quantity, for example, as percentages of the full-scale output. Zero-offset can be corrected by proper calibration.

11.1.4 Reproducibility

Reproducibility refers to the a sensor's ability to repeatedly yield the same output (or comparatively the same) for the same input under the same operational condition. This is an important aspect particularly for:

- magnetic sensors, because of a potential hysteresis effect (discussed below)

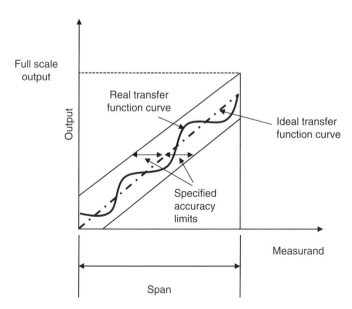

Figure 11.3 The relationship between span, full scale output, and transfer function.

- piezoelectric sensors
- strain gauges, because of the possibility of inelastic characteristics in the sensing elements.

It must be noted, however, that a highly reproducible sensor does not necessarily imply one that is highly accurate. Rather, reproducibility is a quality of consistency. It is sometimes referred to as precision.

11.1.5 Span

The difference between the highest and the lowest magnitudes of a measurand that can be detected by a sensor within an acceptable accuracy and without damaging or significantly affecting its reproducibility is called the sensing span. The span of a sensor is not merely a static value, as it is affected by the frequency response (or transfer function). Figure 11.3 summarises the relationship between span, full-scale output and ideal and real transfer functions.

11.1.6 Stability

Stability (or long-term stability) refers to the physical changes the sensor undergoes over time as a result of which there is a slow drift or shift in its operational condition. For example, some sensors adsorb biological and chemical artefacts on their surfaces in order to sense a measurand. This is, for instance, the case with chemical sensors that employ Raman spectroscopy (for example, a pH sensor). As a result of the accumulation of adsorbent, the sensor's response to the measurand gradually drifts or shifts. This drift can be observed in a shift in the operational frequency or in the expected spectrum of an output for a known input. Stability may also refer to the change in the spectrum of noise to which the sensor is sensitive. Other factors that affect the stability of a sensor are ageing, adsorption or desorption of contaminants, stress, package leaks, and gassing and chemical reactions (Clark Jr et al. 1988; Romain and Nicolas 2010; Spassov et al. 2008).

11.1.7 Resolution

Resolution refers to the minimum change in the measurand that can be detected by the sensor. The resolution of a sensor is affected by the sensing element itself as well as all the subsequent stages, including analogue-to-digital conversion (ADC). At the ADC level, the resolution is affect by the quantisation error, which in turn is affected by the number of bits allocated to digitise the analogue input. If the peak-to-peak output of the sensor is 10 mV, the resolution of a 10-bit ADC is given as:

$$\Delta_r = \frac{10}{2^{10}} = 0.009765625 = 9.8 \ \mu V \tag{11.1}$$

In other words, a variation in measurand values that correspond to an output voltage of less than 9.8 μV will not be detected by the sensor as a whole system, even though the sensing element itself may be able to detect the change.

11.1.8 Selectivity

The selectivity of a sensor is a measure of its capacity to respond only to the measurand while rejecting undesired signals (noise and interference) that may in some respect have overlapping characteristics with the measurand. This is a quality of both the sensing element and the conditioning circuit. When the measurand and the interfering inputs have overlapping spectra, difference amplifiers can be used to improve the selectivity of the sensor. If the magnitude of the measurand is appreciably large, clippers and limiters can be used to suppressed the noise. If the measurand and the interfering signals occupy different spectra, analogue and digital filters can be used to separate them (to selectively amplify or suppress). However, there are also more subtle sources of noise, such as thermal, radiative, and magnetic sources, which cannot be easily suppressed or rejected. In this case, selectivity becomes a part of the structure of the sensing element. For chemical and biochemical sensors, selectivity refers to the ability of a sensor to selectively react to particular molecules or compounds while staying unresponsive to others.

11.1.9 Response Time

The duration a sensor requires to approach its true output when subjected to a step input is referred to as its response time. This parameter is typically affected by the frequency response of the sensor. An ideal sensor will have a constant (flat) speed of response within its operational band.

11.1.10 Self-heating

Self-heating is an important consideration and particularly affects electrical sensing elements. When a current (for example, a biasing current) circulates through an electrical circuit, there will be a voltage drop across the inductive, capacitive, and resistive components, as a result of which there is power dissipation in the form of heat. This phenomenon is called self-heating and it is problematic to sensing because some of the characteristics of the sensors are temperature dependent. The problem is particularly pronounced if the sensing element has a significant resistive characteristic.

11.1.11 Hysteresis

Hysteresis refers to the difficulty of a sensing element (this is specifically the characteristic of the sensing element and not of the conditioning circuits) to faithfully reproduce the same one-to-one relationship between the measurand and the sensor output in the opposite direction of operation (when the magnitude of the measurand is decreasing) as in the forward direction (when the magnitude of the measurand is increasing). A hysteresis-free sensing element faithfully reproduces the same one-to-one relationship in both directions. An instance of the effect of hysteresis on the reproducibility of a sensor is depicted in Figure 11.4. As can be seen, the sensor produces two different output curves for the same values of a measurand, depending on whether the sensor is measuring an increasing or a decreasing measurand.

11.1.12 Ambient Condition

All sensors require ambient conditions to function properly. This may refer to the ambient temperature, the maximum permissible exposure to external magnetic fields or radiation, surrounding vibration, and so on. In order to function properly, sensing devices such as ECGs, EEGs, SQUIDs, and many others require compete shielding from external magnetic fields and from signals produced by power lines. Similarly, the stability of most magnetic and optical sensors can be affected by thermal conditions. In general, the ambient condition of a sensor describes the surrounding or environmental conditions that should be maintained for it to function properly.

11.1.13 Overload Characteristics

The maximum amount of current or power that can be drawn from a sensor or the maximum amount of power that can be dissipated inside it without significantly affecting its operation is referred to as its overload characteristic.

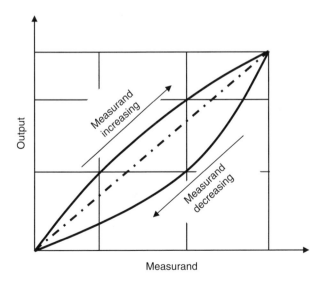

Figure 11.4 The effect of hysteresis on the reproducibility of a sensor.

11.1.14 Operating Life

The operating life of a sensor can be described in many ways, including in terms of its stability and accuracy. Generally it is the expected number of hours it can operate without its stability or accuracy deteriorating beyond a set value. In reality, the operating life of a sensor is not so easy to determine, as it depends on so many factors. For example, the operating life of a chemical sensor depends on the total amount of gas it is exposed to during its lifetime, as well as other environmental conditions, such as temperature, pressure and humidity.

11.1.15 Cost, Size, and Weight

Apart from the functional aspects of a sensor, its non-functional aspects are important too. These include cost, size, and weight. This is particularly true for those sensors that have to be seamlessly embedded into physical bodies or processes. Needless to say, non-functional aspects such as size and weight can have a direct bearing on some functional parameters, such as self-heating, span, and sensitivity. Micro-sized sensors, for example, can only allow a limited amount of power dissipation, which in turn limits their span. The same can be said of weight. Light sensors respond to surrounding heat and radiation more quickly and more accurately than heavy sensors.

11.2 Example: Temperature Sensor Selection

From the list of parameters that measure the quality of a sensor, it is clear that no single sensor can outperform all the others. Sensor selection is a trade-off. To highlight this point, we shall consider the selection of a temperature sensors as an example. A temperature sensor can be realised by using different sensing elements, as already mentioned at the beginning of this chapter. Each element has its own set of merits and demerits.

11.2.1 Resistance Temperature Detectors

An RTD is essentially a resistive sensor. As its name suggests, its resistance varies as a function of temperature. In Chapter 4 we saw that the resistance of a material is directly proportional to its resistivity and length but inversely proportional to its cross-sectional area. We can also express it in terms of the conductivity of the material, as follows (see Figure 11.5):

$$R = \frac{L}{\sigma S} \tag{11.2}$$

where S is the surface area of the material and σ is the conductivity of the resistor. However, the conductivity of the material is temperature dependent (conductivity decreases

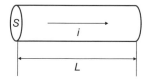

Figure 11.5 The resistance of a resistance temperature detector is temperature dependent.

as the temperature increases because temperature increases the random motion (collision) of electrons). Mathematically, the conductivity is expressed as:

$$\sigma = \frac{\sigma_0}{1 + \alpha(T - T_0)} \tag{11.3}$$

where α is the temperature coefficient of the material and T_0 and σ_0 are its reference temperature and conductivity at the reference temperature, respectively. The resistance of the material as a function of temperature can therefore be expressed as:

$$R(T) = \frac{L}{\sigma_0 S}[1 + \alpha(T - T_0)] \tag{11.4}$$

Among the merits of an RTD is the relative ease with which it can be fabricated. It can be produced as a wire coil (of length from a few centimetres to about 500), which can then be enclosed in a glass, ceramic, or metal housing. Alternatively, a thin-film RTD can be produced by depositing a thin layer of suitable material (such as platinum) on a thermally stable, thermally conducting, and electrically non-conducting material such as a ceramic. Figure 11.6 displays a wire coil and a thin-film temperature sensor. The RTD has a linear response over a wide range of temperatures (approximately from -250 to $700°C$) and can be made small enough to have a response time of the order of a fraction of a second. It is the most stable, most accurate (± 0.01 to $\pm 0.05°C$), and most reproducible temperature sensing element; it is also highly resistant to contamination and corrosion and does not require recalibration after fabrication.

Among its demerits are its relatively high cost, slow response time, poor sensitivity to small temperature changes (low resolution), and high sensitivity to external vibrations (due to the piezoresistive effect). Furthermore, due to the presence of a large resistive component, self-heating and, therefore, power dissipation are high. Finally, all RTDs require a small amount of operating current.

11.2.2 Thermistors

Thermistors are realised by pressing metal oxides into a semiconductor chip, bead, or wafer. Most existing thermistors have negative temperature coefficients (NTC), even though there are also some with positive temperature coefficients. For those with NTC, resistance decreases when temperature increases. Regardless of the type of temperature coefficient, thermistors typically have high temperature coefficients and high resistance.

Figure 11.6 Resistance temperature detectors.
(a) a wire-wound RTD; (b) a thin-film RTD.

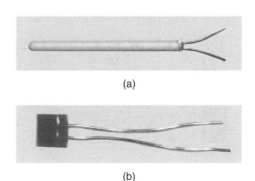

(a)

(b)

For an NTC thermistor, the resistance is approximated by:

$$R(T) = \alpha e^{-\beta/T} \tag{11.5}$$

where α (measured in Ω) and β (measured in Kelvin) are constants. From the equation, it can be seen that the relation between R and T is non-linear, but since β is small, the non-linearity is not significant.

Thermistors have high resolution and can be fabricated in various sizes and shapes. Due to their relatively high resistance, they permit a small amount of current to flow through them and, as a result, self-heating is comparatively small. If copper and nickel extension wires are used, temperature measurements can be stable. However, thermistors are fragile and can measure a relatively limited temperature range (from −50 to 600°C) with accuracy and stability varying along the sensing range.

11.2.3 Thermocouples

Thermocouples measure temperature directly, without requiring a biasing current or voltage. They are simple to fabricate, rugged, and inexpensive. No other sensor technology to date can match the wide sensing range that can be achieved using thermocouples (from approximately −273 to 2700°C). Moreover, they have the fastest response times and enable point temperature sensing and do not exhibit the problems associated with contact resistance (resistance arising from the interface between the sensor and the lead wires connecting the sensor with the conditioning circuit). However, they are the least stable and least reproducible. They have poor resolution and can easily be affected by ambient noise. Moreover, their accuracy is comparatively low. Unlike other temperature-sensing elements, however, thermocouples can be realised from different materials with different sensing parameters. For example, the temperatures that can be sensed by semiconductor thermocouples is typically in the range of −55 to 150°C.

11.2.4 Infrared

A heat-producing object radiates energy in the form of infrared light. A one-to-one relationship between the infrared energy and the temperature of the object can be established by setting up an infrared detector. The detector, for example, can be a photodiode or a phototransistor that generates an electrical voltage proportional to the light it absorbs. A simple infrared setup to measure temperature is displayed in Figure 11.7. This form of temperature measurement is desirable because no contact is needed between the heat-producing body and the sensor. In contrast, all the other

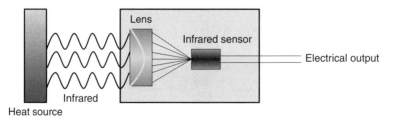

Figure 11.7 The basic set up of an infrared sensing.

temperature sensors we considered above require thermal contact to produce output. Infrared sensors have comparable response times to thermocouples and in some cases, they are even faster. They have good stability and high reproducibility, and because of the absence of direct contact with the object or process they monitor, they are not affected by corrosion or oxidation. As a result, their accuracy does not deteriorate appreciably over time. However, optical sensors in general have high set-up costs. They are more complex than the other technologies and require optoelectronic components to convert optical output to electrical output. Moreover, emissivity variations can affect their accuracy and the field of view of the lens and spot size may restrict their scope and usefulness (the spot size refers to the sensor area on which the infrared beam should focus). Likewise, their accuracy can easily be affected by dust, smoke, and background radiation.

11.3   Sensor Integration

A sensor is rarely a standalone system. Often it is a part of a more complex system, which may include advanced signal processing and actuation units. Consequently, integrating the sensor into the system is a vital step. The integration process has electrical and non-electrical aspects. One of the electrical aspects was considered in Chapter 3, namely impedance matching. Fulfilling the electrical requirements may not be very difficult, because most sensors are produced with standard interfaces, but addressing the non-electrical aspects so that the sensor functions as per its technical specification is the task of the integrator. The ease with which a sensor can be integrated into a system depends on the structure and requirements of both the sensor and the system. Two of the most important integration issues are related to optimal interface of the sensor with the process or the object it monitors and protecting the sensor from internal as well as surrounding interference. In this section, we shall consider these issues in some detail.

11.3.1   Dead Volume

All sensors should be properly interfaced with the process or object they monitor, so that what a sensor perceives is the actual change in the measurand. However, most sensors also require housing for various reasons (we shall discuss some of them shortly). This housing or shield not only minimises the exposure of the sensor to the measurand but also creates a dead volume, which can trap the measurand for some time (see Figure 11.8). The larger the dead volume, the longer the housing will hold a portion of the measurand. As a result, there will be a delay before the sensor perceives a change in the measurand. For example, for a temperature sensor, the dead volume can trap hot air, which may not represent the actual state of the process being monitored. The same can be said of magnetic or humidity sensors. Consequently, this volume has to be minimised.

11.3.2   Self-heating

The heat produced by a sensor due to a small amount of power dissipation in its resistive elements can skew its output in the long run. To reduce the effect of self-heating, a sensor

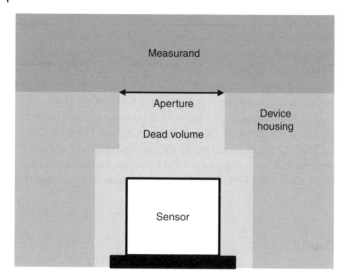

Figure 11.8 The volume created by the housing surrounding a sensor is called the dead volume. The larger the dead volume, the more probable it is the measurand will be trapped for a long time, affecting the sensitivity, accuracy, and response time of the sensor.

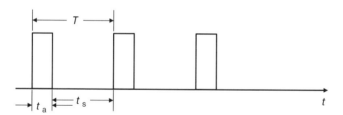

Figure 11.9 The duty cycle of a sensor describes the portion of time the sensor spends in an active state in order to reduce the effect of self-heating.

can operate in a duty cycle, so that during its sleep time, power can be withdrawn and the sensor can cool down. The concept of the duty cycle was defined in Chapter 10 and can be reused here as well:

$$D = \frac{\tau_a}{\tau_a + \tau_s} \times 100\% \tag{11.6}$$

where τ_a is the active time, τ_s is the sleep time, and $T = \tau_a + \tau_s$ is the period of a duty cycle (Figure 11.9). The duty cycle describes the portion of time the sensor spends in an active state. This time has to be very small ($D < 10\%$) for the sensor to cool down effectively. The duty cycle is also useful for reducing the power consumed by the system, particularly when it operates with exhaustible batteries.

Complementary to the duty cycle, and perhaps the most widely used approach to effectively deal with self-heating, is the use of a heat sink. In this approach, the sensor is embedded into material with a high heat conductivity (a metal) and a relatively large volume, so that when the sensor produces heat, the material absorbs the heat quickly, thereby serving as a cooling system to the sensor.

Example 11.1   Minimising the temperature generated in an RTD due to self-heating is of paramount importance for medical instrumentation because it affects the accuracy and response time of the instrument. Self-heating is the result of the biasing current flowing through the RTD, without which it is not possible to establish a relationship between the temperature of the measurand and the change in the resistance of the RTD. Figure 11.10 shows the circuit diagram of a digital RTD provided by Microchip Technology Inc. Determine the maximum power dissipation (self-heating) in the RTD for this set up.

We first begin by explaining the function of the main components. The PIC microcontroller unit (MCU) enables the sensor to be programmed and to be easily integrated with other digital systems. Since the microcontroller unit requires a digital input, the output of the RTD, which is analogue by nature, should first be changed into a digital bit stream by the ADC (MCP 3551). The ADC and the MCU are interfaced with each other by a full-duplex serial bus (SPI) which requires three lines (chip-select, master-in slave-out (MISO), and slave-in master-out (SIMO) pins). Both the microcontroller and the ADC require a DC biasing voltage, V_{DD}. Other than that, the ADC requires a reference voltage, which is the maximum magnitude of the analogue signal it should quantise. Moreover, the RTD requires a biasing voltage, which should be as small as possible to prevent the sensor from self-heating. The LDO component is a voltage regulator, which conditions the ADC reference (V_{REF}) and the RTD biasing voltages.

In order to calculate the voltage across the RTD, we should first calculate the reference voltage:

$$V_{REF} = \frac{R_A + R_{RTD}}{R_A + R_B + R_{RTD}} V_{LDO} \tag{11.7}$$

Then, the voltage across the RTD is given as:

$$V_{RTD} = \frac{R_{RTD}}{R_A + R_{RTD}} V_{REF} \tag{11.8}$$

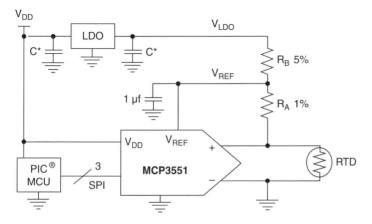

Figure 11.10 The circuit block diagram of a digital RTD. Courtesy of Microchip Technology Inc. (2013).

The nominal current flowing through the RTD is given as:

$$I_{RTD} = \frac{V_{RTD}}{R_{RTD}} \tag{11.9}$$

The temperature produced as a result of this current (self-heating) is equivalent to the power dissipation in the RTD:

$$P_{heat} = I_{RTD}^2 R_{RTD} = \frac{V_{RTD}^2}{R_{RTD}} \tag{11.10}$$

Example 11.2 One of the mechanisms to reduce the effect of self-heating is to use a heat sink to quickly absorb the heat generated by a sensor. Suppose we deposit a thin RTD sensor on an alumina (aluminium oxide, Al_2O_3) substrate, serving as a heat sink. Assuming that the RTD film can be considered as a one-dimensional conduction medium as shown in Figure 11.11, derive an expression relating the heat generated by the self-heating RTD and the temperature of the alumina substrate.

When a temperature difference between two regions exists, a temperature potential is established, in the same way an electric potential can be established as a result of a difference in electric charge concentration between two points. The tendency of this temperature potential is to generate a flow of heat from a higher temperature region to a lower temperature region. Hence, similar to current, which is the flow of electrons, the rate at which heat is transferred by conduction, q, is expressed as:

$$q = \frac{dT}{dR_{th}} \tag{11.11}$$

where R_{th} is the thermal resistance between the two conduction regions. It is a function of the length (l), cross-sectional area (A), and thermal conductivity (k) of the conduction region:

$$R_{th} = \frac{l}{kA} \tag{11.12}$$

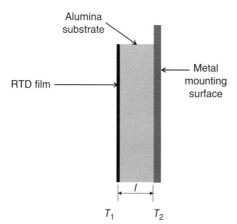

Alumina substrate

RTD film

Metal mounting surface

T_1 l T_2

Figure 11.11 A thin film of RTD is deposited on an alumina substrate to absorb the temperature generated by the self-heating RTD. The relationship between the temperature of the two media (T_1 for the RTD substrate and T_2 for the alumina substrate) can be determined using the heat conduction equation.

where k is the thermal conductivity of the material. If the heat covers the entire surface area and transfers in one direction (along the thickness, l), then, q is proportional to the product of the temperature gradient along the conduction path:

$$q = -kA\frac{dT}{dl} \tag{11.13}$$

More generally, the time-dependent and time-independent components of the temperature gradient are related to one another by the heat conduction equation:

$$\frac{\partial u}{\partial t} - \alpha^2 \left(\frac{\partial u}{\partial x} + \frac{\partial u}{\partial y} + \frac{\partial u}{\partial z} \right) = 0 \tag{11.14}$$

or in short:

$$\frac{\partial u}{\partial t} - \alpha^2 \nabla^2 u = 0 \tag{11.15}$$

where $u(t, x, y, z)$ is the temperature as a spatial and temporal function, ∇ denotes the Laplace operator, and α is a positive constant representing the thermal diffusivity of the material expressed as:

$$\alpha = \frac{k}{\rho c_p} \tag{11.16}$$

where ρ is the density and c_p is the specific heat capacity of the material. The solutions for the heat conduction equation determine the response time and the self-heating of the RTD structure we described by Figure 11.11. Taking into account the fact that the RTD is a very thin film and the temperature distribution can be expressed as $u(x, t)$, the solutions of the differential relations in Eq. (11.15) yield a time-independent final temperature distribution and a summation of exponentially damped orthogonal functions describing the evolution of the temperature distribution from an initial condition to a final condition:

$$u(x, t) = (T_1 - T_2)\frac{x}{l} + T_1 + \sum_{n=1}^{\infty} b_n (e^{-n^2 \pi^2 \alpha t / l^2}) \sin\left(\frac{n\pi x}{l}\right) \tag{11.17}$$

where

$$b_n = \frac{2}{l} \int_0^1 \left(f(x) - (T_1 - T_2)\frac{x}{l} - T_1 \right) \sin\left(\frac{n\pi x}{l}\right) dx \tag{11.18}$$

and $f(x)$ describes the temperature of the system at $t = 0$. When a biasing current flows through the RTD, the power dissipated in the RTD (P) generates self-heating, and this heat flows into the alumina substrate. If we combine Eqs. (11.11) and (11.13) and take into account the single-dimension assumption, the thermal conductivity between the RTD and the alumina substrate as a function of x can be expressed as:

$$q_u = -k\frac{\partial u(T)}{\partial x} \tag{11.19}$$

Furthermore, if we apply Eq. (11.19) to Eq. (11.17) as the boundary condition, we obtain the desired expression:

$$\frac{P}{A} = -\alpha\frac{(T_1 - T_2)}{l} \tag{11.20}$$

where A is the surface area of the RTD. Consequently,

$$T_1 = \frac{lP}{\alpha A} + T_2 = \frac{lV^2}{\alpha A R(T)} + T_2 \tag{11.21}$$

where P and V are the power dissipated and the voltage drop across the RTD, respectively. Remember that T_1 is the temperature of the RTD due to self-heating and T_2 is the temperature of the alumina substrate serving as a heat sink.

Example 11.3   We wish to employ a platinum RTD for monitoring the temperature of a SQUID. The RTD is deposited on an alumina substrate having a thickness of 0.254 mm and a surface area of 4.4×10^{-6} m². If a biasing current of 50 μA flows through the RTD and the temperature of the alumina substrate is kept at 0°C, what will be the self-heating of the RTD? The temperature-dependent resistance of the RTD, $R(T)$, at the specified temperature is 1000 Ω.

Figure 11.12 shows the set up of the biasing condition. The thermal diffusivity of alumina is $\alpha = 20 \times 10^{-5}$ m² s^{-1}. The power dissipation due to the biasing current at the specified temperature can be expressed as $P = I^2 R$. With this, we have all the parameters required by Eq. (11.21) to calculate T_1:

$$T_1 = \frac{0.254 \times 10^{-3}\text{m } ((50 \text{ μA})^2/1000 \text{ Ω})}{(1.2 \times 10^{-5}\text{m}^2 \text{ s}^{-1}) (4.4 \times 10^{-6} \text{ m}^2)} + 0°\text{C} = 12°\text{C}$$

Example 11.4   The maximum permissible junction temperature (T_{JMAX}) is one of the key factors limiting the self-heating of a device. Usually, T_{JMAX} is defined by the manufacturer and takes into consideration specific aspects such as the reliability of the die used in the manufacturing process. Suppose, for a sensor produced by Texas Instruments, the manufacturer defines the following relationship:

$$R_{th} = \frac{T_J - T_A}{P} \tag{11.22}$$

where T_J is the junction temperature (the junction being the interface between the device and the ambient environment), T_A is the ambient temperature, and P the power dissipation in the sensor producing self-heating. Show how the maximum dissipated power of a Texas Instruments synchronous step-down switcher changes when the ambient temperature changes from 25°C to 85°C. The maximum permissible temperature is 125°C and the thermal resistance is 44.5°C W^{-1}.

From Eq. (11.22), we can derive an expression for the permissible dissipated power:

$$P_{MAX} = \frac{T_{JMAX} - T_A}{R_{th}} \tag{11.23}$$

Substituting the given values in the above equation yields 2.25 W for an ambient temperature of 25°C and 0.9 W for 85°C.

Example 11.5   Self-heating in a sensor can be further reduced by identifying a suitable mounting surface during integration. Derive an expression for the power dissipation (self-heating) of an RTD mounted on a metallic surface, as shown in Figure 11.13.

Figure 11.12 An illustration of how self-heating generated by the RTD is absorbed by the alumina substrate in a Honeywell EL-700 platinum RTD. v_{DD} is the biasing voltage.

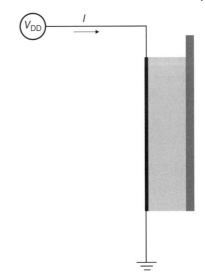

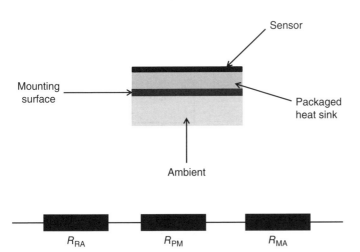

Figure 11.13 The careful choice of a mounting surface during sensor integration may reduce self-heating.

The alumina heat sink we considered above is part and parcel of the packaging process undertaken by the manufacturer and nothing can be done with it. A further improvement, however, can be achieved by using a mounting surface with good heat conduction. The overall thermal resistance of the entire assembly can be regarded as connecting three resistors in series. Subsequently, if we can apply Eq. (11.22) in Figure 11.13, we can rewrite the overall thermal resistance as follows:

$$R_{\text{th}} = R_{\text{RA}} + R_{\text{PM}} + R_{\text{MA}} \tag{11.24}$$

where R_{RA} is the junction thermal resistance between the RTD and the alumina substrate, R_{PM} is the junction thermal resistance between the sensor package and the mounting surface, and R_{MA} the junction thermal resistance between the mounting surface and the ambient temperature. Hence,

$$\frac{T_J - T_A}{P} = R_{th} = R_{RA} + R_{PM} + R_{MA} \tag{11.25}$$

From which we have:

$$P = \frac{T_J - T_A}{R_{RA} + R_{PM} + R_{MA}} \tag{11.26}$$

11.3.3 Internal Heat Sources

The relative position of a sensor with respect to other components, particularly with respect to processing units, power supply units, and on-board voltage-regulation units (power electronics) in a complex system is critical to its proper operation. The reason is that these components cause a significant amount of self-heating, which can affect the sensor's operation. Indeed, not only its placement but also the way it is connected to them is also critical because the connecting wires are heat conductors too. Figure 11.14 shows two possibilities for integrating a sensor with a microcontroller and a power supply unit. In (a), the sensor is placed very close to the two heat-generating units and it is connected to them with thick wires. This is, of course, an example of bad setup. In (b),

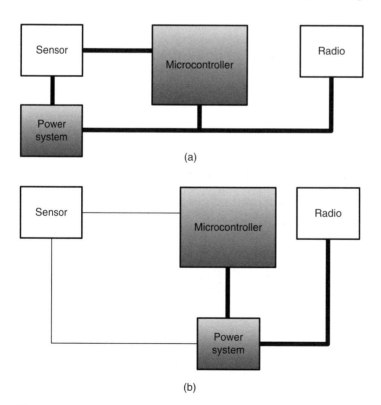

(a)

(b)

Figure 11.14 Sensor placement and wire selection are critical for protecting sensors from internal heat sources.

the sensor is placed as far away from the two heat sources as possible and thin connection wires are used. This may not be always possible, however. For example, the types of wires that can be used to connect a sensor with a processor is constrained by many factors such as the desired bit rate between the sensor and the processor, the type of duplex mechanism required, the space available, and whether or not an ADC internal to

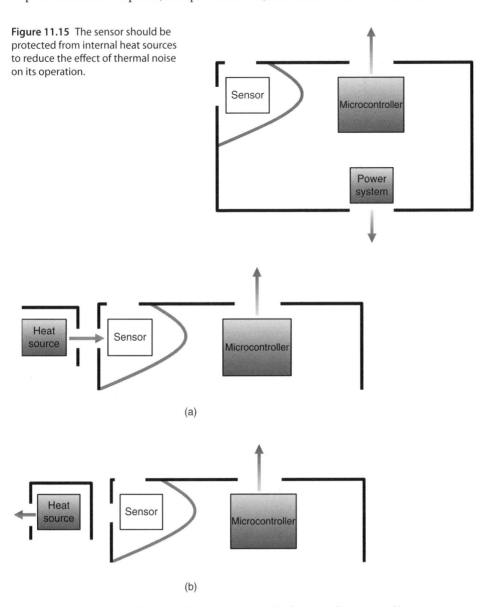

Figure 11.15 The sensor should be protected from internal heat sources to reduce the effect of thermal noise on its operation.

(a)

(b)

Figure 11.16 A mechanism for protecting the sensing path of a sensor from external heat sources: (a) the heat circulation path of an external heat source is not taken into account during sensor placement integration; (b) The heat circulation path of an external heat source is taken into account during sensor placement and integration.

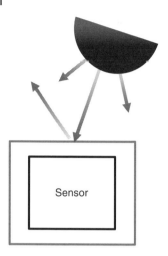

Figure 11.17 Shielding a sensor from the radiation of an external source (the sun) is as important as shielding it from internal and other external heat sources, as this can cause a wide range of interference such as heat, infrared radiation, and ultraviolet radiation.

the microcontroller is used to digitise the sensor signal. Once the placement and wiring issues are addressed, the heat produced by heat sources within the system should still be provided a proper channel to leave the system and the sensor should be shielded from the effect of this internal radiation. Figure 11.15 shows one way of providing an outlet and how a sensor can be shielded from internal infrared radiation. The extra openings at the top left-hand corner prevent the measurand from being trapped by a dead volume, and allow the sensor to make a good determination of the actual state of the measurand.

11.3.3.1 External Heat and Radiation Sources

Lastly, as the sensor or the system integrating it can also be a part of or should work with another system, the contribution of external heat and radiation sources should be taken into consideration during sensor integration. Figure 11.16 compares two scenarios. In the first, the opening through which the sensor is exposed to the measurand is the same as the heat circulation path of an external heat source producing a considerable and undesirable effect on the sensing quality of the sensor. In the second, the two paths are isolated from one another. Likewise, Figure 11.17 shows the way a sensor should be shielded from an external radiation source (in this case, the sun). Similar mechanisms should be adopted to shield a sensor from the influence of external magnetic fields, such as that of the earth.

References

Berruti G, Consales M, Giordano M, Sansone L, Petagna P, Buontempo S, Breglio G and Cusano A 2013 Radiation hard humidity sensors for high energy physics applications using polyimide-coated fiber Bragg gratings sensors. *Sensors and Actuators B: Chemical*, **177**, 94–102.

Carotenuto G, Longo A, Repetto P, Perlo P and Ambrosio L 2007 New polymer additives for photoelectric sensing. *Sensors and Actuators B: Chemical*, **125**(1), 202–206.

Chen CC, Chung TK, Cheng CC and Tseng CY 2014 A novel miniature thermomagnetic energy harvester. *SPIE Smart Structures and Materials+ Nondestructive Evaluation and Health Monitoring*, pp. 90570X–90570X.

Choi HY, Park KS, Park SJ, Paek UC, Lee BH and Choi ES 2008 Miniature fiber-optic high temperature sensor based on a hybrid structured Fabry-Perot interferometer. *Optics Letters*, **33**(21), 2455–2457.

Clark Jr LC, Spokane RB, Homan MM, Sudan R and Miller M 1988 Long-term stability of electroenzymatic glucose sensors implanted in mice. *ASAIO Journal*, **34**(3), 259–265.

Dong S, Li JF and Viehland D 2004 Vortex magnetic field sensor based on ring-type magnetoelectric laminate.

Dürig U 2005 Fundamentals of micromechanical thermoelectric sensors. *Journal of Applied Physics*, **98**(4), 044906.

Duwel A, Gorman J, Weinstein M, Borenstein J and Ward P 2003 Experimental study of thermoelastic damping in MEMS gyros. *Sensors and Actuators A: Physical*, **103**(1), 70–75.

Duwel A, Weinstein M, Gorman J, Borenstein J and Ward P 2002 Quality factors of MEMS gyros and the role of thermoelastic damping. *Micro Electro Mechanical Systems, 2002. The Fifteenth IEEE International Conference on*, pp. 214–219.

Fiebig M 2005 Revival of the magnetoelectric effect. *Journal of Physics D: Applied Physics*, **38**(8), R123.

Giri A, Kirkpatrick E, Moongkhamklang P, Majetich S and Harris V 2002 Photomagnetism and structure in cobalt ferrite nanoparticles. *Applied Physics Letters*, **80** (13), 2341–2343.

Jiles D 1995 Theory of the magnetomechanical effect. *Journal of Physics D: Applied Physics*, **28**(8), 1537.

Nan T, Hui Y, Rinaldi M and Sun NX 2013 Self-biased 215MHz magnetoelectric NEMS resonator for ultra-sensitive DC magnetic field detection. *Scientific Reports*, **3**, #1985.

Romain AC and Nicolas J 2010 Long term stability of metal oxide-based gas sensors for e-nose environmental applications: an overview. *Sensors and Actuators B: Chemical*, **146**(2), 502–506.

Roszhart TV 1990 The effect of thermoelastic internal friction on the Q of micromachined silicon resonators. *Solid-State Sensor and Actuator Workshop, 1990. 4th Technical Digest., IEEE*, pp. 13–16.

Spassov L, Gadjanova V, Velcheva R and Dulmet B 2008 Short-and long-term stability of resonant quartz temperature sensors. *Ultrasonics, Ferroelectrics, and Frequency Control, IEEE Transactions on*, **55**(7), 1626–1631.

Watts MR, Sun J, DeRose C, Trotter DC, Young RW and Nielson GN 2013 Adiabatic thermo-optic mach–zehnder switch. *Optics Letters*, **38**(5), 733–735.

12

Estimation

Designing a good sensor is the first step towards interfacing the physical world with the virtual world. Two additional steps are required before the data obtained from a sensor can be useful. The first step deals with determining whether the data represent the physical reality and the second step deals with understanding the meaning of the sensed data. Erroneous representations or interpretations of sensor data often have detrimental consequences and great care must be taken over the two stages. This chapter deals with the intermediate stage. I will not be dealing with the last stage in this book, as it is application dependent. I shall begin this chapter at the simplest level; you may already know some of the concepts and techniques I will be treating, but my goal is to lay a sound foundation, so that the book is self-contained.

The subjects treated in this chapter are by no means exhaustive. Moreover, I regard *estimation* from a single viewpoint, which is the processing of sensed data. I refer readers wishing to further enrich their knowledge on random variables, stochastic processes, and estimation techniques to the excellent books by Papoulis and Pillai (2002), Ross et al. (1996), Gardiner (1985) and Grewal (2011).

I shall begin this chapter by making a somewhat sensitive statement: we shall never be able to construct a sensor (or any system, for that matter) that captures reality as it is. We can approach reality but never touch it. Even at a quantum level, our approaching of reality is limited by Heisenberg's uncertainty principle. The error with which we perceive reality accumulates as we move away from a quantum reality towards a macro reality. But fortunately for us, this error will never reach a magnitude at which our perception of reality makes existence impossible.

There are two fundamental premises—understood consciously or unconsciously—for relying on sensor data; whether the data come from a biological sensor or a physical sensor constructed by human beings, does not matter. These are:

1) The change in the physical reality (measurand) is a gradual process rather than haphazard and wild; statistically speaking, the measurand is correlated with itself to a certain extent.
2) The state of reality and the output of a sensor are correlated to a certain extent.

The significance of these assertions will become clear when we deal with the mathematics. But for now, consider Figure 12.1, which is a measurement taken from an ordinary temperature sensor having an accuracy (according to the manufacturer) of 1 °C. The sensor was placed outdoors and sampled every second for 30 min. From the reading (and also from intuition) it is clear that even though the measurement fluctuates

Principles and Applications of Ubiquitous Sensing, First Edition. Waltenegus Dargie.
© 2017 John Wiley & Sons, Ltd. Published 2017 by John Wiley & Sons, Ltd.
Companion Website: www.wiley.com/go/dargie2017

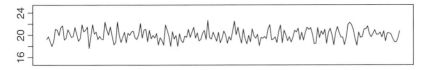

Figure 12.1 The reading of a temperature sensor having an accuracy of 1 °C.

over time, the fluctuation is not haphazard. Secondly, the temperature sensor may not be accurate enough (whatever that means) but it reflects reality. The outside temperature might not have changed at all during the 30 minutes during which we took the reading, or the change might have been quite dissimilar from the one we obtained using the sensor. But there should be certain correlation between the physical reality and the sensor output (unless the sensor is defective). It cannot, for instance, be the case that the temperature fluctuated by around 5 °C much of the time whilst the sensor reading fluctuated by around 20 °C!

12.1 Sensor Error as a Random Variable

The accuracy of a sensor can be quantified but it can also be viewed as a qualitative property. The qualitative and quantitative aspects of a sensor's accuracy can be explained by Figure 12.2. Suppose we expose two sensors having different accuracies to a measurand that does not change; in other words one that is constant over time. For our example, the measurand is a temperature of 20 °C. As can be seen, even if the input is constant, the outputs vary to some extent. Most existing physical sensors share this basic feature. The nearness of the sensor output to the true value of a measurand (as we have already seen in Chapter 11) is what we call accuracy. In Figure 12.2 (top), the output of the sensor appears to be constant, properly reflecting reality, but when we regard it at a fine-grained resolution (in the middle), we see that the samples are different. Indeed, the samples are

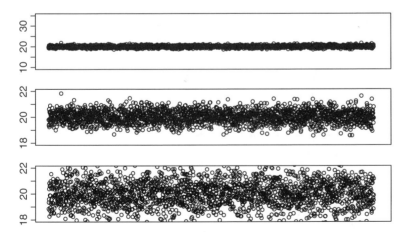

Figure 12.2 A computer generated simulation of the output of two temperature sensors for a fixed (constant) input: (top), (middle) the output of the same sensor at two different granularities; (bottom) the output of a second sensor, having a different accuracy, for the same input.

different not only from the input, but also from each other. The dissimilarity becomes conspicuous in the output of the second sensor (bottom), which has a different accuracy.

Because we cannot be certain of what specific value we will get when we next sample a sensor, we regard its output as a random variable. It is worthwhile emphasising here that the variation in the output of a sensor may have nothing to do with the measurand or the physical process; it can be an inherent property of the sensor itself. The randomness in the output of the sensor nevertheless obeys an underlying probability distribution, because some samples are more likely to occur than others. If we represent the output of a sensor (a random variable) by a small boldface letter (such as $\mathbf{x}$), the function that assigns a probability term to each of the outputs of $\mathbf{x}$ is called a probability density function (pdf), $f(x)$, where x is a real number, representing one of the distinct outcomes of $\mathbf{x}$. Assuming that an infinite number of samples can be obtained if we sample the sensor for long enough, then $f(x)$ provides sufficient information about $\mathbf{x}$.

Figure 12.3 compares the pdfs of the two sensor outputs in Figure 12.2, which happen to be normally distributed. The width of a pdf is an indication of the dissimilarity between the sample outputs of a sensor. The broader it is, the more dissimilar the sample outputs are, and therefore the less reliable the sensor is. If we have the mathematical expression of the pdf, we can ask and answer several questions, such as:

- What is the expected outcome of the random variable (the mean)?
- What is the variance of the random variable (the quantifiable expression of the dissimilarity between the samples, which is also a measure of the error of the sensor)?
- What is the probability that the output of the sensor is between two real numbers, say between 19.5 and 20.5°C, in other words, $P\{19.5 \leq \mathbf{x} \leq 20.5\}$?

The expected outcome or the mean is given by:

$$\eta_x = E[\mathbf{x}] = \int_{-\infty}^{\infty} x f(x) dx \tag{12.1}$$

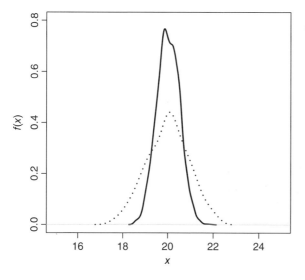

Figure 12.3 The probability density functions of two random variables.

Similarly, the variance of **x** is defined as the square of the expected variance of the sample outputs with respect to the mean:

$$\sigma_x^2 = E[(\mathbf{x} - \eta_x)^2] = \int_{-\infty}^{\infty} (\mathbf{x} - \eta_x)^2 f(x)dx \tag{12.2}$$

The reason we consider $E[(\mathbf{x} - \eta_x)^2]$ instead of $E[(\mathbf{x} - \eta_x)]$ is that the latter will always yield a value of zero, since the summation (integration) results in a net negative value for the samples below the mean and a net positive value for the samples above the mean and both values are equal in magnitude (alternatively, $E[(\mathbf{x} - \eta_x)] = E[\mathbf{x}] - \eta_x = \eta_x - \eta_x = 0$). The variance and mean are related with one another and we shall exploit their relationship to solve some important problems later. Since $\sigma_x^2 = E[(\mathbf{x} - \eta_x)^2]$, we can distribute the right-hand term as follows: $E[\mathbf{x}^2 - 2\eta_x\mathbf{x} + \eta_x^2] = E[\mathbf{x}^2] - 2\eta_x E[\mathbf{x}] + \eta_x^2$, from which we have (as a reminder, the expected value of a constant is the constant itself):

$$\sigma_x^2 = E[\mathbf{x}^2] - \eta_x^2 \tag{12.3}$$

The probability that $\{x_1 \leq \mathbf{x} \leq x_2\}$ can be computed using $f(x)$ alone:

$$P\{x_1 \leq \mathbf{x} \leq x_2\} = \int_{x_1}^{x_2} f(x)dx \tag{12.4}$$

The pdf is also useful to visualise the difference between precision (consistency) and accuracy. Suppose the output of two sensors for a known, fixed input (say, 20 °C), is described by the two pdfs shown in Figure 12.4. As can be seen, the first sensor (the solid line) has a mean that overlaps the input (the true value), while the second sensor (the dotted line) has a mean that is different from the input. Hence we can say that the first sensor is more accurate than the second. On the other hand, the pdf of the second sensor is much wider than that of the first, so we can say that it is more precise than the first, because its output is more consistent and hence more predictable.

Another important function by which a random variable can be described is the cumulative distribution function (CDF) or the probability distribution function (PDF), $F(x)$:

$$F(x) = P\{\mathbf{x} \leq x\} \tag{12.5}$$

Figure 12.4 The probability density functions of the outputs of two sensors are shown to explain the difference between accuracy and precision. Assuming that the known input is 20 °C, the mean of one of the pdfs (solid line) overlaps with the known input, but its variance is big whereas the mean of the other pdf (dotted line) does not overlap with the input but its variance is small. In the first case, we can say that the sensor is more accurate but in the second case we can say that the sensor is more precise.

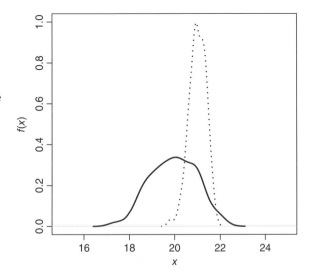

where **x** (boldface type) is the random variable and x (normal font) is a real number. $F(x)$ quantifies the probability that the outcome of a random variable is below a certain value x. For example:

$$F(20) = P\{\mathbf{T} \leq 20\}$$

refers to the probability that the temperature is below $20\,°\text{C}$. Since the probability accumulates as x increases, it is called a cumulative function. Therefore, $F(x)$ is a non-decreasing (monotonically increasing), right-continuous function. For example,

$$F(21) = P\{\mathbf{T} \leq 21\} = F(20) + P\{20 < \mathbf{T} \leq 21\}$$

Figure 12.5 compares the distribution functions of the two simulated temperature sensors of Figure 12.2. The slope of the distribution function indicates the variance of a random variable. The steeper the slope, the smaller the variance, and for our case, the more consistent a sensor is, the gentler the slope, the larger the variance, and the more dissimilar are the outcomes of the sensor. The distribution and the density functions are related with one another:

$$f(x) = \frac{dF(x)}{dx} \tag{12.6}$$

$$F(x) = \int_{-\infty}^{x} f(u)du \tag{12.7}$$

We shall use the two functions alternatively to solve different problems. Sometimes solving problems with one is simpler than solving with the other. It is also worthwhile noticing that since the total probability is always one, we have:

$$\int_{-\infty}^{\infty} f(x)dx = 1 \tag{12.8}$$

and

$$1 = P\{\mathbf{x} \leq x\} + P\{\mathbf{x} > x\} \tag{12.9}$$

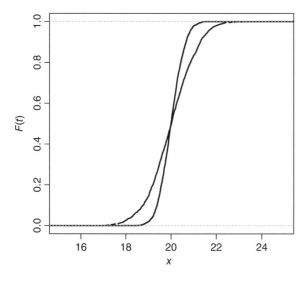

Figure 12.5 A comparison of the cumulative distribution functions of the two simulated temperature sensors, the output of which (for a fixed input) are shown in Figure 12.2.

or

$$1 = F(x) + P\{\mathbf{x} > x\} \tag{12.10}$$

from which we have:

$$P\{\mathbf{x} > x\} = 1 - F(x) \tag{12.11}$$

12.2 Zero-offset Error

One of the inherent errors of physical sensors is the zero-offset error. This error corresponds to the random output of the sensor when there is no input from the measurand. The sources of these outputs can be many, including internal thermal noise, external thermal noise, radiation, and so on. The most significant of these is the internal thermal noise coming from the random vibration of electrons. Most existing sensors have a zero-mean, normal distributed zero-offset error at a given temperature (for example, at room temperature). In other words, the density function of this error can be described as follows:

$$f(e) = \frac{1}{\sqrt{2\pi}\sigma_e} e^{-e^2/2\sigma_e^2} \tag{12.12}$$

where σ_e is the standard deviation or σ_e^2 is the variance of the offset. Figure 12.6 compares the zero-offset errors of two sensors. In general, the inherent error of a sensor due to its internal composition increases as the magnitude and frequency of the measurand increase. You may recall from Chapter 11 how self-heating contributes to the error of a sensor.

Example 12.1 The zero-offset error of a given sensor can be modelled as a uniformly distributed random variable between -2 mV and 2 mV as shown in Figure 12.7. Determine the variance of the random variable. x

Figure 12.6 The zero-offset (the output being a voltage) of two different sensors described by normally distributed random variables. The zero-offset described by the solid line has a standard deviation of 1 mV whilst the zero-offset described by the dashed line has a standard deviation of 0.5 mV.

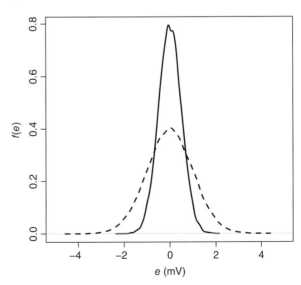

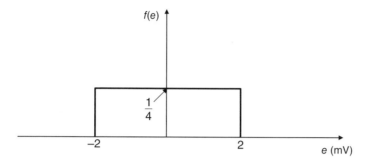

Figure 12.7 The zero-offset of a sensor modelled as a uniformly distributed random variable.

The variance of a random variable is expressed as:

$$\sigma_x^2 = E[(\mathbf{x} - \eta_x)^2] = E[\mathbf{x}^2] - \eta_x^2$$

Since the mean of our random variable is zero, we have:

$$\sigma_x^2 = E[\mathbf{x}^2] = \int_{-2}^{2} x^2 f(x)dx = \frac{1}{12}(2^3 - (-2)^3) = 1.33 \text{ mV}$$

Example 12.2   We wish to build an electronic switch as shown in Figure 12.8. It should respond when the temperature of the measurand crosses a set threshold (specified by V_{REF}). Suppose the RTD has a resistance of 1 kΩ at the reference temperature and a zero-offset voltage distribution as shown in Figure 12.7. Since the sensor contains a zero-offset output, it may trigger the switch erroneously. On the other hand, the sensor may also attenuate an authentic signal due to a negative offset voltage. Assuming the zero-offset is independent of the measurand's temperature and that our priority is reducing false positives, determine a reference voltage that suppresses the contribution of both the biasing voltage and the zero-offset 75% of the time.

At the reference temperature, the voltage across the RTD is a superposition of a portion of the biasing voltage and the zero-offset voltage:

$$V_S = V_B + V_{\text{OFF}}$$

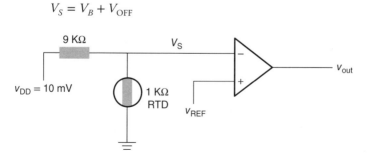

Figure 12.8 The circuit block diagram of a simple comparator. The reference voltage of the comparator is set so that 75 % of the false positive originating from the zero-offset voltage of the sensor can be suppressed.

But,

$$V_B = \left(\frac{1 \, \text{k}\Omega}{9 \, \text{k}\Omega + 1 \, \text{k}\Omega} \right) 10 \, \text{mV} = 1 \, \text{mV}$$

Since V_S contains a random variable, it too is a random variable. As long as V_S is less than V_{REF}, the output of the comparator is negative. If we simply set the reference voltage to be -3 mV, we can be certain that the switch is turned on only as a result of an increment in the measurand's temperature. By doing so, however, we also increase the possibility of a false negative, because the offset voltage might have suppressed an output voltage due to an increase in temperature. In order to satisfy the specified requirement,

$$0.75 = P\{V_S \leq x\} = P\{(V_B + \mathbf{V}_{\text{OFF}}) \leq x\} = P\{\mathbf{V}_{\text{OFF}} \leq (x - V_B)\}$$

Since we have the distribution of $\mathbf{V}_{\text{OFF}}$, the value of x that yields a probability of 0.75 can be determined as follows:

$$0.75 = \frac{1}{4} \int_{-2}^{x - V_B} dx = \frac{1}{4} \int_{-2}^{x-1} dx$$

From which we have:

$$0.75(4) = (x - 1 \, \text{mV}) - (-2 \, \text{mV})$$

Rearranging terms will result in,

$$x = 2 \, \text{mV}$$

12.3 Conversion Error

The output of a sensor passes through many intermediate stages. Therefore, it is important to understand how these stages influence the accuracy and precision of the sensor output. In this section we shall consider how the randomness of the sensor influences the probability distribution of the output. In the next section, we will consider the accumulation of error in more detail. In order to demonstrate how the statistics of the sensor error influence the output voltage, consider Figure 12.9, where we display the electrical circuit diagram of a temperature-to-voltage converter employing a TC1047A temperature sensor manufactured by Microchip Technology Inc. According to the manufacturer, the module has an accuracy of $\pm 2\,^{\circ}\text{C}$ at $25\,^{\circ}\text{C}$ and can measure a change in temperature between -40 and $125\,^{\circ}\text{C}$. For the specified temperature range the output voltage varies from 2.7 V to 4.4 V. Assuming that the sensor has a zero-offset voltage that can be characterised by a zero-mean, normally distributed random variable with a variance of 0.5 mV, we can determine the statistics of the output voltage.

Because the reference voltage is a fixed quantity, the output of the operational amplifier is a square wave, positive when $V_{\text{REF}} > V_S$ and negative when $V_{\text{REF}} < V_S$. Moreover, we have,

$$i_s = -i_f = -\frac{\mathbf{V}_S}{jX_C} = \frac{(\mathbf{V}_S - \mathbf{V}_O)}{RTD}$$

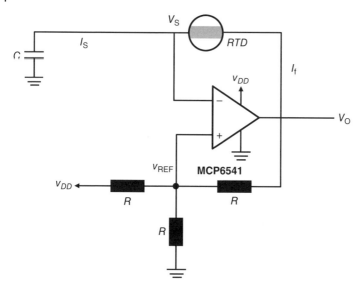

Figure 12.9 A schematic diagram of a temperature-to-voltage converter employing a resistance temperature detector (RTD).

(Recall that because of the high input impedance of the amplifier, we assume that no current flows into the operational amplifier). Hence,

$$\mathbf{V}_O = \mathbf{V}_S \left[1 + \frac{RTD}{jX_C} \right] \tag{12.13}$$

Because the output of the operational amplifier is related to a random variable ($\mathbf{V}_S$), it, too, is a random variable. If we for now disregard the randomness added to the output voltage from the internal noise of the operational amplifier, then the PDF of the output voltage can be expressed in terms of the distribution of the sensor's voltage:

$$F_O(v) = P\{\mathbf{V}_O \leq v\} = P \left\{ \left[1 + \frac{RTD}{jX_C} \right] \mathbf{V}_S \leq v \right\} \tag{12.14}$$

where we use $F_O(v)$ to indicate that the distribution function refers to the output voltage. If we rearrange the terms in Eq. 12.14, we have:

$$F_O(v) = P \left\{ \mathbf{v}_s \leq \left[1 + \frac{jX_C}{RTD} \right] v \right\} = F_S \left(\left[1 + \frac{jX_C}{RTD} \right] v \right) \tag{12.15}$$

where $F_S(.)$ refers to the PDF of $\mathbf{V}_S$. As can be seen, we managed to express the distribution of $\mathbf{V}_O$ in terms of the distribution of $\mathbf{V}_S$.

The labels $\mathbf{V}_O$, $\mathbf{V}_S$ and v should by now be clear; the boldface letters represent random variables whilst the normal font v represents a real number or an instance of the random variables. With the change in the distribution function of the output voltage, its mean and variance change as well. As far as the error of the sensor is concerned, the change in the mean is not much of an issue (it is zero), but the variance is. So how does the variance of the output voltage change?

$$\sigma_{V_o}^2 = E[(\mathbf{V}_o - \eta_{V_O})^2] = E[(\mathbf{V}_O)^2] - \eta_{V_o}^2 \tag{12.16}$$

Substituting Eq. (12.13) into Eq. (12.16) yields,

$$\sigma_{V_O}^2 = \left[1 + \frac{RTD}{jX_C}\right]^2 E(V_S^2) = \left[1 + \frac{RTD}{jX_C}\right]^2 \sigma_{V_S}^2 \tag{12.17}$$

Notice that $\eta_{V_O}^2 = 0$, because V_S is a zero-mean random variable. From Eq. (12.15), it is possible to determine the probability density of the output voltage, because:

$$f_O(v) = \frac{dF_O(v)}{dv} = \left[1 + \frac{jX_C}{RTD}\right] f_S \left(\left[1 + \frac{jX_C}{RTD}\right] v\right) \tag{12.18}$$

where we applied the chain-rule on Eq. (12.15).

Example 12.3   We wish to determine the pdf of the magnitude of the output voltage of the temperature-to-voltage converter in the absence of a measurand. Suppose we are interested in the response of the converter at 100 Hz, with the circuit elements having the values shown in Figure 12.10.

The probability density function of the output voltage is a complex function because of the capacitive reactance. Its magnitude component is given as:

$$|F_O(v)| = \sqrt{\left[1 + \left(\frac{X_C}{RTD}\right)^2\right]} V_S$$

At the specified frequency, the capacitive reactance is:

$$XC = -j\left(\frac{1}{2\pi \times 100 \times 10^{-6}}\right) = -j1.6\,\text{k}\Omega$$

Thus,

$$|F_O(v)| = \sqrt{\left[1 + \left(\frac{1.6\text{k}\Omega}{1\text{k}\Omega}\right)^2\right]} V_S = 1.9\,V_S$$

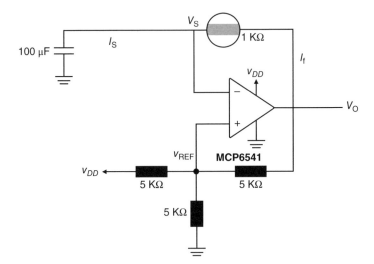

Figure 12.10 The input-output relationship of a temperature-to-voltage converter at 100 Hz.

Moreover, we have:

$$\min(\mathbf{V_O}) = 1.9 \times \min(\mathbf{V_S}) = 1.9 \times -2 = -3.8 \text{ mV}$$

and

$$\max(\mathbf{V_O}) = 1.9 \times \max(\mathbf{V_O}) = 1.9 \times 2 = 3.8 \text{ mV}$$

Figure 12.11 shows the relationship between the two random variables, namely, $\mathbf{V_S}$ and $\mathbf{V_O}$.

In general, if the random variables $\mathbf{y}$ and $\mathbf{x}$ are related to one another, then it is possible to determine the statistics of one of the random variables (the unknown) in terms of the statistics of the other (the known), beginning by describing the PDF of the unknown random variable in terms of the known random variable. For example, if the two random variables are related as follows:

$$\mathbf{y} = a\mathbf{x} + b$$

where a and b are known positive constants, the statistics of $\mathbf{y}$ can be expressed in terms of the statistics of $\mathbf{x}$ and:

$$F(y) = P\{\mathbf{y} \le y\} = P\{(a\mathbf{y} + b) \le y\}$$

$$F(y) = P\left\{ \mathbf{x} \le \frac{(y-b)}{a} \right\}$$

Since

$$\frac{y-b}{a} = C$$

is a constant, then we have:

$$F(y) = P\{\mathbf{x} \le C\} = F_X(C)$$

where $F_X(C) = P\{\mathbf{x} \le C\}$ is the distribution of $\mathbf{x}$. From here on, it is a matter of making the appropriate derivation to determine the statistics pertaining to $\mathbf{y}$.

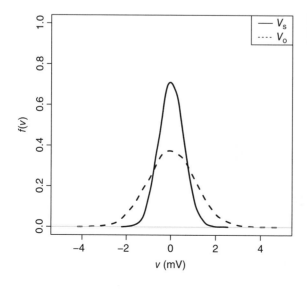

Figure 12.11 The relationship between the PDFs of $\mathbf{V_S}$ and $\mathbf{V_O}$ at 100 Hz frequency.

12.4 Accumulation of Error

A sensing system typically consists of two or more stages. The conditioning circuit, for example, may consist of a Wheatstone bridge, an amplifier, and a filter. Each of these stages will introduce its own error into the signal produced by the sensing stage. These errors should be accounted for in order to meaningfully interpret the sensed signal. Suppose the aggregate error of the conditioning circuit can be modelled by the random variable $\mathbf{c}$, having its own distribution and density functions. The combined error of the sensing and the conditioning stages is given as:

$$\mathbf{e} = \mathbf{s} + \mathbf{c} \tag{12.19}$$

In order to explain the addition of random variables, we shall begin with a simple example. Suppose, in the absence of an input (due to an internal thermal noise), the sensing system produces either -0.5 mV or 0.5 mV with equal probabilities. Likewise, the conditioning circuit, independent of the sensing system, produces either -0.5 mV or +0.5 mV with equal probabilities. If we connect the two systems, as shown in Figure 12.12, and measure the output voltage, we may get the following values with the corresponding probabilities:

$$e = -0.5\,\mathrm{mV} + -0.5\,\mathrm{mV} = -1.0\,\mathrm{mV} \quad (P(-0.5) \times P(-0.5))$$
$$e = -0.5\,\mathrm{mV} + 0.5\,\mathrm{mV} = 0.0\,\mathrm{mV} \quad (P(-0.5) \times P(0.5))$$
$$e = 0.5\,\mathrm{mV} + -0.5\,\mathrm{mV} = 0.0\,\mathrm{mV} \quad (P(0.5) \times P(-0.5))$$
$$e = 0.5\,\mathrm{mV} + 0.5\,\mathrm{mV} = 1.0\,\mathrm{mV} \quad (P(0.5) \times P(0.5))$$

So, we see that $\mathbf{e}$ as a random variable has different outcomes with different probabilities (notice how the new probability distribution should be computed):

$$\mathbf{e} = [-1.0(P = 0.25), 0.0(P = 0.5), 1.0(P = 0.25)]$$

As a result, some values—0.0 mV for example—are more likely to occur—$P\{0.0\,\mathrm{mV}\} = 0.5$—than other values. Figure 12.13 shows the probability mass function of $\mathbf{e}$.

Now suppose, $\mathbf{s}$ and $\mathbf{c}$ each have five discrete outputs (in mV) as follows:

$$\mathbf{s} = \mathbf{c} = [-0.5, -0.25, 0.0, 0.25, 0.5]$$

Then, $\mathbf{e} = \mathbf{s} + \mathbf{c}$ will have 5×5 elements, but some of the elements have equal values. Carrying out the computation as shown above yields nine unique values with corresponding probabilities:

$$\mathbf{e} = [-1.0, -0.75, -0.5, -0.25, 0.0, 0.25, 0.5, 0.75, 1.0]$$

The corresponding probability of occurrence is:

$$f(e) = [0.04, 0.08, 0.12, 0.16, 0.2, 0.16, 0.12, 0.08, 0.04]$$

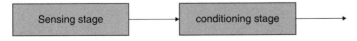

Figure 12.12 The accumulation of error when the sensing and the conditioning systems are connected in series.

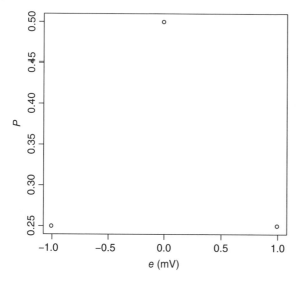

Figure 12.13 The probability mass function of **e** = **s** + **c** when each random variable has only two discrete values with equal probabilities.

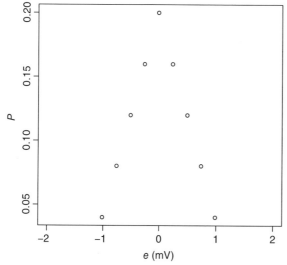

Figure 12.14 The probability mass function(the discrete equivalent of the pdf) of **e** = **s** + **c** when each random variable has only five discrete values with equal probabilities.

Figure 12.14 shows the probability mass function of **e**. In general, if **s** and **c** have I and J discrete values, the elements of **e** can be computed using two `for` loops, as shown in Figure 12.15. Moreover, when some of the elements have equal values, their probabilities should be added. From the above example, it is clear that the computation of the distribution and the density functions for **e**, when **s** and **c** have continuous distributions and density functions, can be carried out as follows:

$$F(e) = P\{\mathbf{s} + \mathbf{c} \leq e\} = \int_{s=-\infty}^{\infty} \int_{c=-\infty}^{e-c} f(s,c)ds \, dc \tag{12.20}$$

where $f(s,c)$ is the joint density function. If **s** and **c** are independent, then $f(s,c) = f(s)f(c)$. You may notice that the two integrations above correspond to the two `for` loops in Figure 12.15 for the case where the two random variables are

```
for(int i; i < sizeof(s);i++){
        for(int j; j < sizeof(c); j++) {
                e[ i * sizeof(c) + j] = s[i] + c[j]
        }
}
```

Figure 12.15 Computing the elements of **e** = **s** + **c** in C. As can be seen, **e** can have sizeof(s) × sizeof(c) distinct elements.

discrete. Similarly, the density function of **e** is given as:

$$f(e) = \int_{-\infty}^{\infty} f(e - c, c)dc \tag{12.21}$$

If the two random variables are independent, then we have:

$$f(e) = \int_{-\infty}^{\infty} f(e - c)f(c)dc \tag{12.22}$$

where $f(e - c)$ is the density of **s** expressed in terms of $s = e - c$. The density function amounts to adding the probabilities of similar values after the elements of **e** are computed in Figure 12.15.

Example 12.4 Suppose both the sensing element and the conditioning circuit of a given sensor generate random outputs **s** and **c**, respectively. In the absence of a measurand, these outputs are described by uniformly distributed voltages, each ranging between 0 mV and 1 mV. Assuming that the two sources of error are independent and their cumulative effect is additive, determine the distribution of the error as **e** = **s** + **c**.

As the sum of two random variables, **e** may have any outcomes between 0 (the minimum value) and 2 (the maximum value). Since both random variables are continuous, the PDF of **e** can be determined by integrating the joint density function. Since we are dealing with definite integrals, we should first determine the boundaries of integration. We shall rely on Figure 12.16 to determine the boundaries. The x- and y-intercepts of the equation **e** = **s** + **c** can be determined by setting one of the random variables to zero. Hence, for the x-intercept we have **c** = 0 and:

$$s = e$$

Similarly, for the y-intercept we have **s** = 0 and,

$$c = e$$

But notice that since **s** and **c** are random variables, the x- and y-intercepts are not constant values as they would be for a deterministic function. Instead, the line **e** = **s** + **c** can be located in different places in the first quadrant bounded by the coordinates $(0, 0)$ and $(1, 1)$. For any positive real value $e \leq 1$, the distribution function,

$$F(e) = P\{e \leq e\} = P\{s + c \leq e\}$$

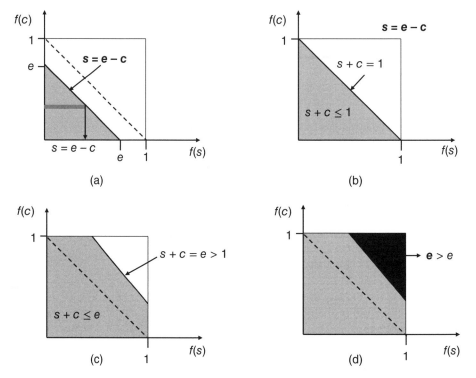

Figure 12.16 Determining the probability distribution function of $e = s + c$: $F(e) = P\{s + c \le e\}$. We are interested in the region where $s + c \le e$. In (a), the shaded region: $s + c \le 1$, and can easily be integrated. Here it is clear that as c varies from zero to e, s varies from zero to the intersection (e - c). In (b), the shaded region subsumes the boundary $s + c = 1$, beyond this boundary, $s + c \le e$ is not simple to integrate on account of the complexity of the geometry of the shaded region, as can be seen in (c). In (d) we can take advantage of the mutual exclusiveness of $\{e \le e\}$ and $\{e > e\}$ and the fact that the probability of the two regions adds to unity.

can be determined by integrating the shaded region of Figure 12.16a, which describes the regions where the two density functions overlap. From the figure, it is apparent that as c varies from 0 to e, s varies from 0 to $e - c$. Alternatively, we can vary s from 0 to e and bind c to vary from 0 to $e - s$. Thus, $F(e)$ can be computed as (remember that both random variables are uniform):

$$
\begin{aligned}
F(e) &= \int_{c=0}^{e} \int_{s=0}^{e-c} f(c)f(e-c)\,dsdc \\
&= \int_{c=0}^{e} \int_{s=0}^{e-c} dsdc \\
&= \int_{c=0}^{e} (e-c)\,dc \\
&= \frac{e^2}{2}
\end{aligned}
$$

But when $\mathbf{e} > 1$, the graph $\mathbf{e} = \mathbf{s} + \mathbf{c}$, which is shown in Figure 12.16c becomes a little complicated to integrate. However, recall from Eq. (12.11) that:

$$F(e) = P\{\mathbf{e} \leq e\} = 1 - P\{\mathbf{e} > e\}$$

Since the black region in Figure 12.16d is $(\mathbf{s} + \mathbf{c}) > e$, it is relatively easy to integrate. Consequently, for $1 < \mathbf{e} \leq 2$, we have:

$$F(e) = 1 - \int_{c=e-1}^{1} \int_{s=e-c}^{1} f(c)f(e-c)\,ds\,dc$$

$$= 1 - \int_{c=e-1}^{1} \int_{s=e-c}^{1} ds\,dc$$

$$= 1 - \int_{c=e-1}^{1} (1 - e + c)\,dc$$

$$= 1 - \frac{(2-e)^2}{2}$$

The pdf of the error can be determined by differentiating $F(e)$ with respect to e. Thus:

$$f(e) = \begin{cases} e & 0 \leq e \leq 1 \\ 2 - e & 1 < e \leq 2 \end{cases}$$

Figure 12.17 displays the probability density function of $\mathbf{e} = \mathbf{s} + \mathbf{c}$.

12.4.1 The Central Limit Theorem

One interesting aspect of an accumulation of error is that, as the number of random variables that should be added increases, their pdf tends to be normally distributed, regardless of the shape of the density functions of the individual stages. If, for example, we divide the conditioning circuit into a Wheatstone bridge ($\mathbf{w}$) and an amplification stage ($\mathbf{a}$) and assume that $\mathbf{s}$, $\mathbf{w}$, and $\mathbf{a}$, are zero-mean, uniformly distributed random

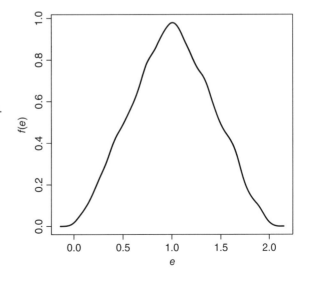

Figure 12.17 The pdf of $\mathbf{e} = \mathbf{s} + \mathbf{c}$ when both random variables are uniformly distributed between 0 and 1. As expected, the plot is similar to the one depicted in Figure 12.14, where we computed the probability mass function for two uniformly distributed discrete random variables.

variables with the following discrete outputs (in mV):

$$\mathbf{s} = \mathbf{w} = \mathbf{a} = [-0.5, -0.25, 0.0, 0.25, 0.5]$$

Then, $\mathbf{e} = \mathbf{s} + \mathbf{w} + \mathbf{a}$ will have $5 \times 5 \times 5 = 125$ entries, but some of the values occur multiple times, as a result of which their frequency of occurrence has to be summed in order to calculate their probability of occurrence. Table 12.1 summarises the statistical parameters of $\mathbf{e}$. The approximated probability mass function of $\mathbf{e}$ is given in Figure 12.18 and reflects a well-studied and statistically well-formulated phenomenon known as the central limit theorem (CLT). The CLT states that the pdf of the sum (or average) of a

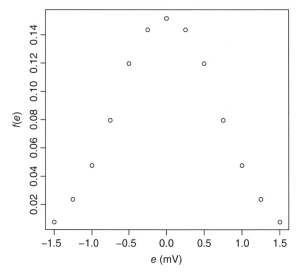

Figure 12.18 The approximate density function of $\mathbf{e} = \mathbf{s} + \mathbf{w} + \mathbf{a}$ is a zero-mean normal distribution function.

Table 12.1 A summary of the statistical parameters of $\mathbf{e} = \mathbf{s} + \mathbf{w} + \mathbf{a}$

Unique entries of e during summing	Frequency of occurrence	Probability of occurrence
−1.5	1	0.08
−1.25	3	0.02
−1.00	6	0.05
−0.75	10	0.08
−0.50	15	0.12
−0.25	18	0.14
0.00	19	0.15
0.25	18	0.14
0.50	15	0.12
0.75	10	0.08
1.00	6	0.05
1.25	3	0.02
1.50	1	0.08

large number of independent random variables with well defined means and variances will be approximately normal, regardless of the underlying distribution of the individual random variables. From Eq. (12.12), a normally distributed density function can be specified by its mean (η) and variance (σ^2). Hence, if we have n independent stages, each of which contributes its own error $\mathbf{e}_i$, having its own mean and variance, then it is possible to sufficiently describe the overall error as $\mathtt{norm}\,(\eta, \sqrt{\sigma^2})$, where:

$$\eta = E\{\mathbf{e}\} = E\{\mathbf{e}_1 + \mathbf{e}_2 + \cdots + \mathbf{e}_n\} = \eta_1 + \eta_2 + \cdots + \eta_n \tag{12.23}$$

To compute the variance (σ^2) of the overall error, $\mathbf{e}$, we shall make use of the relation $\sigma^2 = E[\mathbf{e}^2] - \eta^2$. Moreover,

$$E[\mathbf{e}^2] = E\left[\left(\sum_{i=1}^{n} \mathbf{e}_i\right)^2\right] = \sum_{i=1}^{n}\sum_{j=1}^{n} E[\mathbf{e}_i\mathbf{e}_j] \tag{12.24}$$

Since the individual random variables are independent:

$$E[\mathbf{e}_i\mathbf{e}_j] = \begin{cases} \sigma_i^2 + \eta_i & i = j \\ \eta_i\eta_j & i \neq j \end{cases} \tag{12.25}$$

The double sum in Eq. (12.24) contains n terms for $i = j$ and $n^2 - n$ terms for $i \neq j$. Notice that we have made use of $E[\mathbf{e}_i^2] = \sigma_i^2 + \eta_i^2$. Meanwhile, what happens if all the errors are zero-mean random variables? In that case, the resulting error will also be a zero-mean normally distributed random variable, the variance of which is the sum total of the variance of the individual random variables, because:

$$\eta = \eta_1 + \eta_2 + \cdots + \eta_n = 0$$
$$\eta_i\eta_j = 0$$
$$E[\mathbf{e}_i\mathbf{e}_j] = \begin{cases} \sigma_i^2 & i = j \\ 0 & i \neq j \end{cases}$$

Hence

$$\sigma^2 = \sigma_1^2 + \sigma_2^2 + \cdots + \sigma_n^2$$

12.5 Combining Evidence

So far, we have considered that an error

- is an inherent characteristic of a sensing system
- accumulates as the sensed signal advances towards the processing subsystem (due to the error introduced by the conditioning and additional intermediate stages).

Moreover, we have considered that as the number of independent error sources increases, the overall error assumes a normal pdf. One way of reducing the uncertainty stemming from the inherent sources of error is to employ multiple sensors and combine their evidence. Indeed, arrays of sensors are employed in many practical applications to this end. Hence, the next practical question is determining the appropriate techniques for combining the output of multiple sensors. There can be different combining techniques and we shall examine some of them closely, but one essential aspect to bear in mind when dealing with the combination of evidence is the definition of uncertainty in quantifiable terms. The degree of unreliability of a sensor is directly related to the characteristics of its error. Therefore, some aspects of the error should necessarily be taken into account in the combination equation.

Figure 12.19 shows the outputs (simulated) of three temperature sensors. This figure is similar to Figure 12.2, but there is a slight difference between them. Two of the sensors have the same mean but different variances, whereas two of them have the same variance but different means. A good combination technique is one that takes these aspects into account to minimise the overall error.

12.5.1 Weighted Sum

When all the errors are zero-mean random variables, the simplest way to combine the outputs is as the weighted sum of the individual outputs. The weight given to each sensor output must be inversely proportional to the variance of its error. This is simply because, as we have already seen, the bigger the variance, the bigger our uncertainty. However, uncertainty is a relative term, because it has to be assessed relative to the uncertainty introduced by the other sensors. Suppose we have only two sensors with zero-mean errors and wish to combine their output as follows:

$$\hat{\mathbf{s}} = \alpha_1 \mathbf{s}_1 + \alpha_2 \mathbf{s}_2 \tag{12.26}$$

where $\mathbf{s}_1 = T + \mathbf{e}_1$ and $\mathbf{s}_2 = T + \mathbf{e}_2$ (and T, we assume, is the true value we are seeking to determine). One way to determine the weight for each sensor output is as follows:

$$\hat{\mathbf{s}} = \left(\frac{\sigma_2^2}{\sigma_1^2 + \sigma_2^2} \right) \mathbf{s}_1 + \left(\frac{\sigma_1^2}{\sigma_1^2 + \sigma_2^2} \right) \mathbf{s}_2 \tag{12.27}$$

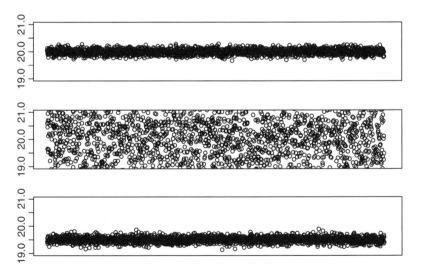

Figure 12.19 The output of three temperature sensors for the same input (20 °C). Two of the sensor outputs have the same mean but different variances (one of them $\sigma_1^2 = 0.5$ and the other $\sigma_2^2 = 1$). Two of the sensor outputs have different means but the same variance. A combination technique should take both aspects into consideration.

Figure 12.20 The density functions of the three simulated temperature sensors shown in Figure 12.19.

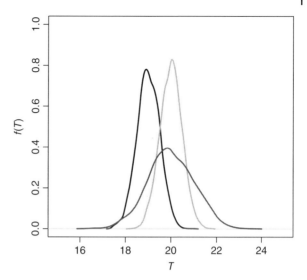

Equation (12.27) fulfils the inverse proportionality requirement as well as the relative significance of the contribution of each sensor (the term $1/(\sigma_1^2 + \sigma_2^2)$, which is a normalisation factor, so that $\alpha_1 + \alpha_2 = 1$, implicitly gives a relative significance to each coefficient). If $\sigma_1^2 > \sigma_2^2$, then we should trust $\mathbf{s}_2$ (which is now multiplied by the larger σ_1^2). If, on the other hand, $\sigma_1^2 < \sigma_2^2$ (which is multiplied by the larger σ_2^2), then we should trust $\mathbf{s}_1$; otherwise, we should trust both equally. We can rewrite Eq. (12.27) as follows:

$$\hat{\mathbf{s}} = \left(\frac{\sigma_2^2}{\sigma_1^2 + \sigma_2^2}\right)\mathbf{s}_1 + \left(\frac{\sigma_1^2}{\sigma_1^2 + \sigma_2^2}\right)\mathbf{s}_2 + (\mathbf{s}_1 - \mathbf{s}_1) \tag{12.28}$$

Taking the $1/(\sigma_1^2 + \sigma_2^2)$ term as the common factor will yield:

$$\hat{\mathbf{s}} = \left(\frac{1}{\sigma_1^2 + \sigma_2^2}\right)[\sigma_2^2\mathbf{s}_1 + \sigma_1^2\mathbf{s}_2 + (\sigma_1^2 + \sigma_2^2)\mathbf{s}_1 - \sigma_1^2\mathbf{s}_1 - \sigma_2^2\mathbf{s}_1] \tag{12.29}$$

Collecting like terms and simplifying and rearranging yields:

$$\hat{\mathbf{s}} = \mathbf{s}_1 + \frac{\sigma_1^2}{\sigma_1^2 + \sigma_2^2}[\mathbf{s}_2 - \mathbf{s}_1] \tag{12.30}$$

If we let

$$K = \frac{\sigma_1^2}{\sigma_1^2 + \sigma_2^2} \tag{12.31}$$

Eq. (12.30) can be expressed in a more compact form:

$$\hat{\mathbf{s}} = \mathbf{s}_1 + K[\mathbf{s}_2 - \mathbf{s}_1] \tag{12.32}$$

The significance of Eq. (12.32) will be clearer later, when we deal with the Kalman filter. For now, it suffices to state the following:

1) Given that we already have evidence from sensor 1, the evidence coming from sensor 2 can be added into the output by properly weighing the difference between sensor 2 and sensor 1.

2) If we trust sensor 1, then the new information (from sensor 2) should not change our belief considerably (K should be small).

3) If, however, we do not trust sensor 1, then, the new evidence should change our belief considerably (K should be large).

If, instead of two, we have n independent sensors from which we can gather evidence, the weighted sum approximating the true value can be described as:

$$\hat{s} = \alpha_1 s_1 + \alpha_2 s_2 + \cdots + \alpha_n s_n \tag{12.33}$$

Notice that $\hat{s}$ is a random variable, because it is the sum of multiple random variables, which means it has its own pdf, mean, and variance. Since our aim is reducing the uncertainty introduced by the error of each sensing element, we must determine the coefficients in such a way that the variance of $\hat{s}$ is a minimum. We can achieve this goal by describing the variance in terms of the coefficients:

$$\sigma_{\hat{s}}^2 = \alpha_1^2 \sigma_1^2 + \alpha_2^2 \sigma_2^2 + \cdots + \alpha_n^2 \sigma_n^2 \tag{12.34}$$

We squared the coefficients merely for reason of convenience. We should select the coefficients such that their sum yields unity:

$$\alpha_1 + \alpha_2 + \cdots + \alpha_n = 1 \tag{12.35}$$

Now we can rewrite Eq. (12.34) as follows:

$$\sigma_{\hat{s}}^2 = \alpha_1^2 \sigma_1^2 + \alpha_2^2 \sigma_2^2 + \cdots + \alpha_n^2 \sigma_n^2 - \lambda(\alpha_1 + \alpha_2 + \cdots + \alpha_n - 1) \tag{12.36}$$

Since $\lambda(\alpha_1 + \alpha_2 + \cdots + \alpha_n - 1)$ is zero, Eq. 12.36 is essentially the same as Eq. 12.34. We call λ a Lagrange multiplier and its significance will be clear shortly. The coefficients minimising the variance in Eq. (12.36) can be determined by partial differentiation:

$$\frac{\partial \sigma_{\hat{s}}^2}{\partial \alpha_i} = 0! \tag{12.37}$$

Consequently, the minimum mean square estimation approach aims to set the expected error to a minimum value:

$$\frac{\partial \sigma_{\hat{s}}^2}{\partial \alpha_i} = 2\alpha_i \sigma_i^2 - \lambda = 0 \tag{12.38}$$

and,

$$\alpha_i = \frac{\lambda}{2\sigma_i^2} \tag{12.39}$$

Equation (12.39) fulfils one of our requirements, namely that our level of confidence in a sensor should be inversely proportional to its variance. However, we have also stressed that our uncertainty in a sensor is a relative characteristic, and should be regarded with respect to the uncertainty the others sensors introduce into the equation, which is why we introduced the λ term in Eq. (12.39). Taking the fact that all the coefficients should add to unity,

$$\alpha_1 + \alpha_2 + \cdots + \alpha_n = \frac{\lambda}{2}\left(\frac{1}{\sigma_1^2} + \frac{1}{\sigma_2^2} + \cdots + \frac{1}{\sigma_n^2}\right) \tag{12.40}$$

from which we have:

$$\frac{\lambda}{2} = \frac{1}{1/\sigma_1^2 + 1/\sigma_2^2 + \cdots + 1/\sigma_n^2} = \sigma_{\hat{s}}^2 \tag{12.41}$$

which now completes our two requirements. We can likewise determine the expected value of $\hat{s}$:

$$\eta_{\hat{s}} = E[\hat{s}] = \alpha_1 E[\mathbf{s}_1] + \alpha_2 E[\mathbf{s}_2] + \cdots + \alpha_n E[\mathbf{s}_n] = \alpha_1 \eta_1 + \alpha_2 \eta_2 + \cdots + \alpha_n \eta_n \tag{12.42}$$

Substituting Eq. (12.39) for each α_i, we have:

$$\eta_{\hat{s}} = \frac{\eta_1/\sigma_1^2 + \eta_2/\sigma_2^2 + \cdots + \eta_n/\sigma_n^2}{1/\sigma_1^2 + 1/\sigma_2^2 + \cdots + 1/\sigma_n^2} \tag{12.43}$$

Similarly:

$$\hat{s} = \frac{\mathbf{s}_1/\sigma_1^2 + \mathbf{s}_2/\sigma_2^2 + \cdots + \mathbf{s}_n/\sigma_n^2}{1/\sigma_1^2 + 1/\sigma_2^2 + \cdots + 1/\sigma_n^2} \tag{12.44}$$

Notice that unlike in Eq. (12.32), we have made no assumption about the mean of the individual random variables ($\mathbf{s}_i$) in Eq. (12.44). They can have any value. But what will happen if all of them have the same mean (η) and variance (σ^2)? In this case,

$$\hat{s} = \frac{1}{n}(\mathbf{s}_1 + \mathbf{s}_2 + \cdots + \mathbf{s}_n) \tag{12.45}$$

and,

$$\sigma_{\hat{s}}^2 = \frac{\sigma^2}{n} \tag{12.46}$$

and,

$$\eta_{\hat{s}} = \eta \tag{12.47}$$

From Eq. (12.46), we can conclude that the more sensors we involve, however imperfect they are when considered individually, the less uncertain we become. Indeed, the variance tends to zero as n tends to infinity.

Example 12.5 An ultrasound scanning system uses an array of 19 microphones arranged as shown in Figure 12.21. Each microphone has a normally distributed, zero-mean, zero-offset output voltage, with a variance of 1 mV. Assuming that the microphones' output voltages are independent of one another and we employ the weighted-sum technique to combine their outputs, determine the pdf of the zero-offset output voltage of the scanning system and compare it with the pdfs of the zero-offset voltage of the individual microphones.

From the CLT we know that the output zero-offset voltage of the entire system will have a normal distribution. Since the output voltages of the sensors are regarded as independent and identically distributed (iid), this suffices for us to determine the variance of $\hat{s}$:

$$\sigma_{\hat{s}}^2 = \frac{\sigma^2}{n} = \frac{1}{19} \approx 0.05 \text{ mV}$$

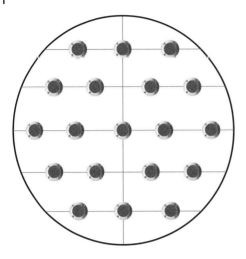

Figure 12.21 An ultrasound scanning system consisting of 19 microphones, which serve as ultrasound receivers. The zero-offset voltages of the microphones are modelled as independent, zero-mean, and normally distributed random variables.

Hence, after inserting the parameters we have into Eq. (12.12), the density function of $\hat{s}$ is given as:

$$f(\hat{s}) = (1.78)e^{-10\hat{s}^2}$$

Figure 12.22 compares the density function of the combined error with the density function of the error of the individual microphones.

Example 12.6 Suppose the manufacturer of the ultrasound scanning system wishes to make its product affordable by mixing two types of microphones. Maintaining that the central seven microphones are more important than the outer 12 microphones, it develops them with an expensive technique, and therefore, each microphone has a normally distributed, zero-mean zero-offset output voltage having a variance of 1 mV. The outer microphones, on the other hand, are cheaper and have a zero-mean, normally distributed zero-offset output voltage with a variance of 5 mV. How can the manufacturer possibly combine the output of the microphones such that the zero-offset error is minimum?

There can be different combination techniques, but one of the plausible approaches is to combine the sensor data in two stages, as shown in Figure 12.23. First the sensors are categorised into two groups, based on their statistical properties, so that we can apply Eq. (12.46). Then we can apply Eq. (12.32) to combine the output of the intermediate stages:

1) *First stage combination (combination based on statistical properties).*

$$\hat{s}_1 = \alpha_{11}s_{11} + \alpha_{12}s_{12} + \cdots + \alpha_{17}s_{17} \tag{12.48}$$

Since the sensors have identical statistics, the variance of $\hat{s}_1$ can be computed as:

$$\sigma_1^2 = \frac{1\,\text{mV}}{7} = 0.14\,\text{mV} \tag{12.49}$$

Similarly,

$$\hat{s}_2 = \alpha_{21}s_{21} + \alpha_{22}s_{22} + \cdots + \alpha_{212}s_{212} \tag{12.50}$$

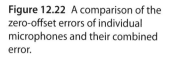

Figure 12.22 A comparison of the zero-offset errors of individual microphones and their combined error.

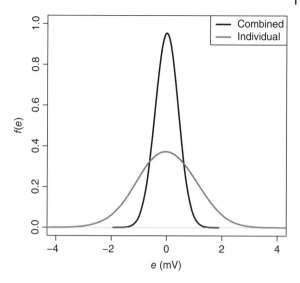

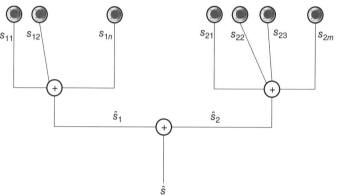

Figure 12.23 Using intermediate stages to systematically combine the outputs of multiple independent sensors having different zero-offset output voltage statistics.

And

$$\sigma_2^2 = \frac{5\,\text{mV}}{12} = 0.42\,\text{mV} \tag{12.51}$$

2) *Second stage combination.*

We can combine the intermediate stages using Eq. (12.32):

$$\hat{\mathbf{s}} = \hat{\mathbf{s}}_1 + K[\hat{\mathbf{s}}_1 - \hat{\mathbf{s}}_2] \tag{12.52}$$

Using Eq. (12.31) we can determine K:

$$K = \frac{0.14\,\text{mV}}{0.14\,\text{mV} + 0.42\,\text{mV}} = 0.25\,\text{mV} \tag{12.53}$$

Hence,

$$\hat{\mathbf{s}} = \hat{\mathbf{s}}_1 + 0.25[\hat{\mathbf{s}}_1 - \hat{\mathbf{s}}_2] \tag{12.54}$$

Likewise, using K, we can compute the variance of $\hat{s}$ as follows:

$$\sigma_{\hat{s}}^2 = \sigma_1^2 - K\sigma_2^2 = 0.14\,\text{mV} - 0.25(0.42\,\text{mV}) = 0.035 \tag{12.55}$$

Figure 12.24 displays how the error of the ultrasound scanning system is improved stage-by-stage through the systematic combination of the different sensor outputs.

12.5.2 Maximum-likelihood Estimation

In Section 12.5.1, we first decided how to combine the outputs of multiple independent sensors, determined the optimal coefficients for minimising our uncertainty, and produced the joint pdf (Figure 12.24). In this section we shall take the reverse order. Our aim is to determine the optimal and unbiased combination technique that optimises the joint density function.

Recall the two important assertions we made in the beginning of this chapter. One of them was stated as follows:

> The state of reality and the output of a sensor are correlated to a certain extent.

This is the assertion upon which the maximum-likelihood estimation (ML) approach is established. Regardless of the quality of the sensors, ML asserts that the output of each sensor has something to do with the real quantity we wish to determine. Take, as an example, the temperature-to-voltage converter we considered previously. Suppose

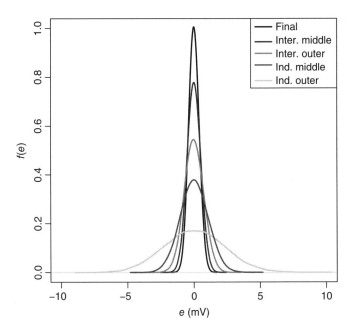

Figure 12.24 Comparison of the pdfs of the different stages during the combination of the zero-offset voltages of different microphones. From bottom to top: The pdfs of the output of the outer microphones, the inner microphones, the intermediate stage combining the outputs of the outer microphones, the intermediate stage combining the outputs of the inner microphones, and the output of the final stage.

between t and $t + dt$ we sample, at a rate of 1 kHz, four independent temperature sensors measuring the temperature of one and the same process and that we obtain the pdfs as shown in Figure 12.25. Assuming that the temperature of the process during this period remains unchanged, we suspect that the most likely temperature the sensors "perceived" should correspond to 2 V. Since the sensors are independent, their joint density function given that they respond to one and the same parameter V – mind you, V is not a random variable, since we assume that the value of the measurand does not change for the time interval (t, dt) – can be computed as:

$$f(\mathbf{s}; V) = f(\mathbf{s}_1; V)f(\mathbf{s}_2; V)\cdots f(\mathbf{s}_n; V) = \prod_{i=1}^{n} f(\mathbf{s}_i; V) \tag{12.56}$$

Since we assume that the output of each sensor has something to do with V, we can express it as follows:

$$\mathbf{s}_1 = V + \mathbf{e}_1$$
$$\mathbf{s}_2 = V + \mathbf{e}_2$$
$$\vdots$$
$$\mathbf{s}_n = V + \mathbf{e}_n$$

If the error introduced by each sensor is a zero-mean, normally distributed random variable, then $\mathbf{s}_i$ is a normally distributed random variable having V as its mean, because:

$$E[\mathbf{s}_i] = V + E[\mathbf{e}_i] = V$$

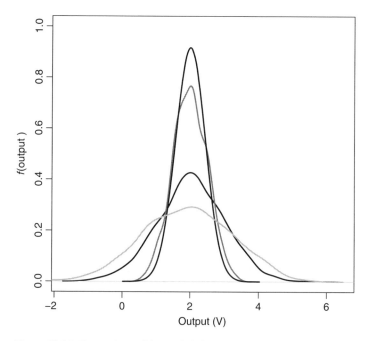

Figure 12.25 Comparison of the probability density functions of four temperature-to-voltage converters measuring the temperature of one and the same process.

and,

$$\sigma_{si}^2 = E[(\mathbf{s}_i - \eta_{si})^2] = E[(V + \mathbf{e}_i - V)^2] = E[(\mathbf{e}_i - \eta_{ei})^2] = \sigma_{ei}^2 = \sigma_i^2$$

Consequently, the density function of the output of each sensor is given as:

$$f(s_i; V) = \frac{1}{\sqrt{2\pi\sigma_i^2}} e^{-(s_i-V)^2/2\sigma_i^2} \tag{12.57}$$

The joint density function as a product of the density functions of the individual sensors is given by:

$$f(s_1, s_2,..., s_n; V) = \frac{1}{(2\pi\sigma^2)^{n/2}} e^{-\sum_{i=1}^{n}((s_i-V)^2/2\sigma_i^2)} \tag{12.58}$$

Now having the joint density function, we can determine the best combination strategy by differentiating Eq. (12.58) with respect to V, because we are interested in the value of V that results in the highest probability or which is the most likely outcome (hence, the name maximum-likelihood estimation). Alternatively, we can differentiate the logarithmic value of Eq. (12.58) (due to the linearity property of logarithms). Thus, we have:

$$\ln\left(f(s_1, s_2,..., s_n; V)\right) = \frac{n}{2} \ln\left(2\pi\sigma^2\right) - \sum_{i=1}^{n} \frac{(s_i - V)^2}{2\sigma_i^2} \tag{12.59}$$

Differentiating Eq. (12.59) with respect to V and setting the result equal to zero to approximate V in terms of $\mathbf{s}_i$ yields:

$$\sum_{i=1}^{n} \frac{(\mathbf{s}_i - V)}{2\sigma_i^2} = 0 \tag{12.60}$$

Consequently,

$$\hat{V}_{ML} = \frac{1}{n} \sum_{i=1}^{n} (\mathbf{s}_i) \tag{12.61}$$

As expected:

$$E[\hat{V}_{ML}(\mathbf{s})] = \frac{1}{n} E\left[\sum_{i=1}^{n} \mathbf{s}_i\right] = \frac{1}{n} \sum_{i=1}^{n} E[\mathbf{s}_i] = V \tag{12.62}$$

The variance of $\hat{V}_{ML}(\mathbf{s})$ can be determined as follows:

$$\sigma_{\hat{V}}^2 = E[(\hat{V}_{ML} - V)^2] = \frac{1}{n^2} E\left[\left(\sum_{i=1}^{n} (\mathbf{s}_i - V)\right)^2\right] \tag{12.63}$$

where V is the true value we wish to approximate and $\hat{V}_{ML}$ is its approximation. Notice that in order to include V into the summation term, we have to divide it by n because it will be added n times as a part of the summation term. The square of the summation term will yield the following:

$$\sigma_{\hat{V}}^2 = \frac{1}{n^2} \left\{ \sum_{i=1}^{n} E[(\mathbf{s}_i - V)^2] + \sum_{i=1}^{n} \sum_{j=1, j\neq i}^{n} E[\mathbf{s}_i - V]E[\mathbf{s}_j - V] \right\} \tag{12.64}$$

The last term of Eq. (12.64) is zero because the sensors are independent and

$$\sum_{i=1}^{n}\sum_{j=1,j\neq i}^{n} E[\mathbf{s}_i - V]E[\mathbf{s}_j - V] = \sum_{i=1}^{n} E[\mathbf{s}_i - V] \sum_{j=1,j\neq i}^{2} E[\mathbf{s}_j - V] = 0 \qquad (12.65)$$

From this we conclude that:

$$\sigma_{\hat{V}}^2 = \frac{1}{n^2}\sum_{i=1}^{n} E[(\mathbf{s}_i - V)^2] = \frac{1}{n^2}\sum_{i=1}^{n}\sigma_i^2 \qquad (12.66)$$

If all the sensors have the same variance, σ^2, then,

$$\sigma_{\hat{V}}^2 = \frac{1}{n^2}\sum_{i=1}^{n}\sigma^2 = \frac{\sigma^2}{n} \qquad (12.67)$$

As a result,

$$\lim_{n\to\infty}\sigma_{\hat{V}}^2 = 0 \qquad (12.68)$$

Under the assumption that all sensors produce errors that can be regarded as iid random variables, Eq. (12.67) produces the same result as Eq. (12.46).

12.5.3 Minimum Mean Square Error Estimation

So far we assumed, at least implicitly, that the error has nothing to do with the magnitude of the measurand or any of its properties. In reality, however, this is not always the case; some of the characteristics of the error may change in response to a change in the characteristics of the measurand. As an example, consider Figure 12.26, where the pdf of the error changes with a change in the magnitude of the measurand. Another assumption we have made so far is that the measurand is a fixed or a constant quantity.

Most measurands we wish to sense in real life are themselves continuously changing and should be regarded as random variables. For instance, the temperature, relative humidity, the intensity of light, the quality of air, and the air pressure change in time, however slowly. Through repeated measurements or knowledge of causes and effects we may have the pdfs of these random variables for particular places or processes. So, if we label the measurand as a random variable $\mathbf{m}$ with its own density function, $f(m)$, the minimum mean square error estimation (MMSE) aims to minimise the mean square error between the real $\mathbf{m}$ and its approximation, $\hat{\mathbf{m}}$:

$$\mathbf{e} = \mathbf{m} - \hat{\mathbf{m}} \qquad (12.69)$$

If you remember an important assertion made at the beginning of this chapter, at this point you may wish to ask how we can ever measure the error between $\mathbf{m}$ and $\hat{\mathbf{m}}$, as we may never be able to measure $\mathbf{m}$. This is correct; we may not be able to measure $\mathbf{m}$. Nevertheless, let us put this question aside for a while and assume that there is a mechanism to measure the error.

Consequently:

$$E[(\mathbf{m} - \hat{\mathbf{m}})^2] \overset{!}{=} minimum \qquad (12.70)$$

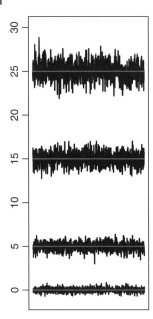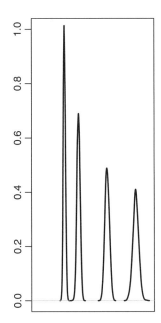

Figure 12.26 An illustration of the change in the pdf of the error of a temperature sensor as the magnitude of the measurand changes.

One way to minimise the error is to employ multiple, independent sensors, just as we did in the previous cases. Thus, if we can approximate the measurand by properly fusing the output of multiple sensors:

$$\hat{\mathbf{m}} = \alpha_1 \mathbf{s}_1 + \alpha_2 \mathbf{s}_2 + \cdots + \alpha_n \mathbf{s}_n \tag{12.71}$$

We can minimise the error by determining the optimal α_i. In other words, we can differentiate Eq. (12.70) with respect to α_i and set the result to zero:

$$E[(\mathbf{m} - \hat{\mathbf{m}})^2] = E[(\mathbf{m} - (\alpha_1 \mathbf{s}_1 + \alpha_2 \mathbf{s}_2 + \cdots + \alpha_n \mathbf{s}_n))^2] \tag{12.72}$$

$$\frac{\partial}{\partial \alpha_i} E[(\mathbf{m} - \hat{\mathbf{m}})^2] = E[(\mathbf{m} - (\alpha_1 \mathbf{s}_1 + \alpha_2 \mathbf{s}_2 + \cdots + \alpha_n \mathbf{s}_n))\mathbf{s}_i] = 0 \tag{12.73}$$

Recalling that,

$$\frac{\partial}{\partial \alpha_i}(\alpha_j \mathbf{s}_j) = 0 \tag{12.74}$$

and, because of our assumption of independence,

$$E[\mathbf{s}_i \mathbf{s}_j] = \eta_i \eta_j \tag{12.75}$$

Listing together all the results, we have,

$$
\begin{aligned}
E[\mathbf{ms}_1] &= \alpha_1 E[\mathbf{s}_1^2] + \alpha_2 \eta_1 \eta_2 + \cdots + \alpha_n \eta_1 \eta_n \\
E[\mathbf{ms}_2] &= \alpha_1 \eta_2 \eta_1 + \alpha_2 E[\mathbf{s}_2^2] \cdots + \alpha_n \eta_2 \eta_n \\
&\ \ \vdots \\
E[\mathbf{ms}_n] &= \alpha_1 \eta_n \eta_1 + \alpha_2 \eta_n \eta_2 + \cdots + \alpha_n E[\mathbf{s}_n^2]
\end{aligned}
\tag{12.76}
$$

If we let $E[\mathbf{ms}_i] = R_{0i}$, $E[\mathbf{s}_i^2] = R_{ii}$ and $E[\mathbf{s}_i\mathbf{s}_j] = \eta_i\eta_j = R_{ij}$, then Eq. (12.76) can be expressed as:

$$
\begin{aligned}
R_{01} &= \alpha_1 R_{11} + \alpha_2 R_{12} + \cdots + \alpha_n R_{1n} \\
R_{02} &= \alpha_1 R_{21} + \alpha_2 R_{22} + \cdots + \alpha_n R_{2n} \\
&\;\;\vdots \\
R_{0n} &= \alpha_1 R_{n1} + \alpha_2 R_{n2} + \cdots + \alpha_n R_{nn}
\end{aligned}
\tag{12.77}
$$

Equation (12.77) can be expressed in matrix form as follows:

$$
\begin{bmatrix} R_{01} \\ R_{02} \\ \vdots \\ R_{0n} \end{bmatrix}
=
\begin{bmatrix}
R_{11} & R_{12} & \cdots & R_{1n} \\
R_{21} & R_{22} & \cdots & R_{2n} \\
& \vdots & & \\
R_{n1} & R_{n2} & \cdots & R_{nn}
\end{bmatrix}
\begin{bmatrix} \alpha_1 \\ \alpha_2 \\ \vdots \\ \alpha_n \end{bmatrix}
\tag{12.78}
$$

At this point, it is imperative to explain some of the variables in Eq. (12.78) and to address the issue of measuring the true value of the measurand. The quantity R_{ii} contains the variance (signifying the error) of the output of the ith sensor ($\mathbf{s}_i$), because $E[\mathbf{s}_i^2] = \sigma_i^2 + \eta_i^2$; σ_i^2 and η_i are usually determined in an environment resembling the typical operating condition of the sensor and by reading the output of the sensor in the absence of a measurand (we have already made reference to this several times). If one can take enough samples, the pdf of the error and, hence, σ_i^2 and η_i, can be determined. The quantity R_{0i} relates the output of the ith sensor to the true value of the measurand. The quantity R_{0i} is determined in a laboratory setting or by the manufacturer of the sensor itself. Here as well, in an environment resembling the typical operation condition and spanning the entire sensing range of the sensor, the sensor is given known inputs ('known' meaning that the controlled input is measured by a highly accurate device, although still an approximation) and for each input, the conditional pdf $f(\mathbf{s}_i|m)$ is carefully determined. Then, the joint pdf can be obtained by multiplying the conditional density function by the density function of the measurand:

$$
f(s_i, m) = f(\mathbf{s}_i|m)f(m)
\tag{12.79}
$$

$f(m)$ is the pdf of $\mathbf{m}$, which we take as a random variable. Our knowledge of $f(m)$ comes from our knowledge of the measurand. Then $E[\mathbf{ms}_i]$ can be determined as:

$$
E[\mathbf{ms}_i] = \int_{-\infty}^{\infty} \int_{-\infty}^{\infty} m\, s_i\, f(s_i, m)\, ds_i\, dm
\tag{12.80}
$$

Finally, the MMSE coefficients can be determined as:

$$
\begin{bmatrix} \alpha_1 \\ \alpha_2 \\ \vdots \\ \alpha_n \end{bmatrix}
=
\begin{bmatrix}
R_{11} & R_{12} & \cdots & R_{1n} \\
R_{21} & R_{22} & \cdots & R_{2n} \\
& \vdots & & \\
R_{n1} & R_{n2} & \cdots & R_{nn}
\end{bmatrix}^{-1}
\begin{bmatrix} R_{01} \\ R_{02} \\ \vdots \\ R_{0n} \end{bmatrix}
\tag{12.81}
$$

Example 12.7   Suppose we use two sensors to estimate the outside temperature. The two sensors have a zero-mean, normally distributed error with different statistical properties and we wish to determine the best way (in the MMSE sense) of combining evidence from these sensors.

The estimated temperature can be expressed as:

$$\hat{T} = \alpha_1 s_1 + \alpha_2 s_2 \tag{12.82}$$

$$\mathbf{e} = \mathbf{T} - \hat{\mathbf{T}} = \mathbf{T} - (u_1 s_1 + u_2 s_2) \tag{12.83}$$

$$\frac{\partial}{\partial \alpha_1} E[\mathbf{e}^2] = E[(\mathbf{T} - (\alpha_1 s_1 + \alpha_2 s_2))(-s_1)] = 0 \tag{12.84}$$

Likewise,

$$\frac{\partial}{\partial \alpha_2} E[\mathbf{e}^2] = E[(\mathbf{T} - (\alpha_1 s_1 + \alpha_2 s_2))(-s_2)] = 0 \tag{12.85}$$

From this we have:

$$\begin{aligned} E[\mathbf{T}s_1] &= \alpha_1 E[s_1^2] \\ E[\mathbf{T}s_2] &= \alpha_2 E[s_2^2] \end{aligned} \tag{12.86}$$

We obtained simplified expressions in Eq. (12.86) because the errors in both sensors have zero means and, as a result, $E[s_1 s_2] = 0$. Moreover, $E[s_1^2] = \sigma_1^2$ and $E[s_2^2] = \sigma_2^2$. Therefore,

$$\alpha_1 = \frac{E[\mathbf{T}s_1]}{\sigma_1^2} \tag{12.87}$$

and,

$$\alpha_2 = \frac{E[\mathbf{T}s_2]}{\sigma_2^2} \tag{12.88}$$

12.5.4 Kalman Filter

So far, even though our evidence combination strategies gradually became more complex, we have nevertheless been entirely dependent on the measurements we got from the sensors in order to determine the values of a measurand. We can reduce our uncertainty about the measurand if we can add knowledge from a different domain. The second assertion made at the beginning of this chapter can serve us towards this end:

> The change in the physical reality (measurand) is a gradual process rather than being haphazard and wild; statistically speaking, the measurand is correlated with itself to a certain extent.

One way of interpreting this assertion is that the future values of a measurand are, to a certain extent, explainable in terms of the present, in the same way its present value is explainable in terms of its past values. Perhaps the poet T.S. Eliot had this in mind when he composed the opening verses of *Burnt Norton in Four Quarters*:

> Time present and time past
> Are both perhaps present in time future,
> And time future contained in time past.
> If all time is eternally present
> All time is unredeemable.
> What might have been is an abstraction

Remaining a perpetual possibility
Only in a world of speculation.
What might have been and what has been
Point to one end, which is always present.

To give a concrete example, the temperature in the city where I am living begins to decline steadily towards the end of September all the way through to the end of March, even though it goes up and down in between. Suppose, based on the knowledge I have up to time $t-1$ (whatever my source of knowledge may be), I predict the temperature for time t with a certain degree of accuracy. Let us label this measurement as $x_p(t)$ (notice the indices p and t; p stands for prediction, because I have not yet made a measurement for time t; t indicates that the prediction is made for time t when I am at time $t-1$). When time t arrives, I make a measurement using a temperature sensor. Let's label this measurement as $x_m(t)$, where the indices m and t represent a measurement taken at time t. Both my prediction and measurement are random variables on account of the uncertainty stemming from prediction and measurement errors. By carefully combining $x_p(t)$ and $x_m(t)$ I can get $\hat{x}_e(t)$, the uncertainty of which is less than if I were to rely on either $x_p(t)$ or $x_m(t)$. Indeed, I can now even improve my prediction of the temperature for time $t+1$ due to my improved estimation of the temperature of t. This is illustrated in Figure 12.27. Formally, the Kalman filter is described by two equations:

$$x_m(t) = x(t) + v(t) \tag{12.89}$$
$$x(t+1) = x(t) + w(t) \tag{12.90}$$

where $x(t)$ is the random variable we are interested in estimating but will never be able to directly measure, $v(t)$ is the measurement error at time t modelled as a random variable, and $w(t)$ is the error made in the prediction as a result of the inherent randomness in the measurand (hence, the measurand is also regarded as a random variable).

To illustrate the above relations by example, suppose we wish to estimate the temperature variation of the city of Dresden for the months between the beginning of September and the end of March; a total of 210 days. Each day exactly at noon we measure the temperature of a particular location and predict the temperature of that same location for the next day (at noon). Our aim is to improve our knowledge of the temperature of each day by carefully combining the values of our prediction and measurement. Suppose that scientific evidence shows that the temperature of Dresden falls about 2% daily in the time period we are concerned with:

$$T(t+1) = 0.98T(t) \tag{12.91}$$

This is illustrated in Figure 12.28a. The scientific claim is, of course, very optimistic, because the future temperature, even though it is to some extent correlated with the present, entails also some randomness; otherwise we need not employ any sensor at all. Suppose the actual temperature variation looks like the trace shown in Figure 12.28b. The good thing about the correlation in the temperature variation is that we can include a process error in Eq. (12.91) to accommodate the randomness in the temperature (process) variation.

$$T(t+1) = 0.98T(t) + w(t) \tag{12.92}$$

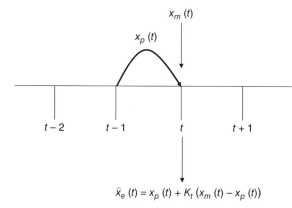

$$\hat{x}_e(t) = x_p(t) + K_t(x_m(t) - x_p(t))$$

Figure 12.27 The basic principle of a Kalman Filter. The Kalman approach combines two types of evidence: one from knowledge of how the measurand evolves in time ($\mathbf{x}_p(t)$), the other from measurement ($\mathbf{x}_m(t)$). The idea is to reduce the uncertainty in the combined evidence $\hat{\mathbf{x}}_e(t)$ by properly weighing $\mathbf{x}_p(t)$ and $\mathbf{x}_m(t)$. As the bottom part illustrates, as our uncertainty decreases, the propagation of our belief into the future further reduces our uncertainty.

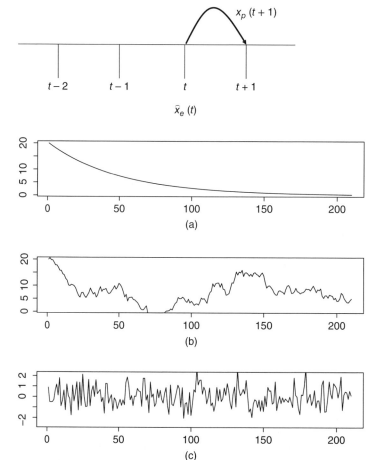

Figure 12.28 An illustration of the temperature variation over time: (a) the temperature at $t+1$ expressed as a function of the temperature at t, in other words, $T(t+1) = 0.98T(t)$; (b) the actual temperature; (c) the process error modelled as a zero-mean, normally distributed random variable.

The process error is shown in Figure 12.28c. At this point it should be noted that Eq. (12.92) should not be confused with the prediction for the time $t + 1$, which we label as $\mathbf{T}_p(t + 1)$. Equation (12.92) stems from the second assertion made at the beginning of the chapter and repeated at the beginning of this section. $\mathbf{T}_p(t + 1)$, on the other hand, depends on the knowledge oft the temperature of t, but this should become clear in the subsequent explanation.

The measurement we take each day using a sensor contains the actual temperature, but this is mixed with the inherent error of the sensor. Thus, we can express it as follows:

$$\mathbf{T}_m(t) = \mathbf{T}(t) + \mathbf{v}(t) \tag{12.93}$$

Notice that even though the temperature of the present as well as the future are scalar quantities, due to $\mathbf{w}$ and $\mathbf{v}$, they should be taken as random variables. If we let $\hat{\mathbf{T}}(t)$ be our estimation of the temperature for the time t, then the mean square error at time t is expressed as follows:

$$P(t) = E[\mathbf{e}^2] = E[(\mathbf{T}(t) - \hat{\mathbf{T}}(t))(\mathbf{T}(t) - \hat{\mathbf{T}}(t))] \tag{12.94}$$

The estimated temperature for time t can be expressed in terms of the predicted temperature for time t and the measured temperature for time t using Eq. (12.32):

$$\hat{\mathbf{T}}(t) = \mathbf{T}_p(t) + K(t)[\mathbf{T}_m(t) - \mathbf{T}_p(t)] \tag{12.95}$$

Since $\mathbf{T}_m(t) = \mathbf{T}(t) + \mathbf{v}(t)$, Eq. (12.95) can be rewritten as:

$$\hat{\mathbf{T}}(t) = \mathbf{T}_p(t) + K(t)[\mathbf{T}(t) + \mathbf{v}(t) - \mathbf{T}_p(t)] \tag{12.96}$$

Substituting Eq. (12.96) into Eq. (12.94) yields,

$$P(t) = E\{([1 - K(t)][\mathbf{T}(t) - \mathbf{T}_p(t)] - K(t)\mathbf{v}(t))^2\} \tag{12.97}$$

The term $E[\mathbf{T}(t) - \mathbf{T}_p(t)]$, denoted as $P_p(t)$, quantifies the prediction error and is sometimes known as the error of the priori estimate. This error does not correlate with the measurement error, since the prediction is made before the measurement is taken. Now we can rewrite Eq. (12.97) in terms of the prediction error:

$$P(t) = (1 - K(t))^2 P_p(t) + K^2(t)R \tag{12.98}$$

where $R = E[\mathbf{v}^2]$. If the measurement error is a zero-mean error, then $R = \sigma_v^2$. If we distribute Eq. (12.98), it yields,

$$P(t) = P_p(t) - 2K(t)P_p(t) - K^2(t)(P_p(t) + R) \tag{12.99}$$

We are now in a position to choose the optimal $K(t)$ that can minimise our estimation error at time t. This can be done by differentiating Eq. (12.99) with respect to $K(t)$ and setting the result to zero. The result is:

$$K(t) = \frac{P_p(t)}{P_p(t) + R} \tag{12.100}$$

Finally, substituting Eq. (12.100) into Eq. (12.99) reduces the expression for $P(t)$ into:

$$P(t) = (1 - K(t))P_p(t) \tag{12.101}$$

One way to continuously improve our ability to predict is to propagate or project the estimation error into the future. Hence our prediction of the temperature for time $t + 1$ based on the evidence we have at time t is:

$$\mathbf{T}_p(t + 1) = 0.98\hat{\mathbf{T}}(t) \tag{12.102}$$

The prediction error for time $t + 1$ is,

$$\mathbf{e}_p(t + 1) = \mathbf{T}(t + 1) - \mathbf{T}_p(t + 1) = 0.98\mathbf{T}(t) + \mathbf{w}(t) - 0.98\hat{\mathbf{T}}(t) = 0.98\mathbf{e}(t) + \mathbf{w}(t) \tag{12.103}$$

The MMSE of the predicted temperature for time $t + 1$, $P_p(t + 1)$, is:

$$P_p(t + 1) = E[\mathbf{e}_p^2(t + 1)] = (0.98)^2 E[\mathbf{e}^2(t)] + E[\mathbf{w}^2(t)] = 0.9604P(t) + Q \tag{12.104}$$

where $Q = E[\mathbf{w}^2(t)]$. If the process error is a zero-mean error, then, $Q = \sigma_w^2$. With $P_p(t)$, $P(t)$, and $P_p(t + 1)$ the Kalman Filter connects the past, the present, and the future, in the same way the poet maintains:

> Time present and time past
> Are both perhaps present in time future,
> And time future contained in time past.

Consequently, we can now estimate step by step the temperature of each day. Suppose we define:

$$\mathbf{T}_p(t) = 0.75\hat{\mathbf{T}}(t - 1) + 0.25\hat{\mathbf{T}}(t - 2)$$

For the first time slot, we shall have no predicted value and therefore the estimated and the measured data are equal:

$$\hat{\mathbf{T}}(1) = \mathbf{T}_m(1)$$

We still do not have sufficient data to make a prediction that satisfies our definition of $\mathbf{T}_p(t)$, but we can make a reasonable estimate using the evidence we have:

$$\mathbf{T}_p(2) = 0.75\hat{\mathbf{T}}(1)$$

With $\mathbf{T}_p(2)$ and $\mathbf{T}_m(2)$ (which we have, because we can always measure them), we can compute $\hat{\mathbf{T}}(2)$, but for that we need $K(2)$, which we do not have. Once again, we can make a reasonable guess; as we don't have any evidence to mistrust either the predicted or the measured values, we can set $K(2) = 0.5$. Thus,

$$\hat{\mathbf{T}}(2) = \mathbf{T}_p(2) + K(2)[\mathbf{T}_m(2) - \mathbf{T}_p(2)]$$

With $K(2)$ determined, we can also determine $P_p(2)$, $P(2)$, and $P_p(2 + 1)$ using Eqs 12.100 -12.102. And with these we can move on to predicting and estimating the temperature for the next time slot, and so on. Figure 12.29 lists the program code in R for computing all the remaining values for each parameter we require for determining prediction and estimations. Figure 12.30 displays the three important temperature values. Clearly, the Kalman estimation is more accurate than either the predicted or measured values.

```
# process noise modelled as a zero-mean normal distribution with a variance of 1
w <- rnorm(210, 0, 1)

# measurement noise modelled as a zero-mean normal distribution with a variance of 9
v <- rnorm(210, 0, 3)

# initialising the vector  for the actual temperature
T <- rep(0, 210)

# The temperature of day one being 20 degree Celsius
T[1] <- 20

# The actual temperature initialised
for(t in 2:length(T) ) {
 T[i] <- 0.98 * T[t - 1] + w[t]
 }

# initialising the vector of the predicted temperature
Tp <- rep(0, 210)
# initialising the vector of the measured temperature
Tm <- rep(0, 210)
# initialising the vector of the estimated temperature
Th <- rep(0, 210)
# initialising the vector of the Kalman constants
K <- rep(0, 210)
# initialising the vector of the minimum mean square prediction error
Pp <- rep(0, 210)
# initialising the vector of the minimum mean square estimation error
P <- rep(0, 210)

# Tm as the addition of the actual temperature and the measurement noise
Tm <- T + v
# Th of day one is set to equal the measured temperature of day one
Th[1] <- Tm[1]
# the predicted temperature of day two is set as 0.75 times Th[1]
Tp[2] <-  0.75 * Th[1]
# setting K[2] = 0.5, because I initially trust Tm[2] and Tp[2] equally
K[2] <- 0.5
# with K[2] = 0.5, Pp[2] will equal to 1
Pp[2] <- 1
# The minimum mean square error of t = 2 is computed
P[2] <- (1-k[2]) * Pp[2]

# the temperature of t = 2 is estimated
Th[2] <- Tp[2] + K[2] * (Tm[2] - Tp[2])

# Computing all the parameters of the Kalman filter

for(t in 3:(length(Tm)-1)) {
 Pp[t] <- P[i-1] * var(w)
 K[t] <- Pp[i] / (Pp[i] + var(v))
# I compute Tp[t] by combining the last two estimated values
 Tp[t] <- 0.75 * Th[t-1] + 0.25 * Th[t-2]
 Th[t] <- Tp[i] + K[t] * (Tm[t] - Tp[t])
 P[t] <- (1- K[t]) * Pp[t]
 }
```

Figure 12.29 Code written in *R* to estimate the temperature fluctuation of 210 days using a Kalman filter. Tp: $\mathbf{T}_p(t)$, Tm: $\mathbf{T}_m(t)$, Th: $\hat{\mathbf{T}}(t)$, Pp: $P_p(t)$, and P: $P(t)$.

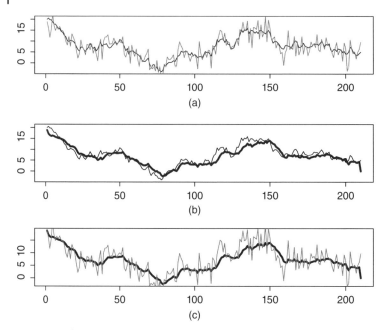

Figure 12.30 Application of the Kalman filter to estimate the temperature variation over time. (a) difference between the actual variation and the measured values; (b) difference between the actual variation and the Kalman estimation (which combines the predicted values and the measured values). (c) difference between the measured and the Kalman estimated values.

12.5.5 The Kalman Filter Formalism

In Eq. (12.89), we assumed (implicitly) that the output of the sensor and the measurand are one and the same type. But this is not generally the case. For the temperature-to-voltage converter sensor, for example, the input is a temperature (**T**) and the output is a voltage (**V**). Therefore, we need a more inclusive approach to formally express the Kalman filter. If **x** is depicted as the measurand which we wish to estimate and **s** is the out put of a sensor (the measurement), the relationship between the present and future values of the measurand are related to one another as follows:

$$\mathbf{x}(t + 1) = \Phi \mathbf{x}(t) + \mathbf{w}(t) \tag{12.105}$$

where Φ is the state transition matrix, from t to $t + 1$, and is assumed to be statistically stationary and **w**(t) is called the process error at time t. Likewise,

$$\mathbf{s}(t) = H\mathbf{m}(t) + \mathbf{v}(t) \tag{12.106}$$

where H establishes an ideal relationship between **s** (t) and **m**(t). With these adjustments in mind, the estimated value of the measurand for time t can be expressed as:

$$\hat{\mathbf{x}}(t) = \mathbf{x}_{p}(t) + K(t)[\mathbf{s}(t) - H\mathbf{x}_{p}(t)] \tag{12.107}$$

where $H\mathbf{x}_{p}(t)$ predicts the sensor output (no measurement is taken yet for time t). The Kalman constant $K(t)$ should have the appropriate unit so as to enable the addition of the two terms on the right side in Eq. (12.107). In the same way, we can modify the expressions for the Kalman constant and the MMSE for time t as well as the prediction error for time $t + 1$ as follows:

$$K(t) = \frac{HP_\text{p}(t)}{H^2P_\text{p}(t) + R} \tag{12.108}$$

$$P(t) = (1 - HK(t))P_\text{p}(t) \tag{12.109}$$

$$P_\text{p}(t + 1) = \Phi^2P(t) + Q \tag{12.110}$$

In conclusion, the estimation of a measurand may deal with past, present, or future values, depending on what we wish to do with the sensor data. In his original publication Kalman (1960) uses the term "estimation" to collectively describe the problem of interpolation, filtering, and prediction. He describes the individual problems as follows (I have made a slight adjustment to the parameter depiction to make the text readable):

> We are given signal $\mathbf{x}_1(t)$ and noise $\mathbf{x}_2(t)$. Only the sum $\mathbf{y}(t) = \mathbf{x}_1(t) + \mathbf{x}_2(t)$ can be observed. Suppose we have observed and know exactly the values of $y(0),..., y(n)$. What can we infer from this knowledge in regard to the (unobservable) value of the signal at [time] t, where t may be less than, equal to, or greater than n? If $t < n$, this is a data-smoothing (interpolation) problem. If $t = n$, this is called filtering. If $t > n$, we have a prediction problem. Since our treatment will be general enough to include these and similar problems, we shall use hereafter the collective term estimation.

References

Gardiner CW 1985 *Handbook of Stochastic Methods*, vol. 3. Springer.

Grewal MS 2011 *Kalman Filtering*. Springer.

Kalman RE 1960 A new approach to linear filtering and prediction problems. *Journal of Basic Engineering*, **82**(1), 35–45.

Papoulis A and Pillai SU 2002 *Probability, Random Variables, and Stochastic Processes*. Tata McGraw-Hill Education.

Ross SM *et al.* 1996 *Stochastic Processes*, vol. **2**. John Wiley & Sons.

Index

Principles and Applications of Ubiquitous Sensing, First Edition. Waltenegus Dargie.
© 2017 John Wiley & Sons, Ltd. Published 2017 by John Wiley & Sons, Ltd.
Companion Website: www.wiley.com/go/dargie2017